A First Course in Ergodic Theory

A First Course in Ergodic Theory

Karma Dajani
Utrecht University
Charlene Kalle
Leiden University

CRC Press
Taylor & Francis Group
Boca Raton London New York

CRC Press is an imprint of the
Taylor & Francis Group, an **informa** business
A CHAPMAN & HALL BOOK

First edition published 2021
by CRC Press
6000 Broken Sound Parkway NW, Suite 300, Boca Raton, FL 33487-2742

and by CRC Press
2 Park Square, Milton Park, Abingdon, Oxon, OX14 4RN

CRC Press is an imprint of Taylor & Francis Group, LLC

Library of Congress Cataloging-in-Publication Data

ISBN: 9780367226206 (hbk)
ISBN: 9781032021843 (pbk)
ISBN: 9780429276019 (ebk)

Typeset in LMR12
by KnowledgeWorks Global Ltd.

To Annabel, Isaac, Noam, and Rafael

Contents

Preface xi

Author Bios xiii

CHAPTER 1 ▪ Measure Preservingness and Basic Examples 1

1.1 WHAT IS ERGODIC THEORY? 1
1.2 MEASURE PRESERVING TRANSFORMATIONS 3
1.3 BASIC EXAMPLES 6

CHAPTER 2 ▪ Recurrence and Ergodicity 19

2.1 RECURRENCE 19
2.2 ERGODICITY 22
2.3 EXAMPLES OF ERGODIC TRANSFORMATIONS 27

CHAPTER 3 ▪ The Pointwise Ergodic Theorem and Mixing 33

3.1 THE POINTWISE ERGODIC THEOREM 34
3.2 NORMAL NUMBERS 43
3.3 IRREDUCIBLE MARKOV CHAINS 46
3.4 MIXING 49

CHAPTER 4 ▪ More Ergodic Theorems 53

4.1 THE MEAN ERGODIC THEOREM 53
4.2 THE HUREWICZ ERGODIC THEOREM 56

CHAPTER 5 ■ Isomorphisms and Factor Maps 65

5.1 MEASURE PRESERVING ISOMORPHISMS 65
5.2 FACTOR MAPS 71
5.3 NATURAL EXTENSIONS 72

CHAPTER 6 ■ The Perron-Frobenius Operator 77

6.1 ABSOLUTELY CONTINUOUS INVARIANT MEASURES 77
6.2 EXACTNESS 82
6.3 PIECEWISE MONOTONE INTERVAL MAPS 91

CHAPTER 7 ■ Invariant Measures for Continuous Transformations 101

7.1 EXISTENCE 101
7.2 UNIQUE ERGODICITY AND UNIFORM DISTRIBUTION 109
7.3 SOME TOPOLOGICAL DYNAMICS 115

CHAPTER 8 ■ Continued Fractions 125

8.1 REGULAR CONTINUED FRACTIONS 126
8.2 ERGODIC PROPERTIES OF THE GAUSS MAP 130
8.3 THE DOEBLIN-LENSTRA CONJECTURE 135
8.4 OTHER CONTINUED FRACTION TRANSFORMATIONS 139

CHAPTER 9 ■ Entropy 147

9.1 RANDOMNESS AND INFORMATION 147
9.2 DEFINITIONS AND PROPERTIES 150
9.3 CALCULATION OF ENTROPY AND EXAMPLES 156
9.4 THE SHANNON-MCMILLAN-BREIMAN THEOREM 160
9.5 LOCHS' THEOREM 168

CHAPTER 10 ■ The Variational Principle 173

10.1 TOPOLOGICAL ENTROPY 173
10.2 PROOF OF THE VARIATIONAL PRINCIPLE 188
10.3 MEASURES OF MAXIMAL ENTROPY 194

CHAPTER 11 ■ Infinite Ergodic Theory 199

11.1 EXAMPLES 199
11.2 CONSERVATIVE AND DISSIPATIVE PART 202
11.3 INDUCED SYSTEMS 207
11.4 JUMP TRANSFORMATIONS 216
11.5 INFINITE ERGODIC THEOREMS 218

CHAPTER 12 ■ Appendix 223

12.1 TOPOLOGY 223
12.2 MEASURE THEORY 224
12.3 LEBESGUE SPACES 229
12.4 LEBESGUE INTEGRATION 231
12.5 HILBERT SPACES 233
12.6 BOREL MEASURES ON COMPACT METRIC SPACES 235
12.7 FUNCTIONS OF BOUNDED VARIATION 237

Bibliography 241

Index 247

Preface

As the title suggests this book is intended as a textbook for an introductory course in ergodic theory. It originated from our own *Ergodic Theory* course that we have been teaching for many years in the Dutch national master program in mathematics called *Mastermath*. Originally we used the lecture notes made by Karma in the 1990's. Over the years we added topics, theorems, examples and explanations from various sources to the course and the list of reference material we provided to the students grew longer and longer. With this book we hoped to create one source that is easy to teach from and that contains all the topics one would like to touch upon in an introductory course in ergodic theory. The text has been put to the test on several groups of students and it contains many worked examples and exercises.

The book is designed for a one semester course. We treat approximately one chapter per week with the exception of Chapters 6, 9 and 11, which take a bit longer. In many places the book combines material that is also covered in several classical textbooks on ergodic theory. We mention in particular in chronological ordering [48] by W. Parry, [12] by I. P. Cornfeld, S. V. Fomin and Ya. G. Sinai, [65] by P. Walters, [50] by K. Petersen, [33] by A. Lasota and M. C. Mackey, [1] by J. Aaronson, [8] by A. Boyarsky and P. Góra, [10] by M. Brin and G. Stuck, [14] by the first author and C. Kraaikamp, [17] by M. Einsiedler and T. Ward, [29] by M. Kesseböhmer, S. Munday and B. O. Stratmann and [63] by M. Viana and K. Oliveira.

The material in this book requires a basic knowledge of measure theory, some topology and a very basic knowledge of functional analysis. We have collected the results from these areas that are most relevant for the content of the book in the appendix for easy reference. For an introduction into these fields, more details and proofs we refer the reader to standard textbooks on the topic as for example [20, 30, 44, 56, 58].

The picture on the cover illustrates the concept of mixing, see Chapter 3, while also referring to the statement by A. Rényi that mathematicians are machines for turning coffee into theorems. And of course, it is a reminder of the many cups that fuelled us while writing the book.

We would like to take this opportunity to thank several people for their help in the process of making this book and for reducing the number of mistakes and typos. Our thanks go to all the students who participated in our Mastermath *Ergodic Theory* course over the past years, in particular the class of 2020, for their careful scrutiny of the text and honest comments. We would also like to thank Jonny Imbierski, Marta Maggioni, Marks Ruziboev and Benthen Zeegers for reading the final versions. Finally, we would like to thank the Hausdorff Research Institute for Mathematics in Bonn for hosting us for a short time during the trimester program *Dynamics: Topology and Numbers.*

Karma and Charlene

Author Bios

Karma Dajani earned her PhD degree from the George Washington University in Washington, DC and is currently an Associate Professor in Mathematics at Utrecht University in the Netherlands. She has over 30 years of teaching experience, close to 60 publications and is a coauthor of the book, *Ergodic Theory of Numbers*. Her research interests are primarily in Ergodic Theory, and its applications to other fields such as Number Theory, Probability Theory, and Symbolic Dynamics. Mathematics is her career but it is also one of her three passions which also includes dance and classical music.

Charlene Kalle obtained her PhD in mathematics at Utrecht University in the Netherlands. After postdoctoral positions in Warwick and Vienna, she moved to Leiden University where she is now an Associate Professor in Mathematics. Her research is in Ergodic Theory with applications to Probability Theory. She is also interested in relations to Number Theory and Fractal Geometry. Besides 20 research articles, she has co-authored a book on extracurricular high school mathematics. She has accumulated 20 years of teaching experience ranging from teaching Italian language to adults, to lecturing master courses in mathematics. She devotes her time not spent on mathematics to her three children and playing bridge.

CHAPTER 1

Measure Preservingness and Basic Examples

1.1 WHAT IS ERGODIC THEORY?

Dynamical systems theory is the area of mathematics dedicated to the study of processes that evolve over time. There are many ways to approach this subject and ergodic theory focuses on describing the asymptotic average behavior of dynamical systems. To do that it uses techniques from many different fields including probability theory, statistical mechanics, number theory, vector fields on manifolds, group actions of homogeneous spaces and fractal geometry.

The word *ergodic* is a mixture of two Greek words: *ergon* (work) and *odos* (path). The word was introduced by L. Boltzmann (1844–1906) regarding his hypothesis in statistical mechanics: *for large systems of interacting particles in equilibrium, the time average along a single trajectory equals the space average.* The hypothesis as he stated it was false, and the search for conditions under which these two quantities are equal led to the birth of ergodic theory as it is known nowadays.

A dynamical system consists of a space X that represents the collection of all states the system can be in, called *state space*, and a way to describe the time evolution. A general setup to represent time evolution is by a (semi-)group G together with a collection of transformations $\{T_g : X \to X\}_{g \in G}$, one for each element of the group, with the property that for any two elements $g, h \in G$ the composition of the corresponding maps satisfies $T_{gh} = T_g \circ T_h$. Continuous time dynamical systems (or flows) are obtained by setting $G = \mathbb{R}$ for example. In this book time will be discrete and time evolution is governed by a single transformation $T : X \to X$. So, if the system is in state $x \in X$ at some moment in time, then it will be in state Tx next. This can be seen as taking $G = \mathbb{N}$ (or $\mathbb{Z}$, depending on whether the dynamics is invertible or not) and associating to each $n \in \mathbb{N}$, the iterate $T^n = \underbrace{T \circ T \circ \cdots \circ T}_{n \text{ times}}$. Our dynamical systems are thus represented by a pair (X, T). This setup is analogous to that of discrete time stochastic processes.

Without further assumptions on the pair (X, T) this setup is too general to lead to any interesting results. The space X usually has a special structure, and we want T to preserve the basic structure on X. For example:

- if X is a measure space, then we like T to be measurable;
- if X is a topological space, then T must be continuous;
- if X has a differentiable structure, then T is a diffeomorphism.

In this book we mostly take the measure theoretic approach: all our spaces are measure spaces $(X, \mathcal{F}, \mu)$ and more specifically finite or infinite measure σ-finite Lebesgue space (see Definition 12.3.2 from the Appendix). The transformations $T : X \to X$ we consider are always measurable (so $T^{-1}A \in \mathcal{F}$ for all $A \in \mathcal{F}$) and also satisfy the following fundamental property, which guarantees that mass cannot appear out of nowhere.

Definition 1.1.1. Let $(X, \mathcal{F}, \mu)$ be a measure space and $T : X \to X$ measurable. Then T is called *non-singular* with respect to μ if for any $A \in \mathcal{F}$ it holds that $\mu(A) = 0$ if and only if $\mu(T^{-1}A) = 0$.

Throughout the book (and without mentioning this any further) we assume the following:

- whenever we write that $(X, \mathcal{F}, \mu)$ is a measure space, we assume it is a **finite or infinite measure σ-finite Lebesgue space**;
- any transformation $T : X \to X$ is always assumed to be **measurable and non-singular**.

In this case we talk about a dynamical system.

Definition 1.1.2. We call a quadruple $(X, \mathcal{F}, \mu, T)$ a *dynamical system* if $(X, \mathcal{F}, \mu)$ is a measure space and $T : X \to X$ is a transformation.

Since any finite measure space can be rescaled to a probability space, the measure spaces under consideration will either be probability spaces or have infinite total measure. In the remainder of this chapter and Chapters 2, 6 and 11 we consider both. Chapters 4, 5, 8 and 9 will be mainly concerned with probability spaces. In Chapters 7 and 10 the underlying state space will be a compact metric space (X, d) and we take the measure structure on X that is compatible with the topology, i.e., the σ-algebra will be the Borel σ-algebra generated by open balls under the metric d. The transformation T will then be continuous on X, which automatically makes it measurable.

1.2 MEASURE PRESERVING TRANSFORMATIONS

Let $(X, \mathcal{F}, \mu)$ be a measure space. Assume it is the state space of a dynamical system for which the time evolution is governed by a measurable transformation $T : X \to X$. The measure structure on X is determined by the σ-algebra as well as the measure, and we would like the measure μ to be compatible with the dynamics of T. This is captured in the definition of measure preservingness.

Definition 1.2.1. Let $(X, \mathcal{F}, \mu)$ be a measure space and $T : X \to X$ a transformation. Then T is said to be *measure preserving* with respect to μ (or equivalently μ is said to be T*-invariant*) if $\mu(T^{-1}A) = \mu(A)$ for all $A \in \mathcal{F}$.

The definition of measure preservingness concerns inverse images of sets, which in contrast to forward images are always measurable. We say T is *invertible* if it is one-to-one and if T^{-1} is measurable. Note that an invertible T is measure preserving if and only if $\mu(TA) = \mu(A)$ for all $A \in \mathcal{F}$. Also note that any measure preserving transformation is automatically non-singular. If a transformation T is not measure preserving with respect to μ, then one way to check the non-singularity of T is by proving that the Radon-Nikodym derivative (see Theorem 12.4.4) $\frac{d\mu \circ T^{-1}}{d\mu}$ is positive μ-a.e. In case T is invertible one can replace T^{-1} by T.

The dynamics of a transformation T is represented by the orbits. For each $x \in X$, the *(forward) orbit* of x under T is the sequence

$$x, Tx, T^2x, \ldots.$$

If T is invertible, then one speaks of the *two-sided orbit*

$$\ldots, T^{-1}x, x, Tx, \ldots.$$

If T is measure preserving, then for any measurable function $f : X \to \mathbb{R}$, the process

$$f, f \circ T, f \circ T^2, \ldots$$

is stationary. This means that for all Borel sets $B_1, \ldots, B_n$, and all integers $r_1 < r_2 < \ldots < r_n$, one has for any $k \geq 1$,

$$\mu(\{x : f(T^{r_1}x) \in B_1, \ldots, f(T^{r_n}x) \in B_n\})$$
$$= \mu(\{x : f(T^{r_1+k}x) \in B_1, \ldots, f(T^{r_n+k}x) \in B_n\}).$$

According to the definitions, to check whether a transformation is measurable and measure preserving, one needs to verify the conditions for **all** measurable sets. It would be convenient of course if it would be sufficient to check the conditions on a more manageable, smaller collection of subsets of X. One possible option is a generating semi-algebra. A collection $\mathcal{S}$ of subsets of X is called a *semi-algebra* if it satisfies

(i) $\emptyset \in \mathcal{S}$,

(ii) $A \cap B \in \mathcal{S}$ whenever $A, B \in \mathcal{S}$, and

(iii) if $A \in \mathcal{S}$, then the complement $A^c = X \setminus A = \cup_{i=1}^n E_i$ is a disjoint union of elements of $\mathcal{S}$.

Recall that a collection $\mathcal{S}$ of subsets of X is *generating* for the σ-algebra $\mathcal{F}$ if $\mathcal{F} = \sigma(\mathcal{S})$, where $\sigma(\mathcal{S})$ denotes the smallest σ-algebra containing all sets from $\mathcal{S}$.

Theorem 1.2.1. *Let $(X, \mathcal{F}, \mu)$ be a measure space and $T : X \to X$ a map. Suppose $\mathcal{S}$ is a generating semi-algebra of $\mathcal{F}$ that contains an exhausting sequence (S_n), i.e., an increasing sequence with $X = \bigcup_{n=1}^{\infty} S_n$. Suppose that for each $A \in \mathcal{S}$ one has $T^{-1}A \in \mathcal{F}$ and $\mu(T^{-1}A) = \mu(A)$. If furthermore, $\mu(S_n) = \mu(T^{-1}S_n) < \infty$ for all n, then T is measurable and measure preserving.*

The proof of this theorem is given in the Appendix, where it is formulated in the slightly more general setting of a transformation $T : X_1 \to X_2$ between two measure spaces, see Theorem 12.3.1.

The next lemma, of which the proof is an easy exercise, says that the notions of measurability and measure preservingness are preserved under taking the completion of the underlying measure space.

Lemma 1.2.1. *Let $(X, \mathcal{F}, \mu)$ be a measure space and $\overline{\mathcal{F}}$ the completion of $\mathcal{F}$. If $T : X \to X$ is measurable and measure preserving on $(X, \mathcal{F}, \mu)$, then it has the same properties on $(X, \overline{\mathcal{F}}, \overline{\mu})$, where $\overline{\mu}$ is the extended measure.*

This lemma is especially useful, since many of the examples of dynamical systems we consider have Borel or Lebesgue σ-algebras and according to the lemma for issues regarding measurability and measure preservingness we do not need to bother with distinguishing between them. Whenever a transformation is measurable and measure preserving with respect to the Borel σ-algebra, the same automatically holds for the Lebesgue σ-algebra.

Another way of verifying whether a given measure is invariant is by using the Koopman operator. If necessary, recall the definition of L^p-spaces from Section 12.2. Let $(X, \mathcal{F}, \mu)$ be a measure space and $T : X \to X$ a transformation. Define the *induced operator* or *Koopman operator* $U_T : L^0(X, \mathcal{F}, \mu) \to L^0(X, \mathcal{F}, \mu)$ by

$$U_T(f) = f \circ T. \tag{1.1}$$

The following properties of U_T are easy to prove.

Proposition 1.2.1. *The operator U_T has the following properties.*

(i) *U_T is linear.*

(ii) *$U_T(fg) = U_T(f)U_T(g)$.*

(iii) *$U_T(c) = c$ for any constant c.*

(iv) *U_T is positive.*

(v) *$U_T(1_B) = 1_B \circ T = 1_{T^{-1}B}$ for all $B \in \mathcal{F}$.*

Exercise 1.2.1. Prove Proposition 1.2.1.

The next proposition helps to check whether a measure is invariant.

Proposition 1.2.2. *Let $(X, \mathcal{F}, \mu, T)$ be a dynamical system. The measure μ is T-invariant if and only if for each $f \in L^0(X, \mathcal{F}, \mu)$ it holds that $\int_X U_T f \, d\mu = \int_X f \, d\mu$ (where if one side doesn't exist or is infinite, then the other side has the same property).*

Proof. First, let $A \in \mathcal{F}$ be given. Then $1_A \in L^0(X, \mathcal{F}, \mu)$ and

$$\int_X 1_A d\mu = \mu(A) = \mu(T^{-1}A) = \int_X 1_{T^{-1}A} \, d\mu = \int_X U_T 1_A \, d\mu,$$

where both sides are infinite precisely if $\mu(A) = \infty$. So the property holds for indicator functions. By linearity of the integral it then also holds for simple functions and by approximation we then get the result for all non-negative measurable functions. By splitting a function in its positive and negative part, we then obtain the statement. For the other direction, let $A \in \mathcal{F}$ be given. Then by assumption $\int_X 1_A \, d\mu = \int_X U_T 1_A \, d\mu = \int_X 1_{T^{-1}A} \, d\mu$, giving the T-invariance of μ. □

A consequence of the previous proposition is the following.

Proposition 1.2.3. *Let $(X, \mathcal{F}, \mu)$ be a measure space and $T : X \to X$ a measure preserving transformation. Let $p \geq 1$. Then, $U_T L^p(X, \mathcal{F}, \mu) \subseteq L^p(X, \mathcal{F}, \mu)$, and $\|U_T f\|_p = \|f\|_p$ for all $f \in L^p(X, \mathcal{F}, \mu)$.*

Exercise 1.2.2. Prove Proposition 1.2.3.

Exercise 1.2.3. Let $(X, \mathcal{F}, \mu)$ be a probability space and $T : X \to X$ a measure preserving transformation. Let $f \in L^1(X, \mathcal{F}, \mu)$. Show that if $U_T f \leq f$ μ-a.e., then $f = U_T f$ μ-a.e.

1.3 BASIC EXAMPLES

Here we collect some examples that will be used throughout the book. In the first couple of examples X is (a subset of) a Euclidean space. To keep the notation simple, in all these cases we use $\mathcal{B}$ for the appropriate Lebesgue σ-algebra and λ for the appropriate Lebesgue measure.

Example 1.3.1 (Translation). For any $t \in \mathbb{R}$ the transformation $T : \mathbb{R} \to \mathbb{R}$ given by $Tx = x + t$ is a translation by t. By the shift invariance of the Lebesgue measure λ it follows immediately from Proposition 1.2.2 that λ is invariant for T. See Figure 1.1(a) for the graph.

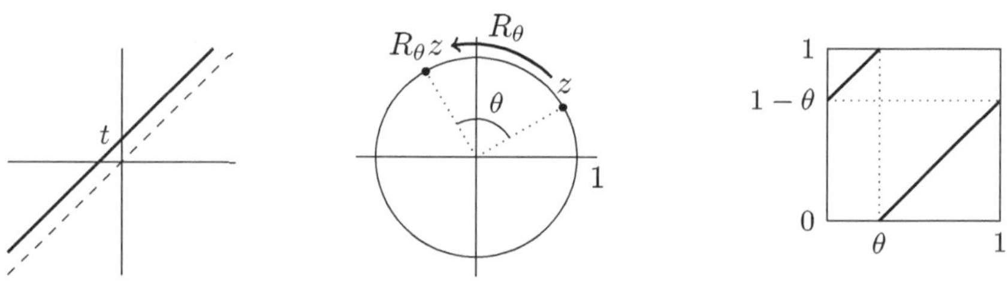

(a) Translation by t (b) Rotation (c) Translation (mod 1)

Figure 1.1 The graphs of the transformation from Example 1.3.1 in (a) and of the transformations from Example 1.3.2 in (b) and (c).

Example 1.3.2 (Rotation). Let $\mathbb{S}^1 \subseteq \mathbb{C}$ denote the complex unit circle. The measure we consider on the Lebesgue σ-algebra $\mathcal{B}$ is the normalized Haar (or Lebesgue) measure λ. Let $0 < \theta < 1$ and define rotation by angle θ on $\mathbb{S}^1$ by $R_\theta : \mathbb{S}^1 \to \mathbb{S}^1$, $z \mapsto e^{2\pi i\theta} z$. One easily verifies that the collection of all half open arcs is a generating semi-algebra for $\mathbb{S}^1$, so by Theorem 1.2.1 it is enough to check measure preservingness only for such arcs. Since R_θ is an isometry, it is clear that λ is R_θ-invariant.

One can also view this transformation additively by defining the transformation T_θ on the unit interval $[0,1)$ by

$$T_\theta x = x + \theta \pmod 1 = x + \theta - \lfloor x + \theta \rfloor.$$

The graphs of R_θ and T_θ are shown in Figures 1.1(b) and (c).

Example 1.3.3 (Doubling map). Let $T : [0,1) \to [0,1)$ be given by

$$Tx = 2x \pmod 1 = \begin{cases} 2x, & \text{if } 0 \le x < \frac{1}{2}, \\ 2x - 1, & \text{if } \frac{1}{2} \le x < 1. \end{cases}$$

T is called the *doubling map*, the graph is shown in Figure 1.2(b). Note that the set of intervals $[a,b)$ form a generating semi-algebra for $([0,1), \mathcal{B})$, so to verify that T is measure preserving with respect to the Lebesgue measure λ, by Theorem 1.2.1 it is enough to only consider such intervals. For any interval $[a,b)$,

$$T^{-1}[a,b) = \left[\frac{a}{2}, \frac{b}{2}\right) \cup \left[\frac{a+1}{2}, \frac{b+1}{2}\right),$$

and

$$\lambda\left(T^{-1}[a,b)\right) = b - a = \lambda\left([a,b)\right).$$

Although this map is very simple, it has in fact many facets. For example, iterations of this map yield the *binary expansion* of points in $[0,1)$. In other words, using T one can associate with each point in $[0,1)$ an infinite sequence $(a_n)_{n\ge1}$ of 0's and 1's, such that $x = \sum_{n\ge1} \frac{b_n}{2^n}$. To do so, define the function b_1 by

$$b_1(x) = \begin{cases} 0, & \text{if } 0 \le x < \frac{1}{2}, \\ 1, & \text{if } \frac{1}{2} \le x < 1, \end{cases} \tag{1.2}$$

so that $Tx = 2x - b_1(x)$. Now, for $n \ge 1$ and $x \in [0,1)$ set $b_n(x) = b_1(T^{n-1}x)$. Fix $x \in [0,1)$. For simplicity we write b_n instead of $b_n(x)$. Then $T^n x = 2T^{n-1}x - b_n$. Rewriting we get $x = \frac{b_1}{2} + \frac{Tx}{2}$. Similarly, $Tx = \frac{b_2}{2} + \frac{T^2x}{2}$. Continuing in this manner, we see that for each $n \ge 1$,

$$x = \frac{b_1}{2} + \frac{b_2}{2^2} + \cdots + \frac{b_n}{2^n} + \frac{T^n x}{2^n}.$$

Since $0 \leq T^n x < 1$, this gives

$$x - \sum_{i=1}^{n} \frac{b_i}{2^i} = \frac{T^n x}{2^n} \to 0 \text{ as } n \to \infty.$$

Thus, we have found the binary expansion of x. We shall later see that the sequence of digits $b_1, b_2, \ldots$ forms an i.i.d. sequence of Bernoulli random variables.

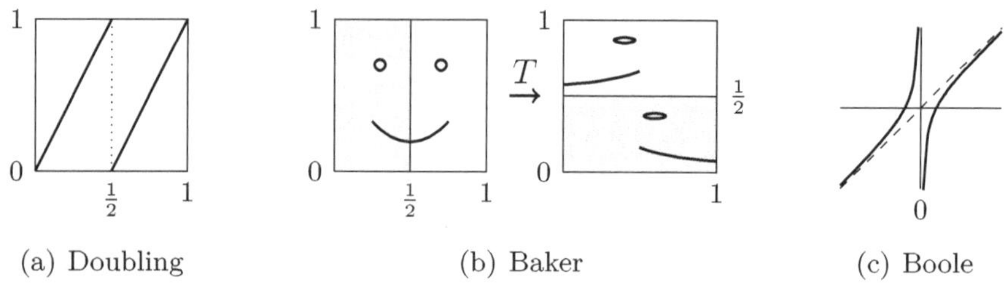

Figure 1.2 The graphs of the transformations from Example 1.3.3, Exercise 1.3.1 and Example 1.3.4.

Exercise 1.3.1 (Baker's transformation). Consider the probability space $([0,1)^2, \mathcal{B}, \lambda)$, where $\mathcal{B}$ is the product Lebesgue σ-algebra and λ is the two-dimensional Lebesgue measure. Define $T : [0,1)^2 \to [0,1)^2$ by

$$T(x,y) = \begin{cases} (2x, \frac{y}{2}), & \text{if } 0 \leq x < \frac{1}{2}, \\ (2x-1, \frac{y+1}{2}), & \text{if } \frac{1}{2} \leq x < 1. \end{cases}$$

T is called the *baker's transformation*, supposedly because the action of the map on the unit square resembles a kneading technique that bakers use when kneading dough. The graph is shown in Figure 1.2(c). Show that T is invertible, measurable and measure preserving.

Example 1.3.4 (Boole's transformation). Let $T : \mathbb{R} \to \mathbb{R}$ be given by $Tx = x - \frac{1}{x}$. This map, the graph of which is shown in Figure 1.2(c), is called *Boole's transformation.* To show that the Lebesgue measure is invariant for T, we use Theorem 1.2.1 again and we consider the inverse image of intervals. Any point $c \in \mathbb{R}$ has two pre-images $c_1 < 0$ and $c_2 > 0$ under T that are the solutions to the equation $x^2 - cx - 1 = 0$. Since these pre-images are real numbers, we can write

$$x^2 - cx - 1 = (x - c_1)(x - c_2) = 0$$

and we see that $c_1 + c_2 = c$. Hence, for any interval $(a, b) \in \mathbb{R}$ it holds that

$$\lambda(T^{-1}(a,b)) = \lambda((a_1,b_1)\cup(a_2,b_2)) = b_1 - a_1 + b_2 - a_2 = b - a = \lambda((a,b)).$$

The same statement holds for any other type of intervals and since the collection of all intervals is a generating semi-algebra for $\mathcal{B}$ on $\mathbb{R}$ and the intervals $(-n, n)$ give an exhausting sequence of finite measure intervals, λ is T-invariant.

Example 1.3.5 (Lüroth series). Besides binary expansions there are many other ways to represent real numbers. In 1883 J. Lüroth introduced in [38] a kind of number expansions that are now known as Lüroth expansions. A *Lüroth (series) expansion* is a representation of a real number x of the form

$$\begin{aligned} x = &\frac{1}{a_1} + \frac{1}{a_1(a_1-1)a_2} + \cdots \\ &+ \frac{1}{a_1(a_1-1)\cdots a_{n-1}(a_{n-1}-1)a_n} + \cdots, \end{aligned}$$

with $a_k \geq 2$ for each $k \geq 1$. The series can be finite or infinite. Lüroth showed that every $x \in (0,1)$ can be written in such a way. How is such a series generated?

The *Lüroth map* $T : [0,1) \to [0,1)$ is defined by

$$Tx = \begin{cases} n(n+1)x - n, & \text{if } x \in [\frac{1}{n+1}, \frac{1}{n}), \\ 0, & \text{if } x = 0. \end{cases} \tag{1.3}$$

The graph is shown in Figure 1.3(a). Let $x \neq 0$. For $k \geq 1$ and $T^{k-1}x \neq 0$ we define the digits $a_k = a_k(x) = n$ if $T^{k-1}x \in [\frac{1}{n}, \frac{1}{n-1})$, $n \geq 2$. Then (1.3) can be written as

$$Tx = \begin{cases} a_1(a_1-1)x - (a_1-1), & \text{if } x \neq 0, \\ 0, & \text{if } x = 0, \end{cases}$$

and for any $x \in (0,1)$ such that $Tx \neq 0$, we have

$$\begin{aligned} x = \frac{1}{a_1} + \frac{Tx}{a_1(a_1-1)} &= \frac{1}{a_1} + \frac{1}{a_1(a_1-1)}\left(\frac{1}{a_2} + \frac{T^2x}{a_2(a_2-1)}\right) \\ &= \frac{1}{a_1} + \frac{1}{a_1(a_1-1)a_2} + \frac{T^2x}{a_1(a_1-1)a_2(a_2-1)}. \end{aligned}$$

After k steps we obtain if $T^{k-1}x \neq 0$ that

$$x = \frac{1}{a_1} + \cdots + \frac{1}{a_1(a_1-1)\cdots a_{k-1}(a_{k-1}-1)a_k} + \frac{T^k x}{a_1(a_1-1)\cdots a_k(a_k-1)}.$$

Notice that, if $T^{k-1}x = 0$ for some $k \geq 1$, and if we assume that k is the smallest positive integer with this property, then

$$x = \frac{1}{a_1} + \cdots + \frac{1}{a_1(a_1-1)\cdots a_{k-1}(a_{k-1}-1)a_k} = \sum_{i=1}^{k}(a_i-1)\prod_{j=1}^{i}\frac{1}{a_j(a_j-1)}.$$

In case $T^{k-1}x \neq 0$ for all $k \geq 1$, one gets an infinite series

$$x = \frac{1}{a_1} + \frac{1}{a_1(a_1-1)a_2} + \cdots + \frac{1}{a_1(a_1-1)\cdots a_{k-1}(a_{k-1}-1)a_k} + \cdots$$
$$= \sum_{i\geq 1}(a_i-1)\prod_{j=1}^{i}\frac{1}{a_j(a_j-1)}.$$

The above infinite series indeed converges to x. To see this, note that

$$\left| x - \sum_{i=1}^{k}(a_i-1)\prod_{j=1}^{i}\frac{1}{a_j(a_j-1)} \right| = \left| \frac{T^k x}{a_1(a_1-1)\cdots a_k(a_k-1)} \right|;$$

since $T^k x \in [0,1)$ and $a_k \geq 2$ for all x and k, we find

$$\left| x - \sum_{i=1}^{k}(a_i-1)\prod_{j=1}^{i}\frac{1}{a_j(a_j-1)} \right| \leq \frac{1}{2^k} \to 0 \text{ as } k \to \infty.$$

From the above we also see that if x and y have the same Lüroth expansion, then, for each $k \geq 1$,

$$|x - y| \leq \frac{1}{2^{k-1}}$$

and it follows that x equals y. So a Lüroth expansion uniquely determines a real number.

Exercise 1.3.2. Show that the Lüroth map T from Example 1.3.5 is measure preserving with respect to Lebesgue measure λ.

Example 1.3.6 (β-transformation). Let $\beta \in (1,2)$. The *β-transformation* $T : [0,1) \to [0,1)$ is defined by

$$Tx = \beta x \pmod 1 = \begin{cases} \beta x, & \text{if } 0 \le x < \frac{1}{\beta}, \\ \beta x - 1, & \text{if } \frac{1}{\beta} \le x < 1. \end{cases}$$

Figure 1.3(b) shows the graph of T in case β equals $G = \frac{1+\sqrt{5}}{2}$, the *golden mean*. The map T is **not** measure preserving with respect to Lebesgue measure λ, but it is non-singular. To see this, note that for any measurable set $A \in \mathcal{B}$ we can write

$$T^{-1}A = \frac{A}{\beta} \cup \frac{(A \cap [0, \beta - 1)) + 1}{\beta},$$

where the union is disjoint. The non-singularity then follows by the translation and scaling invariance of λ. In case $\beta = G$ (notice that $G^2 = G+1$) an invariant measure μ for T that is equivalent to λ is given by

$$\mu(B) = \int_B g\, d\lambda, \quad \text{for all } B \in \mathcal{B},$$

where

$$g(x) = \begin{cases} \frac{5+3\sqrt{5}}{10}, & \text{if } 0 \le x < \frac{1}{\beta}, \\ \frac{5+\sqrt{5}}{10}, & \text{if } \frac{1}{\beta} \le x < 1. \end{cases} \tag{1.4}$$

Exercise 1.3.3. (a) Show that for any $\beta \in (1,2)$ Lebesgue measure is not invariant for T and that (similar to Example 1.3.3) iterations of this map generate expansions for points $x \in [0,1)$ of the form

$$x = \sum_{i=1}^{\infty} \frac{b_i}{\beta^i}, \tag{1.5}$$

where $b_i \in \{0,1\}$ and $b_i b_{i+1} = 0$ for all $i \ge 1$. Expansions as in (1.5) are called *β-expansions.*

(b) Verify that μ from above is an invariant measure for T in case $\beta = G$.

Example 1.3.7 (Continued fractions). The *Gauss map* $T : [0,1) \to [0,1)$ is defined by $T0 = 0$ and for $x \ne 0$,

$$Tx = \frac{1}{x} \pmod 1.$$

Figure 1.3(c) shows the graph.

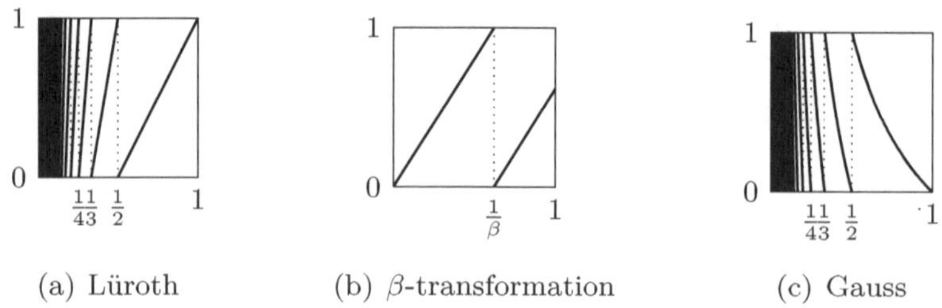

Figure 1.3 The graphs of the transformations from Examples 1.3.5, 1.3.6 and 1.3.7.

Exercise 1.3.4. The *Gauss probability measure* μ is given by

$$\mu(B) = \int_B \frac{1}{\log 2} \frac{1}{1+x}\, d\lambda(x), \quad \text{for all } B \in \mathcal{B}. \tag{1.6}$$

Show that T is **not** measure preserving with respect to Lebesgue measure, but is measure preserving with respect to μ.

An interesting feature of this map is that its iterations generate *continued fraction expansions* for points in $(0,1)$. For if we define for each $k \geq 1$

$$a_k = a_k(x) = \begin{cases} 1, & \text{if } T^{k-1}x \in (\frac{1}{2}, 1), \\ n, & \text{if } T^{k-1}x \in (\frac{1}{n+1}, \frac{1}{n}],\ n \geq 2, \end{cases}$$

then for $x \neq 0$ we can write $Tx = \frac{1}{x} - a_1$ and hence $x = \dfrac{1}{a_1 + Tx}$. After n iterations, if $T^{n-1}x \neq 0$ we see that

$$x = \frac{1}{a_1 + Tx} = \cdots = \cfrac{1}{a_1 + \cfrac{1}{a_2 + \ddots + \cfrac{1}{a_n + T^n x}}}.$$

If $T^n x = 0$ for some n, then we get a finite continued fraction expansion of x. If, on the other hand, $T^n x \neq 0$ for all n, then we can continue this process indefinitely. In fact, in Chapter 8 we show that if we write

$$\frac{p_n}{q_n} = \cfrac{1}{a_1 + \cfrac{1}{a_2 + \ddots + \cfrac{1}{a_n}}},$$

then the sequence (q_n) is monotonically increasing, and $|x - \frac{p_n}{q_n}| < \frac{1}{q_n^2} \to 0$ as $n \to \infty$. The last statement implies that

$$x = \cfrac{1}{a_1 + \cfrac{1}{a_2 + \cfrac{1}{a_3 + \cfrac{1}{\ddots}}}}.$$

Example 1.3.8 (Arnold's cat map). Let $\mathbb{T}^2 = \mathbb{R}^2/\mathbb{Z}^2$ be the two-dimensional torus. The map $T : X \to X$ defined by

$$T(x,y) = (2x + y \pmod 1, x + y \pmod 1) = \begin{pmatrix} 2 & 1 \\ 1 & 1 \end{pmatrix} \begin{pmatrix} x \\ y \end{pmatrix} \pmod{\mathbb{Z}^2}$$

is called *Arnold's cat map.* The name refers to the fact that V. Arnold and A. Avez used a picture of a cat to illustrate the dynamics of this map in [3][1]. The graph of the map is shown in Figure 1.4. The map is invertible, since the matrix $\begin{pmatrix} 2 & 1 \\ 1 & 1 \end{pmatrix}$ has determinant 1. The scaling invariance of the Lebesgue measure together with Proposition 1.2.2 then give that Lebesgue measure λ is T-invariant.

The map T is a specific example of a *hyperbolic toral automorphism.* For any $n \times n$ matrix M, $n \geq 2$, with integer entries and determinant $|det\, M| = 1$ we can define the corresponding *toral automorphism* $T_M : \mathbb{T}^n \to \mathbb{T}^n$ by $T_M x = Mx \pmod{\mathbb{Z}^n}$. Since M maps points in $\mathbb{Z}^n$ to points in $\mathbb{Z}^n$, for any two points $x, y \in \mathbb{R}^n$ that have the same representative in $\mathbb{T}^n$ it holds that $Mx \pmod{\mathbb{Z}^n} = My \pmod{\mathbb{Z}^n}$. Hence, the map T_M is well defined and invertible and Lebesgue measure is invariant. In case none of the eigenvalues of M are on the unit circle, T_M is called a hyperbolic toral automorphism.

The next couple of examples have a symbolic space as a state space, i.e., $X = \{0, 1, \ldots, k-1\}^{\mathbb{Z}}$ or $X = \{0, 1, \ldots, k-1\}^{\mathbb{N}}$ is the set of two-

[1] Another version of the story says that CAT is an abbreviation of Chaotic Automorphism of a Torus (in Russian) and that the picture of a cat was later associated to the cat map due to the appearance of the map in an earlier text by V. Arnold in which the action of another dynamical system was illustrated using the picture of a cat.

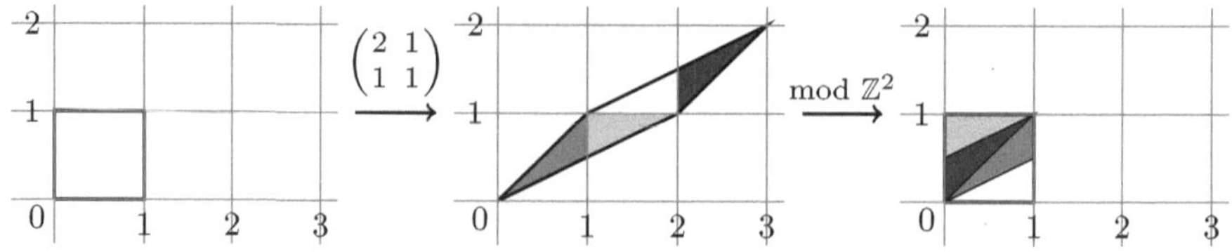

Figure 1.4 An illustration of Arnold's cat map.

or one-sided infinite sequences of symbols $0, 1, \ldots, k-1$. On such spaces we consider the product σ-algebra $\mathcal{C}$ generated by the *cylinder sets*:

$$[a_i \cdots a_j] := \{x \in X : x_i = a_i, \ldots, x_j = a_j\} \tag{1.7}$$

with $a_i, \ldots, a_j \in \{0, 1, \ldots, k-1\}$ and $i \leq j$ (and $i > 0$ in the one-sided case). For one-sided sequences the collection of all cylinders of the form $[a_1 \cdots a_j]$ for $j \geq 0$ (with the empty set for $j = 0$) forms a generating semi-algebra and for two-sided sequences this is obtained from all cylinders of the form $[a_{-i}a_{-i+1} \cdots a_{j-1}a_j]$ with $i, j \geq 0$ together with the empty set. By Theorem 1.2.1 it is enough check measurability and measure preservingness on such sets only.

Example 1.3.9 (Bernoulli shifts). Let $X = \{0, 1, \ldots, k-1\}^{\mathbb{Z}}$. Let $p = (p_0, p_1, \ldots, p_{k-1})$ be a positive probability vector. The measure μ on $\mathcal{C}$ defined by specifying it on cylinder sets by

$$\mu(\{x : x_i = a_i, \ldots, x_j = a_j\}) = p_{a_i} \cdots p_{a_j}$$

is called the *p-Bernoulli measure.* If all p_i are equal and thus equal to $\frac{1}{k}$, then μ is called the *uniform Bernoulli measure.* The transformation $T : X \to X$ given by $Tx = y$, where $y_n = x_{n+1}$ for all n, is called the *left shift.* It is measurable and measure preserving, since

$$T^{-1}\{x : x_i = a_i, \ldots, x_j = a_j\} = \{x : x_{i+1} = a_i, \ldots, x_{j+1} = a_j\}, \tag{1.8}$$

and

$$\mu(\{x : x_{i+1} = a_i, \ldots, x_{j+1} = a_j\}) = p_{a_i} \cdots p_{a_j}.$$

The system $(X, \mathcal{C}, \mu, T)$ is called the *two-sided Bernoulli shift* on k symbols. The *one-sided Bernoulli shift* with $X = \{0, 1, \ldots, k-1\}^{\mathbb{N}}$ is set up in exactly the same way. Note that for the proof of measurability and measure preservingness the inverse image of a cylinder as in (1.8) in this case becomes a union of k disjoint cylinder sets.

Example 1.3.10 (Markov shifts). Let $(X, \mathcal{C}, T)$ be as in Example 1.3.9. We now define a measure ν on $\mathcal{C}$ as follows. Let $P = (p_{ij})$ be a stochastic $k \times k$ matrix, and $q = (q_0, q_1, \ldots, q_{k-1})$ a positive probability vector such that $qP = q$. Define ν on cylinders by

$$\nu(\{x : x_i = a_i, \ldots, x_j = a_j\}) = q_{a_i} p_{a_i a_{i+1}} \cdots p_{a_{j-1} a_j}.$$

The measure ν is called a *Markov measure.* Just as in Example 1.3.9, one sees that T is measurable and measure preserving.

Example 1.3.11 (Binary odometer). On $\{0,1\}^{\mathbb{N}}$ consider the $(p, 1-p)$-Bernoulli measure μ_p for some $0 < p < 1$, that is

$$\mu_p(\{x : x_1 = a_1, \ldots, x_n = a_n\}) = p^{n - \sum_{i=1}^n a_i}(1-p)^{\sum_{i=1}^n a_i}.$$

The transformation $T : \{0,1\}^{\mathbb{N}} \to \{0,1\}^{\mathbb{N}}$ given by $T(1,1,1,\ldots) = (0,0,0,\ldots)$ and for $n \geq 1$,

$$T(\underbrace{1, \ldots, 1}_{n-1}, 0, x_{n+1}, \ldots) = (\underbrace{0, \ldots, 0}_{n-1}, 1, x_{n+1}, \ldots)$$

is called the *binary odometer.* We will show that T is a non-singular transformation. First note that T is invertible, so it is enough to check that μ_p and $\mu_p \circ T$ are equivalent measures. Let $d(x) = \inf\{n \geq 1 : x_n = 0\}$ and set $m(x) = d(x) - 2$. Note that $d(x) = \infty$ only for the constant sequence $(1,1,1,\ldots)$. Consider any cylinder set $A \subseteq \{x \in X : d(x) = k\}$, then

$$A = \{x \in X : x_1 = \cdots = x_{k-1} = 1, x_k = 0, x_{k+1} = \ell_1, \ldots, x_{k+j} = \ell_j\}$$

for some $\ell_1, \ldots, \ell_j \in \{0,1\}$. We have

$$TA = \{x \in X : x_1 = \cdots = x_{k-1} = 0, x_k = 1, x_{k+1} = \ell_1, \ldots, x_{k+j} = \ell_j\},$$

and

$$\begin{aligned}\mu_p(TA) &= p^{k-1}(1-p)\mu_p(\{x : x_{k+1} = \ell_1, \ldots, x_{k+j} = \ell_j\}) \\ &= \left(\frac{p}{1-p}\right)^{k-2} \mu_p(A).\end{aligned}$$

Note that for all $x \in A$ one has $m(x) = k - 2$, hence the last expression can be written as

$$\mu_p(TA) = \int_A \left(\frac{p}{1-p}\right)^{m(x)} d\mu_p(x).$$

This shows that $\frac{d\mu_p \circ T}{d\mu_p}(x) = \left(\frac{p}{1-p}\right)^{m(x)}$ μ_p-a.e. and hence μ_p and $\mu_p \circ T$ are equivalent measures. So T is non-singular with respect to μ_p for all p. Notice that $\frac{d\mu_p \circ T}{d\mu_p} = 1$ μ_p-a.e. and so T is measure preserving with respect to μ_p if and only if $p = \frac{1}{2}$.

We end this section with two examples of symbolic systems of a more probabilistic flavor.

Example 1.3.12 (Stationary stochastic processes). Let $(\Omega, \mathcal{F}, \mathbb{P})$ be a probability space, and

$$\ldots, Y_{-2}, Y_{-1}, Y_0, Y_1, Y_2, \ldots$$

a stationary stochastic process on Ω with values in $\mathbb{R}$. Hence, for each $i \in \mathbb{Z}$, $Y_i : \Omega \to \mathbb{R}$ is a random variable (or measurable function) and for each $k \in \mathbb{Z}$,

$$\mathbb{P}\left(Y_{n_1} \in B_1, \ldots, Y_{n_r} \in B_r\right) = \mathbb{P}\left(Y_{n_1+k} \in B_1, \ldots, Y_{n_r+k} \in B_r\right)$$

for any $n_1 < n_2 < \cdots < n_r$ and any Lebesgue sets $B_1, \ldots, B_r$. We can see this process as coming from a measure preserving transformation in the following way.

Let $X = \mathbb{R}^{\mathbb{Z}} = \{x = (\ldots, x_1, x_0, x_1, \ldots) : x_i \in \mathbb{R}\}$ with the product σ-algebra (i.e., generated by the cylinder sets). Let $T : X \to X$ be the left shift. Define $\Psi : \Omega \to X$ by

$$\Psi(\omega) = (\ldots, Y_{-2}(\omega), Y_{-1}(\omega), Y_0(\omega), Y_1(\omega), Y_2(\omega), \ldots).$$

Then, Ψ is measurable since if $B_1, \ldots, B_r$ are Lebesgue sets in $\mathbb{R}$, then

$$\Psi^{-1}\left(\{x \in X : x_{n_1} \in B_1, \ldots x_{n_r} \in B_r\}\right) = Y_{n_1}^{-1}(B_1) \cap \ldots \cap Y_{n_r}^{-1}(B_r) \in \mathcal{F}.$$

Define a measure μ on X by $\mu(E) = \mathbb{P}\left(\Psi^{-1}(E)\right)$ for any $E \in \mathcal{F}$. On cylinder-type sets μ has the form,

$$\mu\left(\{x \in X : x_{n_1} \in B_1, \ldots, x_{n_r} \in B_r\}\right) = \mathbb{P}\left(Y_{n_1} \in B_1, \ldots, Y_{n_r} \in B_r\right).$$

Since

$$T^{-1}(\{x : x_{n_1} \in B_1, \ldots, x_{n_r} \in B_r\}) \\ = \{x : x_{n_1+1} \in B_1, \ldots, x_{n_r+1} \in B_r\},$$

the stationarity of the process (Y_n) implies that T is measure preserving. Furthermore, if we let $\pi_i : X \to \mathbb{R}$ be the natural projection onto the i-th coordinate, then $Y_i(\omega) = \pi_i(\Psi(\omega)) = \pi_0 \circ T^i(\Psi(\omega))$.

Example 1.3.13 (Random shifts). Let $(X, \mathcal{F}, \mu)$ be a measure space, and $T : X \to X$ an invertible measure preserving transformation. Then, T^{-1} is measurable and measure preserving with respect to μ. Suppose now that at each moment instead of moving forward by T $(x \to Tx)$, we first flip a fair coin to decide whether we will use T or T^{-1}. We can describe this random system by means of a measure preserving transformation in the following way.

Let $\Omega = \{-1, 1\}^{\mathbb{Z}}$ with product σ-algebra $\mathcal{C}$, the uniform Bernoulli measure $\mathbb{P}$ and the left shift $S : \Omega \to \Omega$, which, as we saw in Example 1.3.9, is measure preserving. Now, let $Y = \Omega \times X$ with the product σ-algebra, and product measure $\mathbb{P} \times \mu$. Define $R : Y \to Y$ by

$$R(\omega, x) = (S\omega, T^{\omega_0} x).$$

Then R is invertible (why?), and measure preserving with respect to $\mathbb{P} \times \mu$. To see the latter, for any set $C \in \mathcal{C}$, and any $A \in \mathcal{F}$, we have

$$\begin{aligned}
&(\mathbb{P} \times \mu)\left(R^{-1}(C \times A)\right) \\
&\quad = (\mathbb{P} \times \mu)\left(\{(\omega, x) : R(\omega, x) \in (C \times A)\right) \\
&\quad = (\mathbb{P} \times \mu)\left(\{(\omega, x) : \omega_0 = 1, S\omega \in C, Tx \in A\right) \\
&\qquad + (\mathbb{P} \times \mu)\left(\{(\omega, x) : \omega_0 = -1, S\omega \in C, T^{-1}x \in A\right) \\
&\quad = (\mathbb{P} \times \mu)\left(\{\omega_0 = 1\} \cap S^{-1}C \times T^{-1}A\right) \\
&\qquad + (\mathbb{P} \times \mu)\left(\{\omega_0 = -1\} \cap S^{-1}C \times TA\right) \\
&\quad = \mathbb{P}\left(\{\omega_0 = 1\} \cap S^{-1}C\right) \mu\left(T^{-1}A\right) \\
&\qquad + \mathbb{P}\left(\{\omega_0 = -1\} \cap S^{-1}C\right) \mu\left(TA\right) \\
&\quad = \mathbb{P}\left(\{\omega_0 = 1\} \cap S^{-1}C\right) \mu(A) + \mathbb{P}\left(\{\omega_0 = -1\} \cap S^{-1}C\right) \mu(A) \\
&\quad = \mathbb{P}(S^{-1}C)\mu(A) = \mathbb{P}(C)\mu(A) = (\mathbb{P} \times \mu)(C \times A).
\end{aligned}$$

CHAPTER 2

Recurrence and Ergodicity

2.1 RECURRENCE

One of the most general statements in ergodic theory is the Poincaré Recurrence Theorem. It was formulated and discussed by H. Poincaré in 1890 (see [51]) and it is one of the first results that uses a measure theoretic approach in the study of dynamical systems. For any subset B of the state space X a point $x \in B$ is said to be *B-recurrent* if it eventually returns to B under iterations of T, i.e., if there exists a $k \geq 1$ such that $T^k x \in B$.

Theorem 2.1.1 (Poincaré Recurrence Theorem). *Let $(X, \mathcal{F}, \mu)$ be a probability space and $T : X \to X$ a measure preserving transformation. Let $B \in \mathcal{F}$ with $\mu(B) > 0$. Then μ-a.e. $x \in B$ is B-recurrent.*

Proof. Let N be the subset of B consisting of all elements that are not B-recurrent. Then,

$$N = \{x \in B \, : \, T^k x \notin B \text{ for all } k \geq 1\}.$$

We want to show that $\mu(N) = 0$. First notice that $N \cap T^{-k}N = \emptyset$ for all $k \geq 1$, hence $T^{-\ell}N \cap T^{-m}N = \emptyset$ for all $\ell \neq m$. Thus, the sets $N, T^{-1}N, T^{-2}N, \ldots$ are pairwise disjoint, and $\mu(T^{-n}N) = \mu(N)$ for all $n \geq 1$ (T is measure preserving). If $\mu(N) > 0$, then

$$1 = \mu(X) \geq \mu\Big(\bigcup_{k \geq 0} T^{-k}N\Big) = \sum_{k \geq 0} \mu(N) = \infty,$$

a contradiction. □

The proof of the Poincaré Recurrence Theorem implies that almost every $x \in B$ returns to B infinitely often. In other words, there exist infinitely many integers $n_1 < n_2 < \ldots$ such that $T^{n_i}x \in B$ for all $i \geq 1$. Recall that we can write

$$\{x \in B : T^n x \in B \text{ i.o.}\} = \bigcap_{n \geq 0} \bigcup_{k \geq n} T^{-k}B.$$

To see that $\mu(\{x \in B : T^n x \in B \text{ i.o.}\}) = \mu(B)$, let

$$D = \{x \in B : T^k x \in B \text{ for finitely many } k \geq 1\}.$$

Then,

$$D = \{x \in B : T^k x \in N \text{ for some } k \geq 0\} \subseteq \bigcup_{k=0}^{\infty} T^{-k}N.$$

Thus, $\mu(D) = 0$ since $\mu(N) = 0$ and T is non-singular.

The assumption that the measure of the space $\mu(X)$ is finite is essential in the proof. It is not hard to see that the statement does not necessarily hold for infinite measure systems. Recall the translation $Tx = x+t$ on $\mathbb{R}$ from Example 1.3.1 and assume that $t > 0$. We saw that T is measure preserving with respect to the Lebesgue measure. The interval $[0, t)$ has positive Lebesgue measure, but $\lambda(T^n[0, t) \cap [0, t)) = 0$ for all $n \geq 1$. In other words, no element from $[0, t)$ ever returns to $[0, t)$ under iterations of T.

Definition 2.1.1. Let $(X, \mathcal{F}, \mu)$ be a measure space. A set $W \in \mathcal{F}$ is called *wandering* for a transformation $T : X \to X$ if $\mu(T^{-n}W \cap T^{-m}W) = 0$ for all $0 \leq n < m$. We denote the collection of all wandering sets of T by $\mathcal{W}_T$.

In other words, a set W is wandering for T if the collection $\{T^{-n}W : n \geq 0\}$ is essentially pairwise disjoint. Note that any subset of a wandering set is wandering itself. In the example above the interval $[0, t)$ is wandering and so is any other interval of length at most t.

Definition 2.1.2. Let $(X, \mathcal{F}, \mu)$ be a measure space. A transformation $T : X \to X$ is called *conservative* if $\mu(W) = 0$ for all $W \in \mathcal{W}_T$.

Note that if T is a measure preserving transformation and W is a wandering set, then

$$\mu(X) \geq \mu\Big(\bigcup_{n \geq 0} T^{-n}W\Big) = \sum_{n \geq 0} \mu(T^{-n}W).$$

This shows that any measure preserving transformation on a probability space is automatically conservative.

If T is conservative, then no set $B \in \mathcal{F}$ with $\mu(B) > 0$ can be wandering. So, if $\mu(B) > 0$, then there are $0 \leq n < m$, such that $\mu(T^{-n}B \cap T^{-m}B) > 0$. The non-singularity of T then gives that $\mu(B \cap T^{-(m-n)}B) > 0$. So, if T is conservative, then from any positive measure set at least a positive measure part returns to the set after some time. It is a consequence of the following more general recurrence theorem, which is due to P. R. Halmos [20], that conservativity actually guarantees that a system satisfies the statement from the Poincaré Recurrence Theorem.

Theorem 2.1.2 (Halmos Recurrence Theorem). *Let $(X, \mathcal{F}, \mu)$ be a measure space and $T : X \to X$ a transformation. Then for every $A \in \mathcal{F}$ the following property holds: $\mu(A \cap W) = 0$ for all $W \in \mathcal{W}_T$ if and only if for all measurable sets $B \subseteq A$,*

$$\mu(\{x \in B \,:\, T^n x \in B \ \ i.o.\}) = \mu(B). \tag{2.1}$$

Proof. Assume first that $\mu(A \cap W) = 0$ for all wandering sets W. Fix a measurable set $B \subseteq A$ and let

$$N = \{x \in B \,:\, T^k x \notin B \text{ for all } k \geq 1\} = B \setminus \bigcup_{k \geq 1} T^{-k}B$$

be the set of points in B that never return to B. It suffices to show that

$$\mu\Big(B \setminus \bigcap_{k \geq 0} \bigcup_{n \geq k} T^{-n}B\Big) = 0.$$

From the definition of N it follows that $T^{-n}N \cap T^{-m}N = \emptyset$ for all $n \neq m$, so N is wandering and by assumption $\mu(N) = \mu(N \cap A) = 0$. The non-singularity of T then implies that for each $k \geq 0$,

$$0 = \mu(T^{-k}N) = \mu\Big(T^{-k}B \setminus \bigcup_{n \geq k+1} T^{-n}B\Big),$$

so also $\mu\big(\bigcup_{n \geq k} T^{-n}B \setminus \bigcup_{n \geq k+1} T^{-n}B\big) = 0$. Since $\bigcup_{n \geq k+1} T^{-n}B \subseteq \bigcup_{n \geq k} T^{-n}B$ for all k, from this we obtain inductively, starting from $B \subseteq \bigcup_{n \geq 0} T^{-n}B$, that

$$\mu\Big(B \setminus \bigcup_{n \geq k} T^{-n}B\Big) = 0$$

for all $k \geq 0$ and the result follows.

For the other direction, let W be a wandering set and assume that $\mu(A \cap W) > 0$. Set $B = A \cap W$. Then $B \subseteq A$, so by assumption

$$\mu(\{x \in B : T^n x \in B \text{ i.o.}\}) = \mu(B) > 0.$$

On the other hand B is wandering, so no positive measure set of points can return to B. This gives a contradiction. □

Note that if T is conservative, we can take $A = X$ in Theorem 2.1.2 to get the statement from the Poincaré Recurrence Theorem for every set $B \in \mathcal{F}$. Conversely, if (2.1) holds for all $B \in \mathcal{F}$, then $\mu(W) = 0$ for every wandering set W and T is conservative. Hence, conservativity can be characterized as follows.

Corollary 2.1.1. *Let $(X, \mathcal{F}, \mu)$ be a measure space and $T : X \to X$ a transformation. Then T is conservative if and only if for each $B \in \mathcal{F}$ it holds that μ-almost every $x \in B$ returns to B infinitely often.*

If T is invertible, then we can replace T by T^{-1} in the above results and obtain that for μ-a.e. $x \in B$ there are infinitely many positive and negative integers n such that $T^n x \in B$.

Note that since any measure preserving transformation on a probability space is conservative, the Poincaré Recurrence Theorem follows from this corollary. We stated and proved it separately, because of its historical importance.

2.2 ERGODICITY

The right condition under which Boltzmann's Hypothesis is true, turned out to be ergodicity. Ergodicity is an irreducibility condition defined as follows. Recall that if A and B are measurable sets, then their *symmetric difference* is defined by

$$A \Delta B = (A \cup B) \setminus (A \cap B) = (A \setminus B) \cup (B \setminus A).$$

We use the notation A^c for the *complement* $X \setminus A$ of A in X.

Definition 2.2.1. Let $T : X \to X$ be a transformation on a measure space $(X, \mathcal{F}, \mu)$. The map T is said to be *ergodic* if for every measurable set A satisfying $\mu(A \Delta T^{-1} A) = 0$, we have $\mu(A) = 0$ or $\mu(A^c) = 0$.

We can actually replace essential invariance by total invariance, using the following small exercise.

Exercise 2.2.1. Let $(X, \mathcal{F}, \mu)$ be a measure space.

(a) Show that for any measurable sets A, B, C one has $\mu(A \Delta B) \leq \mu(A \Delta C) + \mu(C \Delta B)$.

(b) Let $T : X \to X$ be a transformation and let $A \in \mathcal{F}$ satisfy $\mu(A \Delta T^{-1}A) = 0$. Prove that then $\mu(A \Delta T^{-k}A) = 0$ for each $k \geq 1$.

Proposition 2.2.1. *Let $T : X \to X$ be a transformation on a measure space $(X, \mathcal{F}, \mu)$. Then T is ergodic if and only if for every measurable set A satisfying $T^{-1}A = A$, we have $\mu(A) = 0$ or $\mu(A^c) = 0$.*

Proof. One direction is immediate. For the other direction, let $A \in \mathcal{F}$ be such that $\mu(A \Delta T^{-1}A) = 0$. Set $B = \{x \in X : T^n x \in A \text{ i.o.}\}$. Then obviously $T^{-1}B = B$, so by assumption $\mu(B) = 0$ or $\mu(B^c) = 0$. Furthermore,

$$\begin{aligned}
\mu(A \Delta B) &= \mu\Big(\bigcap_{n\geq 1}\bigcup_{k\geq n} T^{-k}A \cap A^c\Big) + \mu\Big(\bigcup_{n\geq 1}\bigcap_{k\geq n} T^{-k}A^c \cap A\Big) \\
&\leq \mu\Big(\bigcup_{k\geq 1} T^{-k}A \cap A^c\Big) + \mu\Big(\bigcup_{k\geq 1} T^{-k}A^c \cap A\Big) \\
&\leq \sum_{k\geq 1} \mu(T^{-k}A \Delta A).
\end{aligned}$$

It follows from Exercise 2.2.1 that $\mu\left(T^{-k}A \Delta A\right) = 0$ for each $k \geq 1$. Hence, $\mu(B \Delta A) = 0$ which implies that $\mu(B) = \mu(A)$. Therefore, $\mu(A) = 0$ or $\mu(A^c) = 0$. □

To illustrate the concept, consider the following example of a transformation that is **not** ergodic.

Example 2.2.1. Let $T : [0, 1] \to [0, 1]$ be given by

$$Tx = \begin{cases} 2x, & \text{if } 0 \leq x < \frac{1}{4}, \\ 2x - \frac{1}{2}, & \text{if } \frac{1}{4} \leq x < \frac{3}{4}, \\ 2x - 1, & \text{if } \frac{3}{4} \leq x \leq 1. \end{cases}$$

See Figure 2.1 for the graph. One readily checks that Lebesgue measure λ is invariant for T, but for $A = [0, \frac{1}{2})$ we see that $T^{-1}A = A$ and $0 < \lambda(A) < 1$. The dynamics of T splits into two independent parts: the orbit of any $x \in A$ is completely contained in A and similarly for points in A^c.

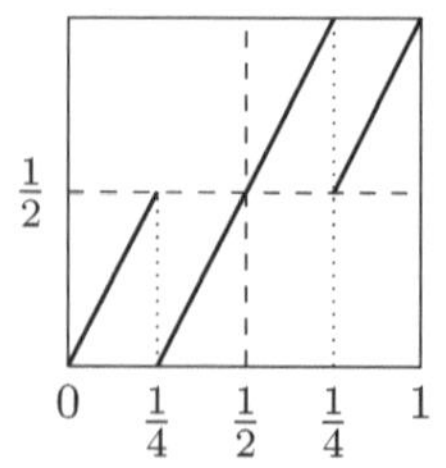

Figure 2.1 The graph of the transformation from Example 2.2.1.

The following proposition says that in a conservative and ergodic system, any positive measure set is eventually visited by almost all points in the space.

Proposition 2.2.2. *Let $(X, \mathcal{F}, \mu)$ be a measure space and let $A \in \mathcal{F}$ with $\mu(A) > 0$ be given. If $T : X \to X$ is a conservative and ergodic transformation, then $\mu(X \setminus \bigcup_{n\geq 0} T^{-n}A) = 0$.*

Proof. Let $\mu(A) > 0$ and set $B = X \setminus \bigcup_{n\geq 0} T^{-n}A = \bigcap_{n\geq 0} T^{-n}A^c$. By definition $B \subseteq T^{-1}B$ and $B = A^c \cap T^{-1}B$. The set $T^{-1}B$ contains precisely those elements of X that do not enter A under any iteration of T. Hence, by conservativity Corollary 2.1.1 gives that $\mu(A \cap T^{-1}B) = 0$. This yields $\mu(T^{-1}B) = \mu(A^c \cap T^{-1}B) = \mu(B)$, so $\mu(B \Delta T^{-1}B) = \mu(T^{-1}B \setminus B) = \mu(A \cap T^{-1}B) = 0$. By ergodicity then $\mu(B) = 0$ or $\mu(B^c) = 0$. Since $A \subseteq B^c$, we get the result. □

Remark 2.2.1. Note that the non-singularity of T implies that if $\mu(X \setminus \bigcup_{n\geq 0} T^{-n}A) = 0$, then

$$\mu(X \setminus \bigcup_{n\geq k} T^{-n}A) = \mu(T^{-k}(X \setminus \bigcup_{n\geq 0} T^{-n}A)) = 0$$

for each $k \geq 0$ and hence,

$$\begin{aligned} 0 \leq \mu(X \setminus \bigcap_{k\geq 0} \bigcup_{n\geq k} T^{-n}A) &= \mu(\bigcup_{k\geq 0}(X \setminus \bigcup_{n\geq k} T^{-n}A)) \\ &\leq \sum_{k\geq 0} \mu(X \setminus \bigcup_{n\geq k} T^{-n}A) = 0. \end{aligned}$$

So μ-a.e. $x \in X$ visits a set A eventually under iterations of T if and only if μ-a.e. $x \in X$ visits A infinitely often under T.

Proposition 2.2.2 can be strengthened.

Proposition 2.2.3. *A transformation $T : X \to X$ on a measure space $(X, \mathcal{F}, \mu)$ is conservative and ergodic if and only if for each $f \in L^1(X, \mathcal{F}, \mu)$ with $f \geq 0$ and $\int_X f \, d\mu > 0$ it holds that $\sum_{n \geq 0} f(T^n x) = \infty$ for μ-almost all $x \in X$.*

Proof. Assume first that T is conservative and ergodic and let $f \in L^1(X, \mathcal{F}, \mu)$ with $f \geq 0$ and $\int_X f \, d\mu > 0$ be given. Then there must be an $\varepsilon > 0$, such that $\mu(\{x \in X : f(x) > \varepsilon\}) > 0$. Set $A = \{x \in X : f(x) > \varepsilon\}$. From Proposition 2.2.2 we obtain as in Remark 2.2.1 that

$$\mu\Big(X \setminus \bigcap_{n \geq 0} \bigcup_{k \geq n} T^{-k} A\Big) = 0,$$

which is equivalent to

$$\sum_{n \geq 0} 1_A(T^n x) = \infty \quad \mu - \text{a.e.}$$

By definition of A this implies the statement. For the other direction, first let $W \in \mathcal{F}$ be a wandering set for T with $\mu(W) > 0$. Then $\int_X 1_W \, d\mu > 0$, but $\sum_{n \geq 0} 1_W(T^n x) \leq 1$ for μ-a.e. $x \in X$ contradicting the assumption made. Hence $\mu(W) = 0$ and T is conservative. Now, let $A \in \mathcal{F}$ be a T-invariant set with $\mu(A) > 0$. Then by assumption

$$\sum_{n \geq 0} 1_A(x) = \sum_{n \geq 0} 1_{T^{-n}A}(x) = \sum_{n \geq 0} 1_A(T^n x) = \infty \quad \mu\text{-a.e.},$$

so μ-a.e. $x \in X$ is an element of A and thus $\mu(X \setminus A) = 0$. Hence, T is ergodic. □

The following theorem contains several equivalent characterizations of ergodicity in case T is conservative.

Theorem 2.2.1. *Let $(X, \mathcal{F}, \mu)$ be a measure space and $T : X \to X$ conservative. The following are equivalent.*

(i) *T is ergodic.*

(ii) *If $A \in \mathcal{F}$ with $\mu(A) > 0$, then $\mu\left(X \setminus \bigcup_{n \geq 1} T^{-n} A\right) = 0$.*

(iii) *If $A, B \in \mathcal{F}$ with $\mu(A)\mu(B) > 0$, then there exists an $n > 0$ such that $\mu(T^{-n} A \cap B) > 0$.*

Remark 2.2.2. (1) In case T is invertible, then in the above characterization one can replace T^{-n} by T^n.

(2) In words, (ii) says that if A is a set of positive measure, almost every $x \in X$ will eventually (in fact infinitely often) visit A.

(3) Item (iii) says that a positive measure set of elements from B will eventually enter A.

Proof of Theorem 2.2.1. (i) $\Rightarrow$ (ii) This follows from Proposition 2.2.2 and the non-singularity of T.

(ii) $\Rightarrow$ (iii) Let $A, B \in \mathcal{F}$ be such that $\mu(A)\mu(B) > 0$. From $\mu\left(X \setminus \bigcup_{n\geq 1} T^{-n}A\right) = 0$ we can conclude that

$$0 < \mu(B) = \mu\Big(B \cap \bigcup_{n\geq 1} T^{-n}A\Big) = \mu\Big(\bigcup_{n\geq 1}(B \cap T^{-n}A)\Big).$$

Hence, there exists an $n \geq 1$ such that $\mu(B \cap T^{-n}A) > 0$.

(iii) $\Rightarrow$ (i) Let $A \in \mathcal{F}$ be such that $\mu(A \Delta T^{-1}A) = 0$ and $\mu(A) > 0$. If $\mu(A^c) > 0$, then by (iii) there exists an $n \geq 1$ such that $\mu(A^c \cap T^{-n}A) > 0$. On the other hand, by Exercise 2.2.1,

$$0 = \mu(A \Delta T^{-n}A) = \mu(A^c \cap T^{-n}A) + \mu(A \cap T^{-n}A^c),$$

a contradiction. Hence, $\mu(A^c) = 0$ and T is ergodic. □

We can obtain several other characterizations of ergodicity using the Koopman operator. Recall its definition from (1.1).

Theorem 2.2.2. *Let $(X, \mathcal{F}, \mu)$ be a measure space, and $T : X \to X$ a transformation. The following are equivalent.*

(i) *T is ergodic.*

(ii) *If $f \in L^0(X, \mathcal{F}, \mu)$ with $U_T f = f$ μ-a.e., then f is a constant μ-a.e.*

In addition, if $(X, \mathcal{F}, \mu)$ is a probability space, then (i) and (ii) are equivalent to

(iii) *If $f \in L^2(X, \mathcal{F}, \mu)$ with $U_T f = f$ μ-a.e., then f is a constant μ-a.e.*

Proof. (i) $\Rightarrow$ (ii) Suppose $f \in L^0(X, \mathcal{F}, \mu)$ satisfies $U_T f = f$ μ-a.e. and assume without any loss of generality that f is real (otherwise we consider separately the real and imaginary parts of f). For each $a \in \mathbb{R}$, let $B_a = \{x \in X : f(x) \leq a\}$. From $f(Tx) = f(x)$ μ-a.e., it follows that $\mu(B_a \Delta T^{-1} B_a) = 0$ and by ergodicity then $\mu(B_a) = 0$ or $\mu(B_a^c) = 0$. Let $a_0 = \inf\{a \in \mathbb{R} : \mu(B_a^c) = 0\}$, and note that $\{x \in X : f(x) = a_0\} = B_{a_0} \setminus \bigcup_{n=1}^{\infty} B_{a_0 - \frac{1}{n}}$. Since $\mu(B_{a_0}^c) = 0$ and $\mu(B_{a_0 - \frac{1}{n}}) = 0$ for all $n \geq 1$, we have $\mu(\{x \in X : f(x) = a_0\}^c) = 0$ and f is a constant μ-a.e.

(ii) $\Rightarrow$ (i) Let $A \in \mathcal{F}$ with $\mu(A \Delta T^{-1} A) = 0$ be given. Then $1_A \in L^0(X, \mathcal{F}, \mu)$ and $1_A = 1_{T^{-1}A} = U_T 1_A$ μ-a.e., so by assumption 1_A is a constant μ-a.e. and $\mu(A) = 0$ or $\mu(A^c) = 0$.

Finally, in case $\mu(X) = 1$ just notice that for any $A \in \mathcal{F}$ the indicator function 1_A is in $L^2(X, \mathcal{F}, \mu)$ and the arguments giving (ii) $\Rightarrow$ (i) also provide (iii) $\Rightarrow$ (i). Since the implication (ii) $\Rightarrow$ (iii) is immediate, this finishes the proof. □

2.3 EXAMPLES OF ERGODIC TRANSFORMATIONS

Example 2.3.1 (Ergodicity of irrational rotations). For $\theta \in (0,1)$, let $T_\theta : [0,1) \to [0,1)$ be the map $T_\theta x = x + \theta \pmod 1$ as in Example 1.3.2. We have seen that T_θ is measure preserving with respect to Lebesgue measure λ. When is T_θ ergodic?

As an example, consider $\theta = \frac{1}{4}$. Then the set

$$A = \left[0, \frac{1}{8}\right) \cup \left[\frac{1}{4}, \frac{3}{8}\right) \cup \left[\frac{1}{2}, \frac{5}{8}\right) \cup \left[\frac{3}{4}, \frac{7}{8}\right)$$

is T_θ-invariant, but $\lambda(A) = \frac{1}{2}$. Hence, $T_{\frac{1}{4}}$ is not ergodic.

Exercise 2.3.1. Suppose $\theta = \frac{p}{q}$ with $\gcd(p,q) = 1$. Prove that the orbit of any point $x \in [0,1)$ is periodic with period q, i.e., show that $T_\theta^q x = x$ and $T_\theta^i x \neq x$ for all $1 \leq i < q$. Find a non-trivial T_θ-invariant set and conclude that T_θ is not ergodic if θ is a rational.

Claim. T_θ is ergodic if and only if θ is irrational.

Proof of claim. Suppose θ is irrational, and let $f \in L^2([0,1), \mathcal{B}, \lambda)$ be T_θ-invariant λ-a.e. Let $A = \{x \in [0,1) : f(x) = f(T_\theta x)\}$. Then $\lambda(A) = 1$. Since $f \in L^2([0,1), \mathcal{B}, \lambda)$, we can write f in its Fourier series[1]:

$$f(x) = \sum_{n \in \mathbb{Z}} a_n e^{2\pi i n x}. \tag{2.2}$$

[1]It was proven by L. Carleson in [11] that for L^2-functions on $[0,1]$ the convergence of the Fourier series from (2.2) holds almost everywhere.

For $x \in A$ we get from $f(T_\theta x) = f(x)$ that

$$\begin{aligned} f(T_\theta x) &= \sum_{n\in\mathbb{Z}} a_n e^{2\pi i n(x+\theta)} = \sum_{n\in\mathbb{Z}} a_n e^{2\pi i n\theta} e^{2\pi i n x} \\ &= f(x) = \sum_{n\in\mathbb{Z}} a_n e^{2\pi i n x}. \end{aligned}$$

Hence, $\sum_{n\in\mathbb{Z}} a_n(1 - e^{2\pi i n\theta})e^{2\pi i n x} = 0$. By the uniqueness of the Fourier coefficients, we have $a_n(1 - e^{2\pi i n\theta}) = 0$, for all $n \in \mathbb{Z}$. For $n \neq 0$ the irrationality of θ implies that $1 - e^{2\pi i n\theta} \neq 0$. Thus, $a_n = 0$ for all $n \neq 0$, and therefore $f(x) = a_0$ for all $x \in A$. By Theorem 2.2.2, T_θ is ergodic. Together with Exercise 2.3.1 this proves the claim. □

Exercise 2.3.2. Consider the probability space $([0,1)^2, \mathcal{B}, \lambda)$, where $\mathcal{B}$ is the Lebesgue σ-algebra on $[0,1)^2$, and λ the two-dimensional Lebesgue measure. Suppose $\theta \in (0,1)$ is irrational, and define $T_\theta \times T_\theta : [0,1)^2 \to [0,1)^2$ by

$$T_\theta \times T_\theta(x,y) = (x + \theta \pmod 1, y + \theta \pmod 1).$$

Show that $T_\theta \times T_\theta$ is measure preserving, but is **not** ergodic.

Exercise 2.3.3. Consider the probability space $([0,1]^2, \mathcal{B}, \lambda)$, where $\mathcal{B}$ is the two-dimensional Lebesgue σ-algebra and λ is the two-dimensional Lebesgue measure. Prove that the transformation $S : [0,1]^2 \to [0,1]^2$ given by

$$S(x,y) = (x + \theta \pmod 1, x + y \pmod 1)$$

with $\theta \in (0,1)$ irrational is measure preserving and ergodic with respect to λ.

(Hint: The Fourier series $\sum_{n,m} c_{n,m} e^{2\pi i(nx+my)}$ of a function $f \in L^2([0,1]^2, \mathcal{B}, \lambda)$ satisfies $\sum_{n,m} |c_{n,m}|^2 < \infty$.)

Example 2.3.2 (Ergodicity of one- (or two-)sided shift). Let $(X, \mathcal{C}, \mu, T)$ be the one-sided Bernoulli shift from Example 1.3.9 with positive probability vector $p = (p_0, p_1, \ldots, p_{k-1})$. To show that the left shift T is ergodic, let E be a measurable subset of X which is T-invariant i.e., $T^{-1}E = E$. For any $\epsilon > 0$ there exists an $A \in \mathcal{C}$ which is a finite disjoint union of cylinders and satisfies $\mu(E\Delta A) < \epsilon$, see Lemma 12.2.1. Then

$$\begin{aligned} |\mu(E) - \mu(A)| &= |\mu(E \setminus A) - \mu(A \setminus E)| \\ &\leq \mu(E \setminus A) + \mu(A \setminus E) = \mu(E\Delta A) < \epsilon. \end{aligned}$$

A depends on finitely many coordinates only, so there exists an $n_0 > 0$ such that $T^{-n_0}A$ depends on different coordinates than A. Since μ is a product measure, we have

$$\mu(A \cap T^{-n_0}A) = \mu(A)\mu(T^{-n_0}A) = \mu(A)^2.$$

Further,

$$\mu(E\Delta T^{-n_0}A) = \mu(T^{-n_0}E\Delta T^{-n_0}A) = \mu(E\Delta A) < \epsilon,$$

and thus

$$\begin{aligned}|\mu(E) - \mu(A \cap T^{-n_0}A)| &\leq \mu\left(E\Delta(A \cap T^{-n_0}A)\right) \\ &\leq \mu(E\Delta A) + \mu(E\Delta T^{-n_0}A) < 2\epsilon.\end{aligned}$$

Hence,

$$\begin{aligned}|\mu(E) - \mu(E)^2| &\leq |\mu(E) - \mu(A)^2| + |\mu(A)^2 - \mu(E)^2| \\ &= |\mu(E) - \mu(A \cap T^{-n_0}A)| \\ &\quad + (\mu(A) + \mu(E))|\mu(A) - \mu(E)| \\ &< 4\epsilon.\end{aligned}$$

Since $\epsilon > 0$ is arbitrary, it follows that $\mu(E) = \mu(E)^2$, hence $\mu(E) = 0$ or 1. Therefore, T is ergodic.

The following lemma provides, in some cases, a useful tool to verify that a measure preserving transformation defined on $([0, 1), \mathcal{B}, \mu)$ is ergodic. Here $\mathcal{B}$ is the Lebesgue σ-algebra, and μ is a probability measure equivalent to Lebesgue measure λ (i.e., $\mu(A) = 0$ if and only if $\lambda(A) = 0$). It was proven by K. Knopp in [31] from 1926.

Lemma 2.3.1 (Knopp's Lemma). *If $E \in \mathcal{B}$ is a Lebesgue set and $\mathcal{S}$ is a class of subintervals of $[0, 1)$, satisfying*

(a) *every open subinterval of $[0, 1)$ is at most a countable union of disjoint elements from $\mathcal{S}$,*

(b) *$\forall A \in \mathcal{S}$, $\lambda(A \cap E) \geq \gamma\lambda(A)$, where $\gamma > 0$ is independent of A,*

then $\lambda(E) = 1$.

Proof. The proof is done by contradiction. Suppose $\lambda(E^c) > 0$. Given $\varepsilon > 0$ there exists by Lemma 12.2.1 a set A_ε that is a finite disjoint union of open intervals such that $\lambda(E^c \triangle A_\varepsilon) < \varepsilon$. Now by conditions (a) and (b) (that is, writing A_ε as a countable union of disjoint elements of $\mathcal{S}$)

one gets that $\lambda(E \cap A_\varepsilon) \geq \gamma\lambda(A_\varepsilon)$. Also from our choice of A_ε and the fact that

$$\lambda(E^c \triangle A_\varepsilon) \geq \lambda(E \cap A_\varepsilon) \geq \gamma\lambda(A_\varepsilon) \geq \gamma\lambda(E^c \cap A_\varepsilon) > \gamma(\lambda(E^c) - \varepsilon),$$

we obtain

$$\gamma(\lambda(E^c) - \varepsilon) < \lambda(E^c \triangle A_\varepsilon) < \varepsilon,$$

implying that $\gamma\lambda(E^c) < \varepsilon + \gamma\varepsilon$. Since $\varepsilon > 0$ is arbitrary, we get a contradiction. □

Example 2.3.3 (Ergodicity of the doubling map). Let $T : [0,1) \to [0,1)$ be the doubling map from Example 1.3.3 given by

$$Tx = 2x \pmod 1) = \begin{cases} 2x, & \text{if } 0 \leq x < \frac{1}{2}, \\ 2x - 1, & \text{if } \frac{1}{2} \leq x < 1. \end{cases}$$

We have already seen that T is measure preserving and we will use Lemma 2.3.1 to show that T is ergodic. Let

$$\mathcal{S} = \left\{ \left[\frac{k}{2^n}, \frac{k+1}{2^n} \right) : n \geq 1 \text{ and } 0 \leq k \leq 2^n - 1 \right\}$$

be the collection of *dyadic intervals*. Notice that the set $\{\frac{k}{2^n} : n \geq 1, 0 \leq k < 2^n - 1\}$ of *dyadic rationals* is dense in $[0,1)$, hence each open interval is an at most countable union of disjoint elements of $\mathcal{S}$. So, $\mathcal{S}$ satisfies the first condition of Knopp's Lemma. Now, T^n maps each dyadic interval of the form $[\frac{k}{2^n}, \frac{k+1}{2^n})$ linearly onto $[0,1)$, (we call such an interval *dyadic of order* n); in fact, $T^n x = 2^n x \pmod 1$. Let E be T-invariant Lebesgue set, and assume $\lambda(E) > 0$. Let $A \in \mathcal{S}$, and assume that A is dyadic of order n. Then, $T^n A = [0,1)$ and

$$\lambda(A \cap E) = \lambda(A \cap T^{-n}E) = \frac{1}{2^n}\lambda(E) = \lambda(A)\lambda(E).$$

Thus, the second condition of Knopp's Lemma is satisfied with $\gamma = \lambda(E) > 0$. Hence, $\lambda(E) = 1$ and T is ergodic.

Example 2.3.4 (Ergodicity of the Lüroth transformation). Consider the map T from Example 1.3.5. In Exercise 1.3.2 we saw that T is measure preserving with respect to Lebesgue measure λ. We now show that T is ergodic with respect to Lebesgue measure λ using Knopp's Lemma and essentially the same approach as for the doubling map from the previous example. We first define the collection $\mathcal{S}$. Recall the definition of the digits $a_i \geq 2$ in Example 1.3.5. For each $i \geq 2$ we set

$$\Delta(i) = \left[\frac{1}{i}, \frac{1}{i-1}\right) = \{x \in [0,1) \,:\, a_1(x) = i\}$$

and for $n \geq 1$ we define the *fundamental intervals* of rank n to be all intervals of the form

$$\begin{aligned}\Delta(i_1, i_2, \ldots, i_n) &= \Delta(i_1) \cap T^{-1}\Delta(i_2) \cap \cdots \cap T^{-(n-1)}\Delta(i_n) \\ &= \{x : a_1(x) = i_1, a_2(x) = i_2, \ldots, a_n(x) = i_n\}.\end{aligned}$$

We choose as $\mathcal{S}$ the collection of all fundamental intervals of all ranks. Notice that $\Delta(i_1, i_2, \ldots, i_n)$ is an interval with endpoints

$$\frac{P_n}{Q_n} \quad \text{and} \quad \frac{P_n}{Q_n} + \frac{1}{i_1(i_1-1)\cdots i_n(i_n-1)},$$

where

$$\frac{P_n}{Q_n} = \frac{1}{i_1} + \frac{1}{i_1(i_1-1)i_2} + \cdots + \frac{1}{i_1(i_1-1)\cdots i_{n-1}(i_{n-1}-1)i_n}.$$

Furthermore, $T^n(\Delta(i_1, i_2, \ldots, i_n)) = [0,1)$, and T^n restricted to $\Delta(i_1, i_2, \ldots, i_n)$ has slope

$$i_1(i_1-1)\cdots i_{n-1}(i_{n-1}-1)i_n(i_n-1) = \frac{1}{\lambda(\Delta(i_1, i_2, \ldots, i_n))}.$$

Since $\lim_{n\to\infty} \lambda(\Delta(i_1, i_2, \ldots, i_n)) = 0$ for any choice of digits $i_1, i_2, \ldots$, the collection $\mathcal{S}$ generates the Lebesgue σ-algebra. Now let E be a T-invariant Lebesgue set of positive Lebesgue measure, and let A be any fundamental interval of rank n. Then

$$\lambda(A \cap E) = \lambda(A \cap T^{-n}E) = \lambda(E)\lambda(A).$$

By Knopp's Lemma with $\gamma = \lambda(E)$ we get that $\lambda(E) = 1$, i.e., T is ergodic with respect to λ.

Remark 2.3.1. The notion of fundamental intervals introduced in the previous example is not specific to the Lüroth map. In fact, they are just the maximal intervals on which the iterates T^n are monotone, and one can define them similarly for any piecewise smooth, monotone interval map. For the doubling map these correspond to the dyadic intervals $\left[\frac{k}{2^n}, \frac{k+1}{2^n}\right)$. In some sense one can think of these intervals as analogues for interval maps of the cylinder sets for symbolic systems.

Exercise 2.3.4. Let λ be the Lebesgue measure on $([0,1),\mathcal{B})$, where $\mathcal{B}$ is the Lebesgue σ-algebra. Consider the transformation $T:[0,1)\to[0,1)$ given by

$$Tx=\begin{cases} 3x, & \text{if } 0\le x<\frac{1}{3},\\ \frac{3}{2}x-\frac{1}{2}, & \text{if } \frac{1}{3}\le x<1.\end{cases}$$

For $x\in[0,1)$ let

$$s_1(x)=\begin{cases} 3, & \text{if } 0\le x<\frac{1}{3},\\ \frac{3}{2}, & \text{if } \frac{1}{3}\le x<1,\end{cases} \quad \text{and} \quad h_1(x)=\begin{cases} 0, & \text{if } 0\le x<\frac{1}{3},\\ \frac{1}{2}, & \text{if } \frac{1}{3}\le x<1.\end{cases}$$

Set $s_n=s_n(x)=s_1(T^{n-1}x)$ and $h_n=h_n(x)=h_1(T^{n-1}x)$.

(a) Show that for any $x\in[0,1)$ one has $x=\displaystyle\sum_{k=1}^{\infty}\frac{h_k}{s_1s_2\cdots s_k}$.

(b) Show that T is measure preserving and ergodic with respect to the measure λ.

Let

$$a_1(x)=\begin{cases} 0, & \text{if } 0\le x<\frac{1}{3},\\ 1, & \text{if } \frac{1}{3}\le x<1,\end{cases}$$

and set $a_n=a_n(x)=a_1(T^{n-1}x)$ for $n\ge 1$.

(c) Show that for each $n\ge 1$ and any sequence $i_1,i_2,\ldots,i_n\in\{0,1\}$ one has

$$\lambda\left(\{x\in[0,1):a_1(x)=i_1,a_2(x)=i_2,\ldots,a_n(x)=i_n\}\right)=\frac{2^k}{3^n},$$

where $k=\#\{1\le j\le n: i_j=1\}$.

Exercise 2.3.5. Consider for $\beta=\frac{1+\sqrt{5}}{2}$, the golden mean, the β-transformation $T_\beta:[0,1)\to[0,1)$ given by $T_\beta x=\beta x \pmod 1$ from Example 1.3.6. Use Lemma 2.3.1 to show that T_β is ergodic with respect to Lebesgue measure λ and thus also with respect to the invariant measure μ given in Example 1.3.6.

CHAPTER 3

The Pointwise Ergodic Theorem and Mixing

In this chapter we let $(X, \mathcal{F}, \mu)$ be a probability space. If $T : X \to X$ is measure preserving and $f \in L^1(X, \mathcal{F}, \mu)$, then the sequence $f, f \circ T, f \circ T^2, \ldots$, when considered as random variables, is identically distributed and $\mathbb{E}_\mu(|f|) = \mathbb{E}_\mu(|f \circ T^n|) = \int_X |f \circ T^n| \, d\mu < \infty$. The Strong Law of Large Numbers in probability theory states that for a sequence $Y_1, Y_2, \ldots$ of i.i.d. random variables on a probability space $(\Omega, \mathcal{G}, \mathbb{P})$, with $\mathbb{E}_\mu(|Y_i|) < \infty$ one has

$$\lim_{n\to\infty} \frac{1}{n} \sum_{i=1}^{n} Y_i = \mathbb{E}_\mu(Y_1) \quad \mathbb{P} - \text{a.e.}$$

Since the sequence $f, f \circ T, f \circ T^2, \ldots$ satisfies two out of three conditions of the Strong Law of Large Numbers (only independence is missing) one could wonder about the asymptotic behavior of the sequence of averages $(\frac{1}{n}\sum_{i=0}^{n-1} f \circ T^i)$. When does it converge? In this chapter we will see one result where independence is replaced by the weaker condition of ergodicity and pointwise convergence of $(\frac{1}{n}\sum_{i=0}^{n-1} f \circ T^i)$ to $\int_X f \, d\mu$ is guaranteed. This is the Pointwise Ergodic Theorem, first proved in 1931 by G. D. Birkhoff, see [5]. The next chapter contains two other ergodic theorems with different conditions on the system and different modes of convergence.

3.1 THE POINTWISE ERGODIC THEOREM

Since 1931 several proofs of the Pointwise Ergodic Theorem have been obtained; here we present a more recent proof given by T. Kamae and M. S. Keane in [28].

Theorem 3.1.1 (Pointwise Ergodic Theorem). *Let $(X, \mathcal{F}, \mu)$ be a probability space and $T : X \to X$ a measure preserving transformation. Then, for any $f \in L^1(X, \mathcal{F}, \mu)$,*

$$\lim_{n\to\infty} \frac{1}{n}\sum_{i=0}^{n-1} f(T^i(x)) = f^*(x)$$

exists μ-a.e., is T-invariant and $\int_X f\, d\mu = \int_X f^*\, d\mu$. If moreover T is ergodic, then f^* is a constant μ-a.e. and $f^* = \int_X f\, d\mu$.

For the proof of the above theorem, we need the following simple lemma.

Lemma 3.1.1. *Let $M > 0$ be an integer, and suppose $(a_n)_{n\geq 0}$, $(b_n)_{n\geq 0}$ are sequences of non-negative real numbers such that for each $n \geq 0$ there exists an integer $1 \leq m \leq M$ with*

$$a_n + \cdots + a_{n+m-1} \geq b_n + \cdots + b_{n+m-1}.$$

Then, for each positive integer $N > M$, one has

$$a_0 + \cdots + a_{N-1} \geq b_0 + \cdots + b_{N-M-1}.$$

Proof of Lemma 3.1.1. Using the hypothesis we recursively find integers $m_0 < m_1 < \cdots < m_k < N$ with the following properties:

$$\begin{aligned} m_0 \leq M,\ m_{i+1} - m_i \leq M \text{ for } i = 0, \ldots, k-1, \text{ and } N - m_k < M,\\ a_0 + \cdots + a_{m_0-1} \geq b_0 + \cdots + b_{m_0-1},\\ a_{m_0} + \cdots + a_{m_1-1} \geq b_{m_0} + \cdots + b_{m_1-1},\\ \vdots\\ a_{m_{k-1}} + \cdots + a_{m_k-1} \geq b_{m_{k-1}} + \cdots + b_{m_k-1}. \end{aligned}$$

Then,

$$\begin{aligned} a_0 + \cdots + a_{N-1} &\geq a_0 + \cdots + a_{m_k-1}\\ &\geq b_0 + \cdots + b_{m_k-1} \geq b_0 + \cdots + b_{N-M-1}. \end{aligned}$$

□

Proof of Theorem 3.1.1. Assume with no loss of generality that $f \geq 0$ (otherwise we write $f = f^+ - f^-$, and we consider each part separately). Write $f_n(x) = f(x) + \cdots + f(T^{n-1}x)$,

$$\overline{f}(x) = \limsup_{n\to\infty} \frac{f_n(x)}{n}, \quad \text{and} \quad \underline{f}(x) = \liminf_{n\to\infty} \frac{f_n(x)}{n}.$$

Then both $\overline{f}$ and $\underline{f}$ are T-invariant, since

$$\begin{aligned}\overline{f}(Tx) &= \limsup_{n\to\infty} \frac{f_n(Tx)}{n} \\ &= \limsup_{n\to\infty} \left(\frac{f_{n+1}(x)}{n+1} \cdot \frac{n+1}{n} - \frac{f(x)}{n} \right) \\ &= \limsup_{n\to\infty} \frac{f_{n+1}(x)}{n+1} = \overline{f}(x)\end{aligned}$$

and a similarly for $\underline{f}$. Now, to prove that f^* exists, is integrable and T-invariant, it is enough to show that

$$\int_X \underline{f}\, \mathrm{d}\mu \geq \int_X f\, \mathrm{d}\mu \geq \int_X \overline{f}\, \mathrm{d}\mu.$$

For since $\overline{f} - \underline{f} \geq 0$, this would imply that $\overline{f} = \underline{f} =: f^*$ μ-a.e.

We first prove that $\int_X \overline{f} d\mu \leq \int_X f\, \mathrm{d}\mu$. Fix any $0 < \epsilon < 1$, and let $L > 0$ be any real number. By definition of $\overline{f}$, for any $x \in X$, there exists an integer $m > 0$ such that

$$\frac{f_m(x)}{m} \geq \min(\overline{f}(x), L)(1 - \epsilon).$$

For $n \geq 1$, let $Y_n = \{x \in X : \exists\, 1 \leq m \leq n \text{ with } f_m(x) \geq m \min(\overline{f}(x), L)(1-\epsilon)\}$. Note that (Y_n) is an increasing sequence with $X = \bigcup_{n=1}^{\infty} Y_n$. Thus, for any $\delta > 0$ there exists an integer $M > 0$ such that the set

$$Y_M = \{x \in X : \exists\, 1 \leq m \leq M \text{ with } f_m(x) \geq m \min(\overline{f}(x), L)(1 - \epsilon)\}$$

has measure at least $1 - \delta$. Set $X_0 = Y_M$ and define F on X by

$$F(x) = \begin{cases} f(x), & \text{if } x \in X_0, \\ L, & \text{if } x \notin X_0. \end{cases}$$

Notice that $f \leq F$ (why?). For any $x \in X$ and $n \geq 0$, let $a_n = a_n(x) = F(T^n x)$, and

$$b_n = b_n(x) = \min(\overline{f}(x), L)(1 - \epsilon)$$

(so b_n is independent of n). We now show that (a_n) and (b_n) satisfy the hypothesis of Lemma 3.1.1 with $M > 0$ as above. For any $n \geq 0$ there are two cases:

- If $T^n x \in X_0$, then there exists an $1 \leq m \leq M$ such that

$$\begin{aligned} f_m(T^n x) &\geq m \min(\overline{f}(T^n x), L)(1 - \epsilon) \\ &= m \min(\overline{f}(x), L)(1 - \epsilon) \\ &= b_n + \cdots + b_{n+m-1}. \end{aligned}$$

  Hence,

$$\begin{aligned} a_n + \cdots + a_{n+m-1} &= F(T^n x) + \cdots + F(T^{n+m-1} x) \\ &\geq f(T^n x) + \cdots + f(T^{n+m-1} x) = f_m(T^n x) \\ &\geq b_n + \cdots + b_{n+m-1}. \end{aligned}$$

- If $T^n x \notin X_0$, then take $m = 1$ since

$$a_n = F(T^n x) = L \geq \min(\overline{f}(x), L)(1 - \epsilon) = b_n.$$

Hence by Lemma 3.1.1 for all integers $N > M$ one has

$$F(x) + \cdots + F(T^{N-1} x) \geq (N - M) \min(\overline{f}(x), L)(1 - \epsilon).$$

Integrating both sides, and using the fact that T is measure preserving, one gets by Proposition 1.2.2 that

$$N \int_X F \, d\mu \geq (N - M) \int_X \min(\overline{f}(x), L)(1 - \epsilon) \, d\mu(x).$$

Since

$$\int_X F \, d\mu = \int_{X_0} f \, d\mu + L\mu(X \setminus X_0),$$

one has

$$\begin{aligned} \int_X f \, d\mu &\geq \int_{X_0} f \, d\mu \\ &= \int_X F \, d\mu - L\mu(X \setminus X_0) \\ &\geq \frac{(N - M)}{N} \int_X \min(\overline{f}(x), L)(1 - \epsilon) \, d\mu(x) - L\delta. \end{aligned}$$

Now letting first $N \to \infty$, then $\delta \to 0$, then $\epsilon \to 0$, and lastly $L \to \infty$ one gets together with the Monotone Convergence Theorem that $\overline{f}$ is integrable, and

$$\int_X f\,d\mu \geq \int_X \overline{f}\,d\mu.$$

We now prove that $\int_X f\,d\mu \leq \int_X \underline{f}\,d\mu$. Fix $\epsilon > 0$ and $\delta_0 > 0$. Since $f \geq 0$, there exists a $\delta > 0$ such that whenever $A \in \mathcal{F}$ with $\mu(A) < \delta$, then $\int_A f\,d\mu < \delta_0$. Note that for any $x \in X$ there exists an integer m such that

$$\frac{f_m(x)}{m} \leq (\underline{f}(x) + \epsilon).$$

Now choose $M > 0$ such that the set

$$Y_0 = \{x \in X \, : \, \exists\, 1 \leq m \leq M \text{ with } f_m(x) \leq m\,(\underline{f}(x) + \epsilon)\}$$

has measure at least $1 - \delta$. Define G on X by

$$G(x) = \begin{cases} f(x), & \text{if } x \in Y_0, \\ 0, & \text{if } x \notin Y_0. \end{cases}$$

Then $G \leq f$. Let $b_n = G(T^n x)$, and $a_n = \underline{f}(x) + \epsilon$ (so a_n is independent of n). One can easily check that the sequences (a_n) and (b_n) satisfy the hypothesis of Lemma 3.1.1 with $M > 0$ as above. Hence for any $N > M$, one has

$$G(x) + \cdots + G(T^{N-M-1}x) \leq N(\underline{f}(x) + \epsilon).$$

Integrating both sides yields

$$(N - M) \int_X G(x)\,d\mu(x) \leq N\left(\int_X \underline{f}(x)\,d\mu(x) + \epsilon\right).$$

Since $\mu(X \setminus Y_0) < \delta$, then $\int_{X \setminus Y_0} f\,d\mu < \delta_0$. Hence,

$$\begin{aligned} \int_X f\,d\mu &= \int_{Y_0} G\,d\mu + \int_{X \setminus Y_0} f\,d\mu \\ &\leq \frac{N}{N - M} \int_X (\underline{f}(x) + \epsilon)\,d\mu(x) + \delta_0. \end{aligned}$$

Now, by letting first $N \to \infty$, then $\delta \to 0$, $\delta_0 \to 0$, and finally $\epsilon \to 0$, one gets

$$\int_X f\,d\mu \leq \int_X \underline{f}\,d\mu.$$

This shows that

$$\int_X \underline{f}\, d\mu \geq \int_X f\, d\mu \geq \int_X \overline{f}\, d\mu,$$

hence, $\overline{f} = \underline{f} = f^*$ μ-a.e., and f^* is T-invariant. In case T is ergodic, then by Theorem 2.2.2 the T-invariance of f^* implies that f^* is a constant μ-a.e. Therefore, for μ-a.e. $x \in X$,

$$f^*(x) = \int_X f^*\, d\mu = \int_X f\, d\mu. \qquad \square$$

Remark 3.1.1. (i) We can say a bit more about the limit f^* in case T is not ergodic. Let $\mathcal{I}$ be the sub-σ-algebra of $\mathcal{F}$ consisting of all T-invariant subsets $A \in \mathcal{F}$. Notice that if $f \in L^1(X, \mathcal{F}, \mu)$, then the *conditional expectation* of f given $\mathcal{I}$ (denoted by $\mathbb{E}_\mu(f|\mathcal{I})$), is the unique μ-a.e. $\mathcal{I}$-measurable $L^1(X, \mathcal{F}, \mu)$-function with the property that

$$\int_A f\, d\mu = \int_A \mathbb{E}_\mu(f|\mathcal{I})\, d\mu$$

for all $A \in \mathcal{I}$, i.e., with $T^{-1}A = A$. We claim that $f^* = \mathbb{E}_\mu(f|\mathcal{I})$. Since the limit function f^* is T-invariant, it follows that f^* is $\mathcal{I}$-measurable. Furthermore, for any $A \in \mathcal{I}$, by the Pointwise Ergodic Theorem and the T-invariance of 1_A,

$$\lim_{n\to\infty} \frac{1}{n} \sum_{i=0}^{n-1} (f 1_A)(T^i x) = 1_A(x) \lim_{n\to\infty} \frac{1}{n} \sum_{i=0}^{n-1} f(T^i x) = 1_A(x) f^*(x) \ \mu - \text{a.e.}$$

and

$$\int_X f 1_A\, d\mu = \int_X f^* 1_A\, d\mu.$$

This shows that $f^* = \mathbb{E}_\mu(f|\mathcal{I})$.

(ii) Suppose T is ergodic and measure preserving with respect to a probability measure μ, and let ν be a probability measure equivalent to μ (i.e., μ and ν have the same sets of measure zero). Then for every $f \in L^1(X, \mathcal{F}, \mu)$ one has ν-a.e. that

$$\lim_{n\to\infty} \frac{1}{n} \sum_{i=0}^{n-1} f(T^i(x)) = \int_X f\, d\mu.$$

This observation is useful in Exercise 3.1.2 below.

Exercise 3.1.1. Show that if T is measure preserving on the probability space $(X, \mathcal{F}, \mu)$ and $f \in L^1(X, \mathcal{F}, \mu)$, then

$$\lim_{n\to\infty} \frac{f(T^n x)}{n} = 0$$

for μ-a.e. $x \in X$.

Exercise 3.1.2. For $\beta = \frac{1+\sqrt{5}}{2}$, the golden mean, consider the β-transformation $T_\beta : [0,1) \to [0,1)$, given by $T_\beta x = \beta x \pmod 1 = \beta x - \lfloor \beta x \rfloor$ as in Example 1.3.6. Define b_n on $[0,1)$ by

$$b_1(x) = \begin{cases} 0, & \text{if } 0 \le x < \frac{1}{\beta}, \\ 1, & \text{if } \frac{1}{\beta} \le x < 1, \end{cases}$$

and $b_n(x) = b_1(T_\beta^{n-1} x)$ for $n \ge 1$. Fix $k \ge 0$. Consider the limit

$$\lim_{n\to\infty} \frac{1}{n} \#\{1 \le i \le n : b_i = 0, b_{i+1} = 0, \ldots, b_{i+k} = 0\}$$

and find its a.e. value (with respect to Lebesgue measure).

Exercise 3.1.3. Let $(X, \mathcal{F}, \mu)$ be a probability space and $f \in L^1(X, \mathcal{F}, \mu)$. Suppose $\{T_t : t \in \mathbb{R}\}$ is a family of transformations $T_t : X \to X$ satisfying

(i) $T_0 = \mathrm{id}_X$ and $T_{t+s} = T_t \circ T_s$,
(ii) T_t is measurable, measure preserving and ergodic with respect to μ,
(iii) The map $G : X \times \mathbb{R} \to X$ given by $G(x,t) = f(T_t(x))$ is measurable, where $X \times \mathbb{R}$ is endowed with the product σ-algebra $\mathcal{F} \otimes \mathcal{B}$ and product measure $\mu \times \lambda$, with $\mathcal{B}$ the Borel σ-algebra on $\mathbb{R}$ and λ the Lebesgue measure.

(a) Show that for all $s \ge 0$,

$$\int_{[0,s]} \int_X f(T_t(x))\, d\mu(x)\, d\lambda(t) = \int_X \int_{[0,s]} f(T_t(x))\, d\lambda(t)\, d\mu(x)$$
$$= s \int_X f\, d\mu.$$

(b) Show that for all $s \ge 0$, $\int_{[0,s]} f(T_t(x))\, d\lambda(t) < \infty$ μ-a.e.
(c) Define $F : X \to \mathbb{R}$ by $F(x) = \int_{[0,1]} f(T_t(x))\, d\lambda(t)$, and consider the

transformation T_1 corresponding to $t = 1$. Show that for any $n \geq 1$ one has

$$\sum_{k=0}^{n-1} F(T_1^k(x)) = \int_{[0,n]} f(T_t(x))\, d\lambda(t),$$

and $\int_X F\, d\mu = \int_X f\, d\mu$.

(d) Show that

$$\lim_{n\to\infty} \frac{1}{n} \int_{[0,n]} f(T_t(x))\, d\lambda(t) = \int_X f\, d\mu$$

holds for μ-a.e. x.

Using the Pointwise Ergodic Theorem, one can give yet another characterization of ergodicity.

Corollary 3.1.1. *Let $(X, \mathcal{F}, \mu)$ be a probability space and $T : X \to X$ a measure preserving transformation. Then, T is ergodic if and only if for all $A, B \in \mathcal{F}$, one has*

$$\lim_{n\to\infty} \frac{1}{n} \sum_{i=0}^{n-1} \mu(T^{-i}A \cap B) = \mu(A)\mu(B). \tag{3.1}$$

Proof. Suppose T is ergodic, and let $A, B \in \mathcal{F}$. Since the indicator function $1_A \in L^1(X, \mathcal{F}, \mu)$, by the Pointwise Ergodic Theorem one has

$$\lim_{n\to\infty} \frac{1}{n} \sum_{i=0}^{n-1} 1_A(T^i x) = \int_X 1_A\, d\mu = \mu(A) \quad \mu\text{-a.e.}$$

Then,

$$\begin{aligned}
\lim_{n\to\infty} \frac{1}{n} \sum_{i=0}^{n-1} 1_{T^{-i}A\cap B}(x) &= \lim_{n\to\infty} \frac{1}{n} \sum_{i=0}^{n-1} 1_{T^{-i}A}(x) 1_B(x) \\
&= 1_B(x) \lim_{n\to\infty} \frac{1}{n} \sum_{i=0}^{n-1} 1_A(T^i x) \\
&= 1_B(x)\mu(A) \qquad \mu\text{-a.e.}
\end{aligned}$$

Since for each n, the function $\frac{1}{n}\sum_{i=0}^{n-1} 1_{T^{-i}A\cap B}$ is dominated by the constant function 1, it follows by the Dominated Convergence Theorem that

$$\begin{aligned}
\lim_{n\to\infty} \frac{1}{n} \sum_{i=0}^{n-1} \mu(T^{-i}A \cap B) &= \int_X \lim_{n\to\infty} \frac{1}{n} \sum_{i=0}^{n-1} 1_{T^{-i}A\cap B}(x)\, d\mu(x) \\
&= \int_X 1_B \mu(A)\, d\mu(x) \\
&= \mu(A)\mu(B).
\end{aligned}$$

Conversely, suppose (3.1) holds for every $A, B \in \mathcal{F}$. Let $E \in \mathcal{F}$ be such that $T^{-1}E = E$. By invariance of E, we have $\mu(T^{-i}E \cap E) = \mu(E)$ for each i, so

$$\lim_{n\to\infty} \frac{1}{n} \sum_{i=0}^{n-1} \mu(T^{-i}E \cap E) = \mu(E).$$

On the other hand, by (3.1)

$$\lim_{n\to\infty} \frac{1}{n} \sum_{i=0}^{n-1} \mu(T^{-i}E \cap E) = \mu(E)^2.$$

Hence, $\mu(E) = \mu(E)^2$, which implies $\mu(E) = 0$ or $\mu(E) = 1$. Therefore, T is ergodic. □

To show ergodicity one needs to verify equation (3.1) for sets A and B belonging to a generating semi-algebra only as the next proposition shows.

Proposition 3.1.1. *Let $(X, \mathcal{F}, \mu)$ be a probability space, and $\mathcal{S}$ a generating semi-algebra of $\mathcal{F}$. Let $T : X \to X$ be a measure preserving transformation. Then, T is ergodic if and only if for all $A, B \in \mathcal{S}$, one has*

$$\lim_{n\to\infty} \frac{1}{n} \sum_{i=0}^{n-1} \mu(T^{-i}A \cap B) = \mu(A)\mu(B). \tag{3.2}$$

Proof. Assume that (3.2) holds for all $A, B \in \mathcal{S}$. Note that it then also holds for all elements in the algebra generated by $\mathcal{S}$. We only need to show that (3.2) holds for all $A, B \in \mathcal{F}$. Let $\epsilon > 0$, and $A, B \in \mathcal{F}$. Then, by Lemma 12.2.1 there exist sets A_0, B_0 each of which is a finite disjoint union of elements of $\mathcal{S}$ such that

$$\mu(A \Delta A_0) < \epsilon, \quad \text{and } \mu(B \Delta B_0) < \epsilon.$$

Since,

$$(T^{-i}A \cap B)\Delta(T^{-i}A_0 \cap B_0) \subseteq (T^{-i}A \Delta T^{-i}A_0) \cup (B \Delta B_0)$$

for any $i \geq 1$, it follows that

$$\begin{aligned} |\mu(T^{-i}A \cap B) - \mu(T^{-i}A_0 \cap B_0)| &\leq \mu((T^{-i}A \cap B)\Delta(T^{-i}A_0 \cap B_0)) \\ &\leq \mu(T^{-i}A \Delta T^{-i}A_0) + \mu(B \Delta B_0) \\ &< 2\epsilon \end{aligned}$$

for any $i \geq 1$. Further,

$$\begin{aligned}|\mu(A)\mu(B) - \mu(A_0)\mu(B_0)| &\leq \mu(A)|\mu(B) - \mu(B_0)| + \mu(B_0)|\mu(A) - \mu(A_0)| \\ &\leq |\mu(B) - \mu(B_0)| + |\mu(A) - \mu(A_0)| \\ &\leq \mu(B\Delta B_0) + \mu(A\Delta A_0) \\ &< 2\epsilon.\end{aligned}$$

Hence, for any n,

$$\begin{aligned}&\Big|\Big(\frac{1}{n}\sum_{i=0}^{n-1}\mu(T^{-i}A\cap B) - \mu(A)\mu(B)\Big) \\ &\qquad - \Big(\frac{1}{n}\sum_{i=0}^{n-1}\mu(T^{-i}A_0\cap B_0) - \mu(A_0)\mu(B_0)\Big)\Big| \\ &\leq \frac{1}{n}\sum_{i=0}^{n-1}|\mu(T^{-i}A\cap B) - \mu(T^{-i}A_0\cap B_0)| + |\mu(A)\mu(B) - \mu(A_0)\mu(B_0)| \\ &< 4\epsilon.\end{aligned}$$

Therefore,

$$\lim_{n\to\infty}\Big(\frac{1}{n}\sum_{i=0}^{n-1}\mu(T^{-i}A\cap B) - \mu(A)\mu(B)\Big) = 0. \qquad \square$$

Theorem 3.1.2. *Suppose μ_1 and μ_2 are probability measures on $(X, \mathcal{F})$, and $T : X \to X$ is a transformation that is measure preserving with respect to both μ_1 and μ_2. Then,*

(i) *if T is ergodic with respect to μ_1, and μ_2 is absolutely continuous with respect to μ_1, then $\mu_1 = \mu_2$, and*

(ii) *if T is ergodic with respect to μ_1 and μ_2, then either $\mu_1 = \mu_2$ or μ_1 and μ_2 are singular with respect to each other.*

Proof. (i) Suppose T is ergodic with respect to μ_1 and μ_2 is absolutely continuous with respect to μ_1. For any $A \in \mathcal{F}$, by the Pointwise Ergodic Theorem for μ_1-a.e. x one has

$$\lim_{n\to\infty}\frac{1}{n}\sum_{i=0}^{n-1}1_A(T^i x) = \mu_1(A).$$

Let

$$C_A = \Big\{x \in X : \lim_{n\to\infty}\frac{1}{n}\sum_{i=0}^{n-1}1_A(T^i x) = \mu_1(A)\Big\},$$

then $\mu_1(C_A) = 1$, and by absolute continuity of μ_2 also $\mu_2(C_A) = 1$. Since T is measure preserving with respect to μ_2, for each $n \geq 1$ one has

$$\frac{1}{n}\sum_{i=0}^{n-1}\int_X 1_A(T^i x)\,d\mu_2(x) = \mu_2(A).$$

On the other hand, by the Dominated Convergence Theorem one has

$$\lim_{n\to\infty}\int_X \frac{1}{n}\sum_{i=0}^{n-1} 1_A(T^i x)\,d\mu_2(x) = \int_X \mu_1(A)\,d\mu_2.$$

This implies that $\mu_1(A) = \mu_2(A)$. Since $A \in \mathcal{F}$ is arbitrary, we have $\mu_1 = \mu_2$.

(ii) Suppose T is ergodic with respect to μ_1 and μ_2. Assume that $\mu_1 \neq \mu_2$. Then, there exists a set $A \in \mathcal{F}$ such that $\mu_1(A) \neq \mu_2(A)$. For $i = 1, 2$, let

$$C_i = \Big\{x \in X : \lim_{n\to\infty}\frac{1}{n}\sum_{j=0}^{n-1} 1_A(T^j x) = \mu_i(A)\Big\}.$$

By the Pointwise Ergodic Theorem $\mu_i(C_i) = 1$ for $i = 1, 2$. Since $\mu_1(A) \neq \mu_2(A)$, then $C_1 \cap C_2 = \emptyset$. Thus μ_1 and μ_2 are supported on disjoint sets, and hence μ_1 and μ_2 are mutually singular. □

Exercise 3.1.4. Imagine a monkey typing on a laptop with a keyboard that has only 26 keys, one key for each letter of the alphabet and no space bar, numbers or other keys. Suppose that the monkey hits one key every second, that he hits each key with equal probability (independently of the preceding letter) and that he goes on forever.

(a) Show how to model this using a shift space on 26 symbols with an appropriate Bernoulli measure.

(b) Use the Pointwise Ergodic Theorem to show that with probability 1 the monkey will eventually type the word "BANANA".

3.2 NORMAL NUMBERS

The Pointwise Ergodic Theorem can be used to compute the frequencies of digits in various number expansions. Recall the doubling map $T_2 : [0, 1) \to [0, 1)$ defined by $T_2 x = 2x \pmod 1$ from Example 1.3.3 and its association to binary expansions of numbers in $[0, 1)$.

Except for the dyadic rationals each number $x \in [0,1)$ has a unique binary expansion and this can be obtained from iterations of T_2 as in (1.2). For a number x the n-th binary digit $b_n(x)$ equals 0 precisely if $T_2^{n-1}x \in [0, \frac{1}{2})$. So the frequency of the digit 0 in the sequence $(b_n(x))$ representing the binary expansion of x is therefore given by

$$\lim_{n\to\infty} \frac{1}{n} \sum_{i=0}^{n-1} 1_{[0,\frac{1}{2})}(T_2^i x).$$

T_2 is measure preserving and ergodic with respect to the Lebesgue measure. The Pointwise Ergodic Theorem then implies that this limit is equal to $\lambda([0, \frac{1}{2})) = \frac{1}{2}$ for Lebesgue almost every $x \in [0,1)$. Obviously for those numbers x the frequency of the digit 1 in their binary expansion is also equal to $\frac{1}{2}$.

Instead of single digits, we could also consider arbitrary blocks of digits $a_1, a_2, \ldots, a_k \in \{0,1\}$. The block $a_1 a_2 \cdots a_k$ occurs at position n of the binary expansion of x precisely if

$$T_2^{n-1}x \in \left[\frac{a_1}{2} + \frac{a_2}{2^2} + \cdots + \frac{a_k}{2^k}, \frac{a_1}{2} + \frac{a_2}{2^2} + \cdots + \frac{a_k+1}{2^k}\right).$$

By the Pointwise Ergodic Theorem the block $a_1 a_2 \cdots a_k$ occurs in the binary expansion of Lebesgue almost every $x \in [0,1)$ with frequency

$$\lambda\left(\left[\frac{a_1}{2} + \frac{a_2}{2^2} + \cdots + \frac{a_k}{2^k}, \frac{a_1}{2} + \frac{a_2}{2^2} + \cdots + \frac{a_k+1}{2^k}\right)\right) = \frac{1}{2^k}.$$

So typically any block of digits of length k occurs with frequency $\frac{1}{2^k}$ in the binary expansion of x. A number with this property is called *normal in base 2*.

Now consider another integer $N \geq 2$ and define the $\times N$ *transformation* $T_N : [0,1) \to [0,1)$ by $T_N x = Nx - \lfloor Nx \rfloor$. Iterations of T_N generate the *base N expansion* of points in the unit interval much like how binary expansions were obtained from the doubling map. For $x \in [0,1)$ and $i \geq 1$ the i-th base N digit is given by $c_i = c_i(x) = k$ if $T_N^{i-1}x \in [\frac{k}{N}, \frac{k+1}{N})$. Then

$$x = \sum_{i=1}^{\infty} \frac{c_i}{N^i}.$$

This representation is unique if $x \neq \frac{k}{N^i}$ for some $i \geq 1$ and $0 \leq k \leq N^i - 1$.

Exercise 3.2.1. Prove that T_N is measure preserving and ergodic on the probability space $([0,1), \mathcal{B}, \lambda)$, where $\mathcal{B}$ is the Lebesgue σ-algebra and λ the Lebesgue measure.

From Exercise 3.2.1 we see that we can again use the Pointwise Ergodic Theorem to obtain the frequency of digits and blocks of digits in typical base N expansions. Take any $a_1, \ldots, a_k \in \{0, 1, \ldots, N-1\}$. For Lebesgue almost all $x \in [0,1)$ the frequency of the block $a_1 a_2 \cdots a_k$ in the base N expansion of x is equal to

$$\begin{aligned}\lim_{n\to\infty} \frac{1}{n} \sum_{i=0}^{n-1} 1_{[\frac{a_1}{N}+\frac{a_2}{N^2}+\cdots+\frac{a_k}{N^k}, \frac{a_1}{N}+\frac{a_2}{N^2}+\cdots+\frac{a_k+1}{N^k})}(T_N^i x) &= \int_{[0,1)} 1_{[\frac{a_1}{N}+\frac{a_2}{N^2}+\cdots+\frac{a_k}{N^k}, \frac{a_1}{N}+\frac{a_2}{N^2}+\cdots+\frac{a_k+1}{N^k})}\, d\lambda \\ &= \frac{1}{N^k}.\end{aligned}$$

Hence, Lebesgue almost every $x \in [0,1)$ is *normal in base* N.

Numbers in $[0,1)$ that are normal in all integer bases $N \geq 2$ are called *(absolutely) normal.* If we denote the exceptional set of numbers that are not normal in base N by $\mathcal{E}_N$, then $\lambda(\mathcal{E}_N) = 0$ for all $N \geq 2$ and thus also $\lambda(\bigcup_{N\geq 2} \mathcal{E}_N) = 0$. So we see that Lebesgue almost every $x \in [0,1)$ is normal. This result was first obtained using the Borel-Cantelli Lemma by E. Borel in [6] from 1909. Unfortunately his proof was not constructive (nor is the proof above) and even though nowadays it is possible to construct a normal number up to any desired precision, it is still not known whether famous constants like π (or $\pi - 3$), e (or $e-2$) and $\frac{1}{2}\sqrt{2}$ are normal or not.

Exercise 3.2.2. For $x \in [0,1]$, let $(a_n(x))_{n\geq 1}$ denote the finite or infinite sequence of digits of x as produced by the Lüroth map from Example 1.3.5.

(a) Let $\mathcal{E}$ denote the set of $x \in [0,1]$ for which the sequence $(a_n(x))_{n\geq 1}$ contains only even integers. Prove that $\mathcal{E}$ has zero Lebesgue measure.

(b) Prove that for Lebesgue almost every $x \in [0,1]$ we have

$$\lim_{n\to\infty} \frac{\#\{1 \leq k \leq n \,:\, a_k(x) \text{ is even}\}}{n} = \ln 2.$$

What is the frequency of the odd digits in the Lüroth expansion of Lebesgue almost every $x \in [0,1]$?

3.3 IRREDUCIBLE MARKOV CHAINS

Consider the Markov shift from Example 1.3.10, i.e., $X = \{0, 1, \ldots, N-1\}^{\mathbb{Z}}$, $\mathcal{C}$ is the σ-algebra generated by the cylinders, $T : X \to X$ is the left shift, and μ the Markov measure defined by a stochastic $N \times N$ matrix $P = (p_{ij})$, and a positive probability vector $\pi = (\pi_0, \pi_1, \ldots, \pi_{N-1})$ satisfying $\pi P = \pi$ through

$$\mu(\{x : x_l = i_l, x_{l+1} = i_{l+1}, \ldots, x_n = i_n\}) = \pi_{i_l} p_{i_l i_{l+1}} p_{i_{l+1} i_{l+2}} \cdots p_{i_{n-1} i_n}.$$

We want to find necessary and sufficient conditions for T to be ergodic. To achieve this, we first set

$$Q = \lim_{n\to\infty} \frac{1}{n} \sum_{k=0}^{n-1} P^k,$$

where $P^k = (p_{ij}^{(k)})$ is the k-th power of the matrix P, and P^0 is the $N \times N$ identity matrix. More precisely, $Q = (q_{ij})$, where

$$q_{ij} = \lim_{n\to\infty} \frac{1}{n} \sum_{k=0}^{n-1} p_{ij}^{(k)}.$$

Lemma 3.3.1. *For each $i, j \in \{0, 1, \ldots, N-1\}$, the limit $\lim_{n\to\infty} \frac{1}{n} \sum_{k=0}^{n-1} p_{ij}^{(k)}$ exists, i.e., q_{ij} is well-defined.*

Proof. For each n,

$$\frac{1}{n} \sum_{k=0}^{n-1} p_{ij}^{(k)} = \frac{1}{\pi_i} \frac{1}{n} \sum_{k=0}^{n-1} \mu(\{x \in X : x_0 = i, x_k = j\}).$$

Since T is measure preserving, by the Pointwise Ergodic Theorem it holds for μ-a.e. $y \in X$ that

$$\lim_{n\to\infty} \frac{1}{n} \sum_{k=0}^{n-1} 1_{\{x : x_k = j\}}(y) = \lim_{n\to\infty} \frac{1}{n} \sum_{k=0}^{n-1} 1_{\{x : x_0 = j\}}(T^k y) = f^*(y),$$

where f^* is T-invariant and integrable. Then,

$$\begin{aligned} \lim_{n\to\infty} \frac{1}{n} \sum_{k=0}^{n-1} 1_{\{x : x_0 = i, x_k = j\}}(y) &= 1_{\{x : x_0 = i\}}(y) \lim_{n\to\infty} \frac{1}{n} \sum_{k=0}^{n-1} 1_{\{x : x_0 = j\}}(T^k y) \\ &= f^*(y) 1_{\{x : x_0 = i\}}(y). \end{aligned}$$

Since $\frac{1}{n}\sum_{k=0}^{n-1} 1_{\{x:x_0=i,x_k=j\}}(y) \leq 1$ for all n, by the Dominated Convergence Theorem,

$$\begin{aligned} q_{ij} &= \frac{1}{\pi_i} \lim_{n\to\infty} \frac{1}{n}\sum_{k=0}^{n-1} \mu(\{x \in X : x_0 = i, x_k = j\}) \\ &= \frac{1}{\pi_i} \int_X \lim_{n\to\infty} \frac{1}{n}\sum_{k=0}^{n-1} 1_{\{x:x_0=i,x_k=j\}}(y)\, d\mu(y) \\ &= \frac{1}{\pi_i} \int_X f^*(y) 1_{\{x:x_0=i\}}(y)\, d\mu(y) \\ &= \frac{1}{\pi_i} \int_{\{x:x_0=i\}} f^*(y)\, d\mu(y), \end{aligned}$$

which is finite since f^* is integrable. Hence q_{ij} exists. □

Exercise 3.3.1. Show that the matrix Q has the following properties:
(a) Q is stochastic,
(b) $Q = QP = PQ = Q^2$,
(c) $\pi Q = \pi$.

We now give a characterization for the ergodicity of T. Recall that the matrix P is said to be *irreducible* if for every $i, j \in \{0, 1, \ldots, N-1\}$, there exists an $n \geq 1$ such that $p_{ij}^{(n)} > 0$.

Theorem 3.3.1. *The following are equivalent.*

(i) *T is ergodic.*

(ii) *All rows of Q are identical.*

(iii) *$q_{ij} > 0$ for all i, j.*

(iv) *P is irreducible.*

(v) *1 is a simple eigenvalue of P.*

Proof. (i) ⇒ (ii) By the Pointwise Ergodic Theorem it holds for each i, j, that

$$\lim_{n\to\infty} \frac{1}{n}\sum_{k=0}^{n-1} 1_{\{x:x_0=i,x_k=j\}}(y) = 1_{\{x:x_0=i\}}(y)\pi_j.$$

By the Dominated Convergence Theorem,

$$\lim_{n\to\infty} \frac{1}{n}\sum_{k=0}^{n-1} \mu(\{x \in X : x_0 = i, x_k = j\}) = \pi_i \pi_j.$$

Hence,

$$q_{ij} = \frac{1}{\pi_i} \lim_{n\to\infty} \frac{1}{n} \sum_{k=0}^{n-1} \mu(\{x \in X : x_0 = i, x_k = j\}) = \pi_j,$$

i.e., q_{ij} is independent of i. Therefore, all rows of Q are identical.

(ii) ⇒ (iii) If all the rows of Q are identical, then all the columns of Q are constants. Thus, for each j there exists a constant c_j such that $q_{ij} = c_j$ for all i. Since $\pi Q = \pi$, it follows that $q_{ij} = c_j = \pi_j > 0$ for all i, j.

(iii) ⇒ (iv) For any i, j,

$$\lim_{n\to\infty} \frac{1}{n} \sum_{k=0}^{n-1} p_{ij}^{(k)} = q_{ij} > 0.$$

Hence, there exists an n such that $p_{ij}^{(n)} > 0$ and therefore P is irreducible.

(iv) ⇒ (iii) Suppose P is irreducible. For any state $i \in \{0, 1, \ldots, N-1\}$, let $S_i = \{j : q_{ij} > 0\}$. Since Q is a stochastic matrix, it follows that $S_i \neq \emptyset$. Let $l \in S_i$, then $q_{il} > 0$. Since $Q = QP = QP^n$ for all n, then for any state j

$$q_{ij} = \sum_{m=0}^{N-1} q_{im} p_{mj}^{(n)} \geq q_{il} p_{lj}^{(n)}$$

for any n. Since P is irreducible, there exists an n such that $p_{lj}^{(n)} > 0$. Hence, $q_{ij} > 0$ for all i, j.

(iii) ⇒ (ii) Suppose $q_{ij} > 0$ for all $j \in \{0, 1, \ldots, N-1\}$. Fix any state j, and let $q_j = \max_{0\leq i\leq N-1} q_{ij}$. Suppose that not all the q_{ij}'s are the same. Then there exists a $k \in \{0, 1, \ldots, N-1\}$ such that $q_{kj} < q_j$. Since Q is stochastic and $Q^2 = Q$, then for any $i \in \{0, 1, \ldots, N-1\}$ we have,

$$q_{ij} = \sum_{l=0}^{N-1} q_{il} q_{lj} < \sum_{l=0}^{N-1} q_{il} q_j = q_j.$$

This implies that $q_j = \max_{0\leq i\leq N-1} q_{ij} < q_j$, a contradiction. Hence, the columns of Q are constants and all the rows are identical.

(ii) ⇒ (i) Suppose all the rows of Q are identical. We have shown above that this implies $q_{ij} = \pi_j$ for all $i, j \in \{0, 1, \ldots, N-1\}$. Hence $\pi_j = \lim_{n\to\infty} \frac{1}{n} \sum_{k=0}^{n-1} p_{ij}^{(k)}$. Let

$$A = \{x : x_r = i_0, \ldots, x_{r+l} = i_l\}, \text{ and } B = \{x : x_s = j_0, \ldots, x_{s+m} = j_m\}$$

be any two cylinder sets of X. By Proposition 3.1.1 it is enough to show that

$$\lim_{n\to\infty}\frac{1}{n}\sum_{i=0}^{n-1}\mu(T^{-i}A\cap B)=\mu(A)\mu(B).$$

Since T is the left shift, for all n sufficiently large, the cylinders $T^{-n}A$ and B depend on different coordinates. Hence, for n sufficiently large,

$$\mu(T^{-n}A\cap B)=\pi_{j_0}p_{j_0j_1}\cdots p_{j_{m-1}j_m}p_{j_mi_0}^{(n+r-s-m)}p_{i_0i_1}\cdots p_{i_{l-1}i_l}.$$

Thus,

$$\begin{aligned}
&\lim_{n\to\infty}\frac{1}{n}\sum_{k=0}^{n-1}\mu(T^{-k}A\cap B)\\
&=\pi_{j_0}p_{j_0j_1}\cdots p_{j_{m-1}j_m}p_{i_0i_1}\cdots p_{i_{l-1}i_l}\lim_{n\to\infty}\frac{1}{n}\sum_{k=0}^{n-1}p_{j_mi_0}^{(k)}\\
&=(\pi_{j_0}p_{j_0j_1}\cdots p_{j_{m-1}j_m})(\pi_{i_0}p_{i_0i_1}\cdots p_{i_{l-1}i_l})\\
&=\mu(B)\mu(A).
\end{aligned}$$

Therefore, T is ergodic.

(ii) ⇒ (v) If all the rows of Q are identical, then $q_{ij}=\pi_j$ for all i,j. If $vP=v$, then $vQ=v$. This implies that for all j, $v_j=\left(\sum_{i=0}^{N-1}v_i\right)\pi_j$. Thus, v is a multiple of π. Therefore, 1 is a simple eigenvalue.

(v) ⇒ (ii) Suppose 1 is a simple eigenvalue. For any i, let $q_i^*=(q_{i0},\ldots,q_{i(N-1)})$ denote the i-th row of Q. Then, q_i^* is a probability vector. From $Q=QP$, we get $q_i^*=q_i^*P$. By hypothesis π is the only probability vector satisfying $\pi P=\pi$, hence $\pi=q_i^*$, and all the rows of Q are identical. □

3.4 MIXING

As a corollary to the Pointwise Ergodic Theorem we found a new definition of ergodicity; namely, asymptotic average independence. Based on the same idea, we now define other notions of *weak* independence that are stronger than ergodicity.

Definition 3.4.1. Let $(X,\mathcal{F},\mu)$ be a probability space, and $T:X\to X$ a measure preserving transformation. Then,

(i) T is *weakly mixing* if for all $A, B \in \mathcal{F}$, one has

$$\lim_{n\to\infty} \frac{1}{n} \sum_{i=0}^{n-1} \left| \mu(T^{-i}A \cap B) - \mu(A)\mu(B) \right| = 0; \tag{3.3}$$

(ii) T is *strongly mixing* if for all $A, B \in \mathcal{F}$, one has

$$\lim_{n\to\infty} \mu(T^{-n}A \cap B) = \mu(A)\mu(B). \tag{3.4}$$

Notice that strongly mixing implies weakly mixing, and weakly mixing implies ergodicity. This follows from the simple fact that if (a_n) is a sequence of real numbers such that $\lim_{n\to\infty} a_n = 0$, then $\lim_{n\to\infty} \frac{1}{n}\sum_{i=0}^{n-1} |a_i| = 0$, and hence $\lim_{n\to\infty} \frac{1}{n}\sum_{i=0}^{n-1} a_i = 0$. Furthermore, if (a_n) is a bounded sequence of non-negative real numbers, then the following are equivalent (see [65] for the proof):

(i) $\lim_{n\to\infty} \frac{1}{n}\sum_{i=0}^{n-1} |a_i| = 0$,

(ii) $\lim_{n\to\infty} \frac{1}{n}\sum_{i=0}^{n-1} |a_i|^2 = 0$,

(iii) there exists a subset J of the integers of *density zero*, i.e.,

$$\lim_{n\to\infty} \frac{1}{n} \# \left(\{0, 1, \ldots, n-1\} \cap J\right) = 0,$$

such that $\lim_{n\to\infty, n\notin J} a_n = 0$.

Using this one can give three equivalent characterizations of weakly mixing transformations. Can you state them?

Exercise 3.4.1. Let $(X, \mathcal{F}, \mu)$ be a probability space, and $T : X \to X$ a measure preserving transformation. Let $\mathcal{S}$ be a generating semi-algebra of $\mathcal{F}$.

(a) Show that if equation (3.3) holds for all $A, B \in \mathcal{S}$, then T is weakly mixing.

(b) Show that if equation (3.4) holds for all $A, B \in \mathcal{S}$, then T is strongly mixing.

Exercise 3.4.2. Consider the one- or two-sided Bernoulli shift T as given in Examples 1.3.9 and 2.3.2. Show that T is strongly mixing.

Exercise 3.4.3. Let $(X, \mathcal{F}, \mu)$ be a probability space, and let $T : X \to X$ be measure preserving and strongly mixing. Consider the probability space $(Y, \mathcal{G}, \nu)$, where

$$Y = X \times \{0\} \cup X \times \{1\} \cup X \times \{2\},$$

$\mathcal{G}$ is the σ-algebra generated by sets of the form $A \times \{i\}$ with $A \in \mathcal{F}$ and $i = 0, 1, 2$, and ν the measure given by $\nu(A \times \{i\}) = \frac{1}{3}\mu(A)$. Define $S : Y \to Y$ by $S(x, 0) = (x, 1)$, $S(x, 1) = (x, 2)$ and $S(x, 2) = (Tx, 0)$.

(a) Show that S is measure preserving and ergodic with respect to ν.

(b) Show that S^2 is ergodic, but S^3 is not ergodic.

(c) Show that S is not strongly mixing.

Exercise 3.4.4. Let $(X, \mathcal{F}, \mu)$ be a probability space, and $T : X \to X$ a measure preserving transformation. Consider the transformation $T \times T$ defined on $(X \times X, \mathcal{F} \otimes \mathcal{F}, \mu \times \mu)$ by $T \times T(x, y) = (Tx, Ty)$.

(a) Show that $T \times T$ is measure preserving with respect to $\mu \times \mu$.

(b) Show that $T \times T$ is ergodic, if and only if T is weakly mixing.

Exercise 3.4.5. Let $(X, \mathcal{F}, \mu)$ be a probability space and $T : X \to X$ measure preserving.

(a) Show that the following are equivalent.

(i) T is weakly mixing.

(ii) For any dynamical system $(Y, \mathcal{G}, \nu, S)$ consisting of a probability space $(Y, \mathcal{G}, \nu)$ and a measure preserving and ergodic transformation $S : Y \to Y$ the measure preserving dynamical system $(X \times Y, \mathcal{F} \otimes \mathcal{G}, \mu \times \nu, T \times S)$ is ergodic, where $\mathcal{F} \otimes \mathcal{G}$ is the product σ-algebra, $\mu \times \nu$ is the product measure, and $T \times S(x, y) = (Tx, Sy)$.

(b) Show that if $T(x) = T_\theta(x) = x + \theta \pmod 1$ is an irrational rotation on $[0, 1)$ as in Example 2.3.1, then T_θ is **not** weakly mixing with respect to λ, where λ is the normalized Lebesgue measure on $[0, 1)$.

Exercise 3.4.6. Given a measure preserving dynamical system $(X, \mathcal{F}, \mu, T)$ with μ a probability measure, show that $T \times T$ is ergodic with respect to $\mu \times \mu$ if and only if $T \times T \times T$ is ergodic with respect to $\mu \times \mu \times \mu$.

CHAPTER 4

More Ergodic Theorems

In the previous chapter, we studied for $f \in L^1(X, \mathcal{F}, \mu)$ the pointwise behavior of the ergodic averages $\frac{1}{n}\sum_{i=0}^{n-1} f(T^i x)$. Here we see two other ergodic theorems. In Section 4.1 below we restrict our attention to $f \in L^2(X, \mathcal{F}, \mu)$, and study the L^2-convergence of the ergodic averages. The result we prove, the Mean Ergodic Theorem, is due to J. Von Neumann and was published in [64] in 1932, but already announced in 1931 even before the appearance of the Pointwise Ergodic Theorem. In Section 4.2 we consider the Hurewicz Ergodic Theorem, which considers ergodic averages for transformations that are invertible and not necessarily measure preserving. W. Hurewicz presented this result in 1944 in [24].

Throughout the chapter $(X, \mathcal{F}, \mu)$ is a probability space.

4.1 THE MEAN ERGODIC THEOREM

Let $(X, \mathcal{F}, \mu)$ be a probability space and $T : X \to X$ a measure preserving transformation. Recall that $L^2(\mu) = L^2(X, \mathcal{F}, \mu)$ is a Hilbert space with inner product

$$(f, g) = \int_X f\,\overline{g}\,d\mu,$$

where $\overline{g}$ is the complex conjugate of g. See also Section 12.5 for more background on Hilbert spaces. Consider the Koopman operator U_T from (1.1), but restricted to $L^2(\mu)$:

$$U_T(f) = f \circ T, \quad \text{for } f \in L^2(\mu).$$

From the fact that T is measure preserving and Proposition 1.2.3 with $p = 2$ it follows that $(U_T(f), U_T(f)) = (f, f)$, so U_T is an isometry. The adjoint $U_T^* : L^2(\mu) \to L^2(\mu)$ is defined by

$$(U_T(f), g) = (f, U_T^*(g)), \quad \text{for all } f, g \in L^2(\mu).$$

Since U_T is an isometry, one gets that $U_T^* U_T = I_{L^2(\mu)}$, where $I_{L^2(\mu)}$ is the identity operator.

Consider the set

$$\mathcal{I}_2 = \{f \in L^2(\mu) : f = U_T(f) = f \circ T\}$$

of T-invariant functions. It is easy to check that $\mathcal{I}_2$ is a closed linear subspace of $L^2(\mu)$. By Theorem 12.5.1, we can write each element $f \in L^2(\mu)$ in a unique way as $f = g + h$, where $g = \Pi_T(f) \in \mathcal{I}_2$ is the projection of f onto the subspace $\mathcal{I}_2$ and $h \in \mathcal{I}_2^{\perp}$ is in the orthogonal complement. In the following lemma, we identify explicitly the elements of $\mathcal{I}_2^{\perp}$.

Lemma 4.1.1. *Let $B = \{U_T(g) - g : g \in L^2(\mu)\}$. Then $\overline{B} = \mathcal{I}_2^{\perp}$, where $\overline{B}$ is the closure of B in the $L^2(\mu)$-norm.*

Proof. Equivalently, we will show that $\overline{B}^{\perp} = \mathcal{I}_2$. Let $f \in \mathcal{I}_2$, then $U_T(f) = f$ and for any $U_T(g) - g \in B$, we have

$$(f, U_T(g) - g) = (U_T(f), U_T(g)) - (f, g) = (f, g) - (f, g) = 0.$$

So f is orthogonal to every element of B. We show this is true for elements of $\overline{B}$ as well. Let $h \in \overline{B}$, and let $(h_n)_n = (U_T(g_n) - g_n)_n$ be a sequence in B converging in L^2 to h. Then, $(f, h_n) = 0$ for all n, and by the Cauchy Schwartz inequality, $(f, h) = \lim_{n\to\infty}(f, h_n) = 0$. Thus $f \in \overline{B}^{\perp}$. Conversely, suppose $f \in \overline{B}^{\perp}$. For every $g \in L^2(\mu)$ we have $(f, U_T(g) - g) = 0$, so

$$(f, g) = (f, U_T(g)) = (U_T^*(f), g).$$

This implies that $f = U_T^*(f)$. We now show that $f = U_T(f)$. To this end consider

$$\begin{aligned}\|U_T(f) - f\|_2^2 &= (U_T(f) - f, U_T(f) - f) \\ &= \|U_T(f)\|_2 - (U_T(f), f) - (f, U_T(f)) + \|f\|_2 \\ &= \|f\|_2 - (f, U_T^*(f)) - (U_T^*(f), f) + \|f\|_2 = 0.\end{aligned}$$

Thus, $f \in \mathcal{I}_2$. □

With this lemma we are ready to prove the Mean Ergodic Theorem.

Theorem 4.1.1 (Mean Ergodic Theorem). *Let $(X, \mathcal{F}, \mu)$ be a probability space and $T : X \to X$ a measure preserving transformation. Let Π_T denote the orthogonal projection of $L^2(\mu)$ onto the closed subspace $\mathcal{I}_2$. Then, for any $f \in L^2(\mu)$ the sequence $\left(\frac{1}{n}\sum_{i=0}^{n-1} f \circ T^i\right)$ converges in L^2 to $\Pi_T(f)$.*

Proof. By Theorem 12.5.1 and Lemma 4.1.1, any $f \in L^2(\mu)$ can be written as $f = \Pi_T(f) + h$ with $h \in \overline{B}$. From $\Pi_T(f) \in \mathcal{I}_2$ it follows that $\Pi_T(f) \circ T = U_T(\Pi_T(f)) = \Pi_T(f)$. This implies that $\frac{1}{n}\sum_{i=0}^{n-1} \Pi_T(f) \circ T^i$ converges in L^2 to $\Pi_T(f)$. Let $(h_j) = (U_T(g_j) - g_j)$ be a sequence in B converging to h in $L^2(\mu)$. Note that for any $n \geq 1$,

$$\frac{1}{n}\sum_{i=0}^{n-1} h_j \circ T^i = \frac{1}{n}(g_j \circ T^n - g_j) = \frac{1}{n}(U_T^n(g_j) - g_j).$$

Thus,

$$\left\|\frac{1}{n}\sum_{i=0}^{n-1} h_j \circ T^i\right\|_2 = \frac{1}{n}\|U_T^n(g_j) - g_j\|_2 \leq \frac{2}{n}\|g_j\|_2.$$

This shows that $\frac{1}{n}\sum_{i=0}^{n-1} h_j \circ T^i$ converges in L^2 to 0 for all j. Finally, for any $\varepsilon > 0$ there is an $N \geq 1$, such that if $j \geq N$,

$$\begin{aligned}
\left\|\frac{1}{n}\sum_{i=0}^{n-1} h \circ T^i\right\|_2 &\leq \frac{1}{n}\sum_{i=0}^{n-1} \|(h - h_j) \circ T^i\|_2 + \|\frac{1}{n}\sum_{i=0}^{n-1} h_j \circ T^i\|_2 \\
&= \frac{1}{n}\sum_{i=0}^{n-1} \|(h - h_j)\|_2 + \left\|\frac{1}{n}\sum_{i=0}^{n-1} h_j \circ T^i\right\|_2 \\
&= \|(h - h_j)\|_2 + \left\|\frac{1}{n}\sum_{i=0}^{n-1} h_j \circ T^i\right\|_2 \\
&< \varepsilon + \left\|\frac{1}{n}\sum_{i=0}^{n-1} h_j \circ T^i\right\|_2.
\end{aligned}$$

Taking the limit as $n \to \infty$ we see that $\displaystyle\frac{1}{n}\sum_{i=0}^{n-1} h \circ T^i$ converges to 0 in $L^2(\mu)$. Since

$$\frac{1}{n}\sum_{i=0}^{n-1} f \circ T^i = \frac{1}{n}\sum_{i=0}^{n-1} \Pi_T(f) \circ T^i + \frac{1}{n}\sum_{i=0}^{n-1} h \circ T^i,$$

we get the required result. □

Exercise 4.1.1. Let $(X, \mathcal{F}, \mu)$ be a probability space and $T : X \to X$ a measure preserving transformation. Prove that for any $f \in L^1(X, \mathcal{F}, \mu)$ the sequence $\left(\frac{1}{n}\sum_{i=0}^{n-1} f \circ T^i\right)$ converges in L^1 to some T-invariant function $f^* \in L^1(X, \mathcal{F}, \mu)$. (Hint: use Scheffé's Lemma, Lemma 12.4.1).

Exercise 4.1.2. Let $(X, \mathcal{F}, \mu)$ be a probability space, $T : X \to X$ a measure preserving transformation and $1 < p < \infty$, prove that for any $f \in L^p(X, \mathcal{F}, \mu)$ the sequence $\left(\frac{1}{n}\sum_{i=0}^{n-1} f \circ T^i\right)$ converges in L^p to some T-invariant function $f^* \in L^p(X, \mathcal{F}, \mu)$.

4.2 THE HUREWICZ ERGODIC THEOREM

In this section we will assume that $T : X \to X$ is an invertible transformation that is not necessarily measure preserving. The invertibility and the non-singularity of T imply that for any $n \in \mathbb{Z}$,

$$\mu(A) = 0 \quad \text{if and only if} \quad \mu(T^n A) = 0.$$

Hence, for any n the measure $\mu \circ T^n$ defined by $\mu \circ T^n(A) = \mu(T^n A)$ for any measurable set A is equivalent to μ (and hence also equivalent to any other measure $\mu \circ T^k$). By the Radon-Nikodym Theorem there then exists for each n, a non-negative measurable function $\omega_n = \frac{d\mu\circ T^n}{d\mu}$ such that

$$\mu(T^n A) = \int_A \omega_n \, d\mu, \qquad A \in \mathcal{F}.$$

(So $\omega_0 = 1$.) We have the following propositions.

Proposition 4.2.1. *Suppose $(X, \mathcal{F}, \mu)$ is a probability space, and $T : X \to X$ is an invertible transformation. Then for every $f \in L^1(X, \mathcal{F}, \mu)$ and $n \in \mathbb{Z}$,*

$$\int_X f \, d\mu = \int_X f(T^n x)\omega_n(x) \, d\mu(x).$$

Proof. We show the result for indicator functions only, the rest of the proof is left to the reader. Let $A \in \mathcal{F}$ and $n \in \mathbb{Z}$. Then

$$\begin{aligned}\int_X 1_A \, d\mu &= \mu(A) = \mu(T^n(T^{-n}A)) \\ &= \int_{T^{-n}A} \omega_n(x) \, d\mu(x) \\ &= \int_X 1_A(T^n x)\omega_n(x) \, d\mu(x).\end{aligned}$$

□

Proposition 4.2.2. *Under the assumptions of Proposition 4.2.1, one has for all $n, m \geq 1$, that the set*

$$Y := \{x \in X : \omega_{n+m}(x) = \omega_n(x)\omega_m(T^n x)\} \tag{4.1}$$

satisfies $\mu(Y) = 1$.

Proof. By Proposition 4.2.1, for any $A \in \mathcal{F}$,

$$\begin{aligned}
\int_A \omega_n(x)\omega_m(T^n x)\, d\mu(x) &= \int_X 1_{T^n A}(T^n x)\omega_m(T^n x)\omega_n(x)\, d\mu(x) \\
&= \int_X 1_{T^n A}(x)\omega_m(x)\, d\mu(x) \\
&= \int_X 1_{T^{n+m} A}(T^m x)\omega_m(x)\, d\mu(x) \\
&= \int_X 1_{T^{m+n} A}(x)\, d\mu(x) \\
&= \int_A \omega_{n+m}(x)\, d\mu(x).
\end{aligned}$$

Hence, $\omega_{n+m}(x) = \omega_n(x)\omega_m(T^n x)$ for μ-a.e x. □

Exercise 4.2.1. Let $(X, \mathcal{F}, \mu)$ be a probability space, and $T : X \to X$ an invertible transformation. For any measurable function f, set $f_n(x) = \sum_{i=0}^{n-1} f(T^i x)\omega_i(x)$, $n \geq 1$. Show that for all $n, m \geq 1$

$$f_{n+m}(x) = f_n(x) + \omega_n(x) f_m(T^n x)$$

for every element x of the set Y from (4.1).

Proposition 4.2.3. *Suppose T is invertible and conservative. Then*

$$\sum_{n=1}^{\infty} \omega_n(x) = \infty, \quad \mu\text{-a.e.}$$

Proof. Let $A = \{x \in X : \sum_{n=1}^{\infty} \omega_n(x) < \infty\}$. Note that

$$A = \bigcup_{M=1}^{\infty} \Big\{x \in X : \sum_{n=1}^{\infty} \omega_n(x) < M\Big\}.$$

If $\mu(A) > 0$, then there exists an $M \geq 1$ such that the set

$$B = \Big\{x \in X : \sum_{n=1}^{\infty} \omega_n(x) < M\Big\}$$

has positive measure. Then, $\int_B \sum_{n=1}^{\infty} \omega_n \, d\mu < M\mu(B) < \infty$. However, by the Monotone Convergence Theorem,

$$\begin{aligned}\int_B \sum_{n=1}^{\infty} \omega_n \, d\mu &= \sum_{n=1}^{\infty} \int_B \omega_n \, d\mu \\ &= \sum_{n=1}^{\infty} \mu(T^n B) \\ &= \sum_{n=1}^{\infty} \int_X 1_{T^n B} \, d\mu \\ &= \int_X \sum_{n=1}^{\infty} 1_B \circ T^{-n} \, d\mu.\end{aligned}$$

Hence, $\int_X \sum_{n=1}^{\infty} 1_B \circ T^{-n} \, d\mu < \infty$, which implies that

$$\sum_{n=1}^{\infty} 1_B(T^{-n}x) < \infty \quad \mu\text{-a.e.}$$

Therefore, for μ-a.e. x one has $T^{-n}x \in B$ for only finitely many $n \geq 1$, contradicting Corollary 2.1.1. Thus $\mu(A) = 0$, and

$$\sum_{n=1}^{\infty} \omega_n(x) = \infty, \quad \mu\text{-a.e.}$$

□

The following theorem by W. Hurewicz is a generalization of the Pointwise Ergodic Theorem to the setting of non-measure preserving transformations; see also W. Hurewicz' original paper [24]. We give a new proof similar to the proof presented for the Pointwise Ergodic Theorem in Section 3.1, see also [28].

Theorem 4.2.1 (Hurewicz Ergodic Theorem). *Let $(X, \mathcal{F}, \mu)$ be a probability space, and $T : X \to X$ an invertible and conservative transformation. For any $f \in L^1(X, \mathcal{F}, \mu)$ the limit*

$$\lim_{n\to\infty} \frac{\sum_{i=0}^{n-1} f(T^i x)\omega_i(x)}{\sum_{i=0}^{n-1} \omega_i(x)} = f_*(x)$$

exists μ-a.e., is T-invariant and $\int_X f \, d\mu = \int_X f_* \, d\mu$. Furthermore, if T is ergodic, then $f_* = \int_X f \, d\mu$ is a constant.

Proof. By the invertibility and non-singularity of T,

$$\mu(\{x \in X : \omega_n(T^k x) > 0 \quad \text{for all } n \geq 1 \text{ and } k \in \mathbb{Z}\}) = 1.$$

In this proof we only consider points x in this set and that moreover satisfy the conclusions of Propositions 4.2.2 and 4.2.3, which leaves a subset of X of full μ-measure.

Assume with no loss of generality that $f \geq 0$ (otherwise we write $f = f^+ - f^-$, and we consider each part separately). For $n \geq 1$ let

$$\begin{aligned} f_n(x) &= f(x) + f(Tx)\omega_1(x) + \cdots + f(T^{n-1}x)\omega_{n-1}(x), \\ g_n(x) &= \omega_0(x) + \omega_1(x) + \cdots + \omega_{n-1}(x). \end{aligned}$$

Moreover, set

$$\overline{f}(x) = \limsup_{n\to\infty} \frac{f_n(x)}{g_n(x)}, \quad \text{and} \quad \underline{f}(x) = \liminf_{n\to\infty} \frac{f_n(x)}{g_n(x)}.$$

By Proposition 4.2.2, one has $g_{n+m}(x) = g_n(x) + \omega_n(x) g_m(T^n x)$. Using Exercise 4.2.1 and Proposition 4.2.3, we will show that $\overline{f}$ and $\underline{f}$ are T-invariant. To this end,

$$\begin{aligned} \overline{f}(Tx) &= \limsup_{n\to\infty} \frac{f_n(Tx)}{g_n(Tx)} \\ &= \limsup_{n\to\infty} \frac{\dfrac{f_{n+1}(x) - f(x)}{\omega_1(x)}}{\dfrac{g_{n+1}(x) - g_1(x)}{\omega_1(x)}} \\ &= \limsup_{n\to\infty} \frac{f_{n+1}(x) - f(x)}{g_{n+1}(x) - g_1(x)} \\ &= \limsup_{n\to\infty} \left[\frac{f_{n+1}(x)}{g_{n+1}(x)} \cdot \frac{g_{n+1}(x)}{g_{n+1}(x) - g_1(x)} - \frac{f(x)}{g_{n+1}(x) - g_1(x)} \right] \\ &= \limsup_{n\to\infty} \frac{f_{n+1}(x)}{g_{n+1}(x)} \\ &= \overline{f}(x). \end{aligned}$$

Similarly $\underline{f}$ is T-invariant.

Now, to prove that f_* exists, is integrable and T-invariant, it is enough to show that

$$\int_X \underline{f}\, d\mu \geq \int_X f\, d\mu \geq \int_X \overline{f}\, d\mu.$$

For since $\overline{f} - \underline{f} \geq 0$, this would imply that $\overline{f} = \underline{f} = f_*$ μ-a.e. We first prove that $\int_X \overline{f}\, d\mu \leq \int_X f\, d\mu$. Fix any $0 < \epsilon < 1$, and let $L > 0$ be any positive real number. By definition of $\overline{f}$, for any x there exists an integer $m > 0$ such that

$$\frac{f_m(x)}{g_m(x)} \geq \min(\overline{f}(x), L)(1-\epsilon).$$

Now, for any $\delta > 0$ there exists an integer $M > 0$ such that the set

$$X_0 = \{x : \exists\, 1 \leq m \leq M \text{ with } f_m(x) \geq g_m(x) \min(\overline{f}(x), L)(1-\epsilon)\}$$

has measure at least $1 - \delta$. Define F by

$$F(x) = \begin{cases} f(x), & \text{if } x \in X_0, \\ L, & \text{if } x \notin X_0. \end{cases}$$

Notice that $f \leq F$ (why?). For any x and $n \geq 0$, let $a_n = a_n(x) = F(T^n x)\omega_n(x)$, and $b_n = b_n(x) = \min(\overline{f}(x), L)(1-\epsilon)\omega_n(x)$. We now show that (a_n) and (b_n) satisfy the hypothesis of Lemma 3.1.1 with $M > 0$ as above. For any $n \geq 0$ the following holds.

- If $T^n x \in X_0$, then there exists an $1 \leq m \leq M$ such that

  $$f_m(T^n x) \geq \min(\overline{f}(x), L)(1-\epsilon) g_m(T^n x).$$

  Hence,

  $$\omega_n(x) f_m(T^n x) \geq \min(\overline{f}(x), L)(1-\epsilon) g_m(T^n x)\omega_n(x).$$

  Now, with Proposition 4.2.2 we get

  $$\begin{aligned} b_n + \cdots + b_{n+m-1} &= \min(\overline{f}(x), L)(1-\epsilon) g_m(T^n x)\omega_n(x) \\ &\leq \omega_n(x) f_m(T^n x) \\ &= f(T^n x)\omega_n(x) + f(T^{n+1}x)\omega_{n+1}(x) \\ &\quad + \cdots + f(T^{n+m-1}x)\omega_{n+m-1}(x) \\ &\leq F(T^n x)\omega_n(x) + F(T^{n+1}x)\omega_{n+1}(x) \\ &\quad + \cdots + F(T^{n+m-1}x)\omega_{n+m-1}(x) \\ &= a_n + a_{n+1} + \cdots + a_{n+m-1}. \end{aligned}$$

- If $T^n x \notin X_0$, then take $m = 1$ since

  $$a_n = F(T^n x)\omega_n(x) = L\omega_n(x) \geq \min(\overline{f}(x), L)(1-\epsilon)\omega_n(x) = b_n.$$

Hence by T-invariance of $\overline{f}$ and Lemma 3.1.1, for all integers $N > M$ one has

$$F(x) + F(Tx)\omega_1(x) + \cdots + F(T^{N-1}x)\omega_{N-1}(x) \\ \geq \min(\overline{f}(x), L)(1-\epsilon)g_{N-M}(x).$$

Integrating both sides, and using Proposition 4.2.1 together with the T-invariance of $\overline{f}$ one gets

$$N \int_X F\,d\mu \geq \int_X \min(\overline{f}(x), L)(1-\epsilon)g_{N-M}(x)\,d\mu(x) \\ = (N-M)\int_X \min(\overline{f}(x), L)(1-\epsilon)\,d\mu(x).$$

Since

$$\int_X F\,d\mu = \int_{X_0} f\,d\mu + L\mu(X \setminus X_0),$$

one has

$$\int_X f\,d\mu \geq \int_{X_0} f\,d\mu \\ = \int_X F\,d\mu - L\mu(X \setminus X_0) \\ \geq \frac{(N-M)}{N}\int_X \min(\overline{f}(x), L)(1-\epsilon)\,d\mu(x) - L\delta.$$

Now letting first $N \to \infty$, then $\delta \to 0$, then $\epsilon \to 0$, and lastly $L \to \infty$ one gets together with the Monotone Convergence Theorem that $\overline{f}$ is integrable, and

$$\int_X f\,d\mu \geq \int_X \overline{f}\,d\mu.$$

We now prove that

$$\int_X f\,d\mu \leq \int_X \underline{f}\,d\mu.$$

Fix $\epsilon > 0$. Then for any x there exists an integer m such that

$$\frac{f_m(x)}{g_m(x)} \leq (\underline{f}(x) + \epsilon).$$

Fix $\delta_0 > 0$. Then there is a $\delta > 0$ such that for any measurable set A with $\mu(A) < \delta$ it holds that $\int_A f d\mu < \delta_0$. Fix such a δ. Then there exists an integer $M > 0$ such that the set

$$Y_0 = \{x : \exists\, 1 \leq m \leq M \text{ with } f_m(x) \leq (\underline{f}(x) + \epsilon)g_m(x)\}$$

has measure at least $1 - \delta$. Define G by

$$G(x) = \begin{cases} f(x), & \text{if } x \in Y_0, \\ 0, & \text{if } x \notin Y_0. \end{cases}$$

Notice that $G \leq f$. For any $n \geq 0$ let $b_n = G(T^n x)\omega_n(x)$, and $a_n = (\underline{f}(x) + \epsilon)\omega_n(x)$. We now check that the sequences (a_n) and (b_n) satisfy the hypothesis of Lemma 3.1.1 with $M > 0$ as above.

- If $T^n x \in Y_0$, then there exists an $1 \leq m \leq M$ such that

$$f_m(T^n x) \leq (\underline{f}(x) + \epsilon) g_m(T^n x).$$

  Hence,

$$\begin{aligned} \omega_n(x) f_m(T^n x) &\leq (\underline{f}(x) + \epsilon) g_m(T^n x)\omega_n(x) \\ &= (\underline{f}(x) + \epsilon)(\omega_n(x) + \cdots + \omega_{n+m-1}(x)). \end{aligned}$$

  By Proposition 4.2.2, and the fact that $f \geq G$, one gets

$$\begin{aligned} b_n + \cdots + b_{n+m-1} &= G(T^n x)\omega_n(x) + \cdots + G(T^{n+m-1}x)\omega_{n+m-1}(x) \\ &\leq f(T^n x)\omega_n(x) + \cdots + f(T^{n+m-1}x)\omega_{n+m-1}(x) \\ &= \omega_n(x) f_m(T^n x) \\ &\leq (\underline{f}(x) + \epsilon)(\omega_n(x) + \cdots + \omega_{n+m+1}(x)) \\ &= a_n + \cdots + a_{n+m-1}. \end{aligned}$$

- If $T^n x \notin Y_0$, then take $m = 1$ since

$$b_n = G(T^n x)\omega_n(x) = 0 \leq (\underline{f}(x) + \epsilon)\omega_n(x) = a_n.$$

Hence by Lemma 3.1.1 one has for all integers $N > M$,

$$\begin{aligned} G(x) + G(Tx)\omega_1(x) + \cdots + G(T^{N-M-1}x)\omega_{N-M-1}(x) \\ \leq (\underline{f}(x) + \epsilon) g_N(x). \end{aligned}$$

Integrating both sides yields

$$(N - M) \int_X G \, d\mu \leq N \Big(\int_X \underline{f} \, d\mu + \epsilon \Big).$$

Since $\mu(X \setminus Y_0) < \delta$, then $\int_{X\setminus Y_0} f\,d\mu < \delta_0$. Hence,

$$\begin{aligned}\int_X f\,d\mu &= \int_X G\,d\mu + \int_{X\setminus Y_0} f\,d\mu \\ &\leq \frac{N}{N-M}\int_X (\underline{f}(x)+\epsilon)\,d\mu(x) + \delta_0.\end{aligned}$$

Now, letting first $N \to \infty$, then $\delta \to 0$, $\delta_0 \to 0$ and finally $\epsilon \to 0$, one gets

$$\int_X f\,d\mu \leq \int_X \underline{f}\,d\mu.$$

This shows that

$$\int_X \underline{f}\,d\mu \geq \int_X f\,d\mu \geq \int_X \overline{f}\,d\mu,$$

hence, $\overline{f} = \underline{f} = f_*$ μ-a.e., and f_* is T-invariant. The last statement follows from Theorem 2.2.2 and the fact that μ is a probability measure.

□

Example 4.2.1. Consider the binary odometer from Example 1.3.11. There we have seen that T is non-singular with respect to the product measure μ_p on $X = \{0,1\}^{\mathbb{N}}$ and that $\omega_1(x) = \frac{d\mu_p \circ T}{d\mu_p}(x) = \left(\frac{p}{1-p}\right)^{m(x)}$, where $m(x) = \inf\{n \geq 1 : x_n = 0\} - 2$. By Proposition 4.2.2, one has

$$\omega_n(x) = \omega_1(x)\omega_1(Tx)\cdots\omega_1(T^{n-1}x) = \left(\frac{p}{1-p}\right)^{m(x)+m(Tx)+\cdots+m(T^{n-1}x)}$$

for μ_p-a.e. $x \in X$. The map T is in fact ergodic (see Example 6.2.1) and conservative (by Exercise 11.2.2). Consider the function f on X given by $f((x_1, x_2, \ldots)) = x_1$, then f is integrable with $\int_X f\,d\mu_p = 1 - p$. Furthermore, for each $n \geq 1$,

$$\sum_{i=0}^{n-1} f(T^i x)\omega_i(x) = \begin{cases} \displaystyle\sum_{i=0}^{\lfloor (n-1)/2 \rfloor} \omega_{2i}(x), & \text{if } x_1 = 1, \\ \displaystyle\sum_{i=1}^{\lfloor n/2 \rfloor} \omega_{2i-1}(x), & \text{if } x_1 = 0. \end{cases}$$

In either case, by the Hurewicz Ergodic Theorem we have for μ_p-a.e. x,

$$\lim_{n\to\infty} \frac{\sum_{i=0}^{n-1} f(T^i x)\omega_i(x)}{\sum_{i=0}^{n-1} \omega_i(x)} = 1 - p.$$

CHAPTER 5

Isomorphisms and Factor Maps

In this chapter we only consider dynamical systems on probability spaces. Given two such dynamical systems $(X, \mathcal{F}, \mu, T)$ and $(Y, \mathcal{G}, \nu, S)$, it is natural to wonder when these are actually two versions of the same thing. What would it mean for two systems to be the same? And when could one system be considered as a subsystem of the other? In this chapter we define the notions that make this precise.

5.1 MEASURE PRESERVING ISOMORPHISMS

Let $(X, \mathcal{F}, \mu, T)$ and $(Y, \mathcal{G}, \nu, S)$ be two dynamical systems on probability spaces with T and S measure preserving. On each space there are two important structures:

(1) the measure structure given by the σ-algebra and the probability measure. Note that in this context sets of measure zero can be ignored;

(2) the dynamical structure given by the measure preserving transformation.

So our notion of *being the same* must mean that we have a map

$$\psi : (X, \mathcal{F}, \mu, T) \to (Y, \mathcal{G}, \nu, S)$$

that preserves both these structures, i.e., that satisfies

(i) ψ is one-to-one and onto,

(ii) ψ is measurable ($\psi^{-1}(G) \in \mathcal{F}$, for all $G \in \mathcal{G}$) and so is ψ^{-1},

(iii) ψ preserves the measures, i.e., $\nu = \mu \circ \psi^{-1}$, and the same holds for ψ^{-1},

(iv) ψ preserves the dynamics of T and S, so $\psi \circ T = S \circ \psi$, which is the same as saying that the diagram from Figure 5.1 commutes,

and for all these properties it is sufficient that they hold almost everywhere.

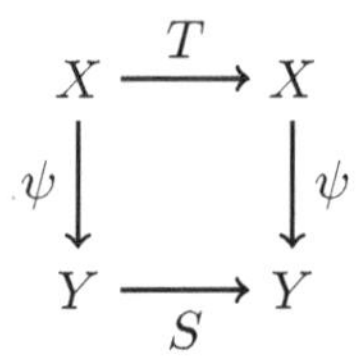

Figure 5.1 A commuting diagram.

The last property means that T-orbits are mapped to S-orbits:

$$\begin{array}{ccccccccccc} x & \to & Tx & \to & T^2x & \to & \cdots & \to & T^nx & \to & \cdots \\ \downarrow & & \downarrow & & \downarrow & & \downarrow & & \downarrow & & \\ \psi(x) & \to & S\psi(x) & \to & S^2\psi(x) & \to & \cdots & \to & S^n\psi(x) & \to & \cdots \end{array}$$

This leads to the following definition.

Definition 5.1.1. Two dynamical systems $(X, \mathcal{F}, \mu, T)$ and $(Y, \mathcal{G}, \nu, S)$ on probability spaces are *(measure preservingly or metrically) isomorphic* if there exist measurable sets $N \subseteq X$ and $M \subseteq Y$ with $\mu(N) = 0 = \nu(M)$ and $T(X \setminus N) \subseteq X \setminus N$, $S(Y \setminus M) \subseteq Y \setminus M$, and finally if there exists a measurable map $\psi : X \setminus N \to Y \setminus M$ such that (i)–(iv) are satisfied for the systems restricted to $X \setminus N$ and $Y \setminus M$. The map ψ is called a *(measurable or metric) isomorphism.*

Exercise 5.1.1. Suppose $(X, \mathcal{F}, \mu, T)$ and $(Y, \mathcal{G}, \nu, S)$ are two isomorphic dynamical systems. Show that

(a) T is ergodic if and only if S is ergodic,

(b) T is weakly mixing if and only if S is weakly mixing,

(c) T is strongly mixing if and only if S is strongly mixing.

Example 5.1.1. Let $\mathbb{S}^1 = \{z \in \mathbb{C} : |z| = 1\}$ be equipped with the Lebesgue σ-algebra $\hat{\mathcal{B}}$ on $\mathbb{S}^1$, and Haar measure (i.e., normalized Lebesgue measure on the unit circle). Define $S : \mathbb{S}^1 \to \mathbb{S}^1$ by $Sz = z^2$; equivalently $Se^{2\pi i\theta} = e^{2\pi i(2\theta)}$. Figure 5.2(a) shows the graph. One can easily check that S is measure preserving. In fact, the map S is isomorphic to the doubling map T on $([0,1),\mathcal{B},\lambda)$ given by $Tx = 2x \pmod 1$ (see Examples 1.3.3 and 2.3.3). Define the map $\psi : [0,1) \to \mathbb{S}^1$ by $\psi(x) = e^{2\pi i x}$. We leave it to the reader to check that ψ is an isomorphism.

Example 5.1.2. Let $T : [0,1) \to [0,1)$ be the $\times N$ transformation given by $Tx = Nx - \lfloor Nx \rfloor$ for some integer $N \geq 2$ as in Section 3.2. See Figure 5.2(b) for the graph. As we have seen iterations of T provide for each $x \in [0,1)$ that is not of the form $\frac{k}{N^i}$ a unique base N expansion $x = \sum_{k=1}^{\infty} \frac{a_k}{N^k}$ with $a_k \in \{0,1,\ldots,N-1\}$ for all $k \geq 1$ and with the property that there is no k such that $a_j = N-1$ for all $j \geq k$. Let $Y := \{0,1,\ldots,N-1\}^{\mathbb{N}}$. We construct an isomorphism between $([0,1),\mathcal{B},\lambda,T)$ and the Bernoulli shift $(Y,\mathcal{C},\mu,S)$ (see Example 1.3.9) with μ the uniform Bernoulli measure defined on cylinders by

$$\mu(\{(y_i)_{i\geq 1} \in Y : y_1 = a_1, y_2 = a_2, \ldots, y_n = a_n\}) = \frac{1}{N^n}.$$

Define $\psi : [0,1) \to Y$ by

$$\psi : x = \sum_{k=1}^{\infty} \frac{a_k}{N^k} \mapsto (a_k)_{k\geq 1},$$

where $\sum_{k=1}^{\infty} \frac{a_k}{N^k}$ is the base N expansion of x produced by T. Let

$$C(i_1,\ldots,i_n) = \{(y_i)_{i\geq 1} \in Y : y_1 = i_1, \ldots, y_n = i_n\}.$$

In order to see that ψ is an isomorphism we verify measurability and measure preservingness on such cylinders:

$$\psi^{-1}(C(i_1,\ldots,i_n)) = \left[\frac{i_1}{N} + \frac{i_2}{N^2} + \cdots + \frac{i_n}{N^n}, \frac{i_1}{N} + \frac{i_2}{N^2} + \cdots + \frac{i_n+1}{N^n}\right)$$

and

$$\lambda(\psi^{-1}(C(i_1,\ldots,i_n))) = \frac{1}{N^n} = \mu(C(i_1,\ldots,i_n)).$$

Note that

$$\mathcal{N} = \{(y_i)_{i\geq 1} \in Y : \exists k \geq 1,\ y_i = N-1 \text{ for all } i \geq k\}$$

is a subset of Y of measure 0. Then $\psi : [0,1) \to Y \setminus \mathcal{N}$ is a bijection. Finally, it is easy to see that $\psi \circ T = S \circ \psi$.

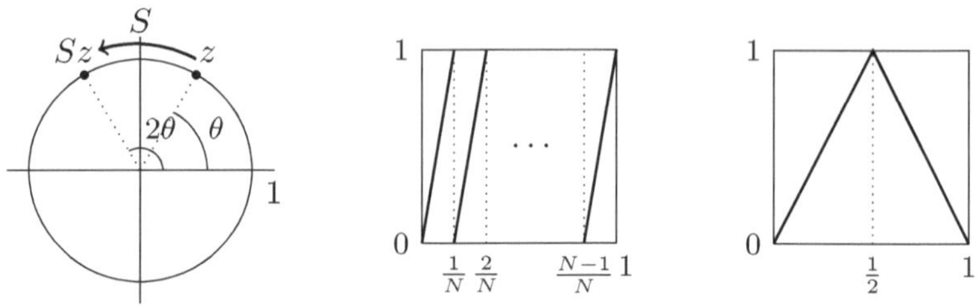

(a) Doubling (b) $\times N$ (c) Tent

Figure 5.2 The transformations from Examples 5.1.1, 5.1.2 and 5.1.3.

Example 5.1.3. Consider the doubling map $Tx = 2x \pmod 1$ and the *tent map* $F : [0,1] \to [0,1]$ defined by $Fx = \min\{2x, 2-2x\}$, see Figure 5.2(c), both on the space $([0,1], \mathcal{B}, \lambda)$. (We can take $T1 = 1$ for example.) As in Example 1.3.3 we can associate to each $x \in [0,1]$ an infinite sequence $(c_n) \in \{0,1\}^{\mathbb{N}}$ that codes the orbit of x under F by setting for each $n \geq 1$,

$$c_n = c_n(x) = \begin{cases} 0, & \text{if } 0 \leq F^{n-1}x < \frac{1}{2}, \\ 1, & \text{if } \frac{1}{2} \leq F^{n-1}x \leq 1. \end{cases}$$

Very similarly to Example 5.1.2 one can show that the map $\phi : [0,1] \to \{0,1\}^{\mathbb{N}}$ defined by $\phi(x) = (c_n)_{n\geq 1}$ is an isomorphism between $([0,1], \mathcal{B}, \lambda, F)$ and the uniform Bernoulli shift $(\{0,1\}^{\mathbb{N}}, \mathcal{C}, \mu, S)$. The composition $\psi \circ \phi^{-1} : [0,1] \to [0,1]$ of ϕ with the isomorphism from Example 5.1.2 then becomes an isomorphism between $([0,1], \mathcal{B}, \lambda, T)$ and $([0,1], \mathcal{B}, \lambda, F)$.

To show that two systems are isomorphic it is enough to identify one isomorphism. Proving that two systems are not isomorphic requires different techniques.

Example 5.1.4 (Positive and negative β-transformations). The β-transformation was defined in Example 1.3.6 for $\beta = \frac{1+\sqrt{5}}{2}$, the golden mean. In Exercise 1.3.3 the relation between this map and number expansions in base β was identified. One can define a β-transformation for any value $\beta \in (1,2)$ on the interval $[0,1]$ by $T_\beta = \beta x \pmod 1$. Each of these transformations has an invariant probability measure μ_β that is equivalent to the Lebesgue measure (i.e., has the same null sets). The density of this measure was given in Example 1.3.6 for the golden mean and we will give it for general $\beta \in (1,2)$ in Example 6.3.4. One can also consider the corresponding *negative β-transformation* $S_\beta : [0,1] \to [0,1]$ defined by

$$S_\beta x = \begin{cases} 1 - \beta x, & \text{if } 0 \le x < \frac{1}{\beta}, \\ 2 - \beta x, & \text{if } \frac{1}{\beta} \le x \le 1. \end{cases}$$

This map is related to β-expansions of numbers in $[0,1]$ with a negative base $-\beta$ and digits $1,2$, i.e., each $x \in [0,1]$ can be expressed as a sum $x = \sum_{i=1}^{\infty} \frac{-b_i}{(-\beta)^i}$ with $b_i \in \{1,2\}$ for each $i \ge 1$. We show that if $1 < \beta < \frac{1+\sqrt{5}}{2}$, then T_β and S_β are not isomorphic.

Assume that $\phi : [0,1] \to [0,1]$ is a map that satisfies $\phi \circ T_\beta = S_\beta \circ \phi$. Then for each $n \ge 1$ it must hold that $\phi \circ T_\beta^n = S_\beta^n \circ \phi$. Figure 5.3 shows the first three iterates of the maps T_β and S_β for some value $\beta < \frac{1+\sqrt{5}}{2}$. Let I be the set of $x \in [0,1)$ that have precisely two pre-image under T_β^3. As one can see from Figure 5.3(c) $I = [T^2(\beta-1), 1)$, so $\mu_\beta(I) > 0$. On the other hand, there are no points x that have precisely two pre-images under S_β, so $\phi(I) = \emptyset$. This shows that no matter what invariant probability measure ν_β we would consider for S_β, we would always get that $0 < \mu_\beta(I) \neq \nu_\beta(\phi(I)) = 0$. So, no map ϕ can satisfy both conditions (iii) and (iv) from Definition 5.1.1, which implies that $([0,1), \mathcal{B}, \mu_\beta, T_\beta)$ cannot be isomorphic to $([0,1], \mathcal{B}, \nu_\beta, S_\beta)$ for any probability space $([0,1], \mathcal{B}, \nu_\beta)$. In fact in [26] it was shown that for any $\beta \in (1,2)$ that is not a so-called *multinacci number*, i.e., the positive real root of a polynomial of the form

$$x^k - x^{k-1} - \cdots - x - 1 = 0, \quad k \ge 2, \tag{5.1}$$

the system $([0,1), \mathcal{B}, \mu_\beta, T_\beta)$ cannot be isomorphic to $([0,1], \mathcal{B}, \nu_\beta, S_\beta)$ for any probability space $([0,1], \mathcal{B}, \nu_\beta)$.

Exercise 5.1.2. Let $T : [0,1)^2 \to [0,1)^2$ be the baker's transformation from Exercise 1.3.1 given by

$$T(x,y) = \begin{cases} (2x, \frac{y}{2}), & \text{if } 0 \le x < \frac{1}{2}, \\ (2x-1, \frac{y+1}{2}), & \text{if } \frac{1}{2} \le x < 1. \end{cases}$$

Show that T is isomorphic to the two-sided Bernoulli shift S on $(\{0,1\}^{\mathbb{Z}}, \mathcal{C}, \mu)$ with μ the uniform Bernoulli measure.

Exercise 5.1.3. Consider the measurable space $(\mathbb{R}, \mathcal{B})$, where $\mathbb{R}$ is the

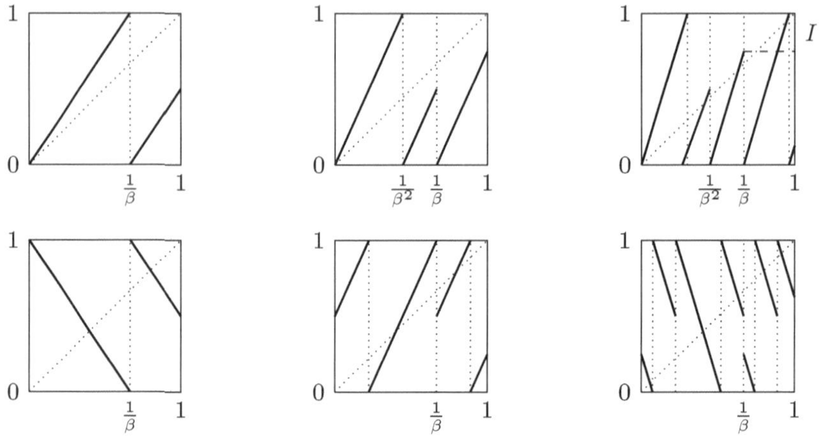

(a) T_β (top) and S_β (bottom)

(b) T_β^2 (top) and S_β^2 (bottom)

(c) T_β^3 (top) and S_β^3 (bottom)

Figure 5.3 The first three iterates of the positive and negative β-transformations for $\beta = \frac{3}{2}$ from Example 5.1.4. The interval I of points that have precisely two pre-images under T_β^3 is indicated in the top picture of (c).

real line and $\mathcal{B}$ is the Lebesgue σ-algebra. Define a transformation $T : \mathbb{R} \to \mathbb{R}$ by $T0 = 0$ and

$$Tx = \frac{1}{2}\Big(x - \frac{1}{x}\Big)$$

for $x \neq 0$. Consider the measure μ on $\mathcal{B}$ given by

$$\mu(A) = \int_A \frac{1}{\pi(1+x^2)}\, d\lambda(x), \qquad \text{for all } A \in \mathcal{B}.$$

(a) Show that T is measure preserving with respect to the probability measure μ.

(b) Let $\psi : \mathbb{R} \to [0,1)$ be defined by

$$\psi(x) = \frac{1}{\pi}\arctan x + \frac{1}{2}.$$

Show that ψ is a measurable isomorphism between the dynamical systems $(\mathbb{R}, \mathcal{B}, \mu, T)$ and $([0,1), \mathcal{B} \cap [0,1), \lambda, S)$, where $Sx = 2x \pmod 1$ is the doubling map.

(c) Show that T is strongly mixing with respect to μ.

5.2 FACTOR MAPS

In the above section, we discussed the notion of isomorphism which describes when two dynamical systems are considered the same. Now, we give a precise definition of what it means for a dynamical system to be a subsystem of another one.

Definition 5.2.1. Let $(X, \mathcal{F}, \mu, T)$ and $(Y, \mathcal{G}, \nu, S)$ be two dynamical systems. We say that S is a *factor* of T if there exist measurable sets $N \in \mathcal{F}$ and $M \in \mathcal{G}$, such that $\mu(N) = 0 = \nu(M)$ and $T(X \setminus N) \subseteq X \setminus N$ and $S(Y \setminus M) \subseteq Y \setminus M$, and finally if there exists a measurable and measure preserving map $\psi : X \setminus N \to Y \setminus M$ that is surjective, and satisfies $\psi(Tx) = S\psi(x)$ for all $x \in X \setminus N$. We call ψ a *factor map.*

Remark 5.2.1. Notice that if ψ is a factor map, then $\mathcal{E} = \psi^{-1}\mathcal{G}$ is a T-invariant sub-σ-algebra of $\mathcal{F}$, since

$$T^{-1}\mathcal{E} = T^{-1}\psi^{-1}\mathcal{G} = \psi^{-1}S^{-1}\mathcal{G} \subseteq \psi^{-1}\mathcal{G} = \mathcal{E}.$$

Example 5.2.1. As in Exercise 5.1.2 let T be the baker's transformation on $([0,1)^2, \mathcal{B}, \lambda)$. Let $(X = \{0,1\}^{\mathbb{N}}, \mathcal{C}, \mu, S)$ be the uniform Bernoulli shift (see Example 1.3.9). Define $\psi : [0,1)^2 \to X$ by

$$\psi(x, y) = (a_1, a_2, \ldots),$$

where $x = \sum_{n=1}^{\infty} \frac{a_n}{2^n}$ is the binary expansion of x. It is easy to check that ψ is a factor map.

Exercise 5.2.1. Let T be the left shift on $X = \{0,1,2\}^{\mathbb{N}}$ which is endowed with the σ-algebra $\mathcal{C}$ generated by the cylinder sets, and the uniform Bernoulli measure μ, giving each symbol probability $\frac{1}{3}$, i.e.,

$$\mu\left(\{x \in X : x_1 = i_1, x_2 = i_2, \ldots, x_n = i_n\}\right) = \frac{1}{3^n},$$

for all $i_1, i_2, \ldots, i_n \in \{0,1,2\}$. Let S be the left shift on $Y = \{0,1\}^{\mathbb{N}}$ which is endowed with the σ-algebra $\mathcal{D}$ generated by the cylinder sets, and the product measure ν giving the symbol 0 probability $\frac{1}{3}$ and the symbol 1 probability $\frac{2}{3}$, i.e.,

$$\mu(\{y \in Y : y_1 = j_1, y_2 = j_2, \ldots, y_n = j_n\}) \\ = \left(\frac{2}{3}\right)^{j_1+\cdots+j_n} \left(\frac{1}{3}\right)^{n-(j_1+\cdots+j_n)},$$

where $j_1, j_2, \ldots, j_n \in \{0,1\}$. Show that S is a factor of T.

Exercise 5.2.2. Show that a factor of an ergodic (weakly mixing/strongly mixing) transformation is also ergodic (weakly mixing/strongly mixing).

5.3 NATURAL EXTENSIONS

Exercise 5.2.2 shows that many properties of a dynamical system carry over to a factor. This fact can also be put to use in the other direction; sometimes the properties of a system can be described better by embedding it in a larger system that is easier to analyze, for example by choosing the larger system to be invertible. A natural extension is such a larger system. By putting a minimality condition on it, it is guaranteed that many properties of the original system also hold for the natural extension.

Definition 5.3.1. Let $(Y, \mathcal{G}, \nu, S)$ be a **non-invertible** measure preserving dynamical system. An **invertible** measure preserving dynamical system $(X, \mathcal{F}, \mu, T)$ is called a *natural extension* of $(Y, \mathcal{G}, \nu, S)$ if S is a factor of T and the factor map ψ satisfies

$$\bigvee_{n=0}^{\infty} T^n \psi^{-1}(\mathcal{G}) = \mathcal{F},$$

where $\bigvee_{n=0}^{\infty} T^n \psi^{-1}(\mathcal{G})$ is the smallest σ-algebra containing the σ-algebras $T^k \psi^{-1}(\mathcal{G})$ for all $k \geq 0$.

Example 5.3.1. Let T on $(\{0,1\}^{\mathbb{Z}}, \mathcal{C}, \mu)$ be the two-sided Bernoulli shift, and S on $(\{0,1\}^{\mathbb{N}}, \mathcal{G}, \nu)$ the one-sided Bernoulli shift, both spaces are endowed with the uniform Bernoulli measure. Notice that T is invertible, while S is not. Define $\psi : \{0,1\}^{\mathbb{Z}} \to \{0,1\}^{\mathbb{N}}$ by

$$\psi(\ldots, x_{-1}, x_0, x_1, \ldots) = (x_1, x_2, \ldots).$$

Then, ψ is a factor map. We claim that

$$\bigvee_{n=0}^{\infty} T^n \psi^{-1}(\mathcal{G}) = \mathcal{C}.$$

To prove this, we show that $\bigvee_{n=0}^{\infty} T^n \psi^{-1}(\mathcal{G})$ contains all cylinders generating $\mathcal{C}$.

Let $\Delta = \{x \in X : x_{-k} = a_{-k}, \ldots, x_\ell = a_\ell\}$ be an arbitrary cylinder in $\mathcal{C}$, and let $D = \{y \in Y : y_1 = a_{-k}, \ldots, y_{k+\ell+1} = a_\ell\}$ which is a cylinder in $\mathcal{G}$. Then,

$$\psi^{-1}(D) = \{x \in X : x_1 = a_{-k}, \ldots, x_{k+\ell+1} = a_\ell\}$$

and $T^{k+1}\psi^{-1}(D) = \Delta$. This shows that

$$\bigvee_{n=0}^{\infty} T^n \psi^{-1}(\mathcal{G}) = \mathcal{C}.$$

Thus, T is a natural extension of S.

Natural extensions were extensively studied by V. A. Rohlin in [55]. In 1961 he gave a canonical construction of a natural extension that resembles the idea of converting a one-sided shift to a two-sided shift. We will outline the construction and end with a theorem summarizing some important results. Start with a non-invertible measure preserving system $(Y, \mathcal{G}, \nu, S)$ with ν a probability measure. As the underlying space of the natural extension we take the set

$$X = \{x = (x_0, x_1, \ldots) : x_i \in Y \text{ and } Sx_{i+1} = x_i,\ i \geq 0\}.$$

The sequence $(x_1, x_2, \ldots)$ can be seen as a possible past of the point x_0 and is typically not unique due to the non-invertibility of S. On X we consider the product σ-algebra generated by sets of the form

$$A_{i,C} = \{x = (x_0, x_1, \ldots) : x_i \in C\},$$

with $C \in \mathcal{G}$ and $i \geq 0$. Note that any cylinder set can be written as a set of this form, since if $C_0, C_1, \ldots, C_n \in \mathcal{G}$, then

$$\{x = (x_0, x_1, \ldots) : x_j \in C_j,\ 0 \leq j \leq n\} = A_{n, S^{-n}C_0 \cap S^{-(n-1)}C_1 \cap \cdots \cap C_n}.$$

On $(X, \mathcal{F})$ we consider the measure μ defined on the generating sets by $\mu(A_{i,C}) = \nu(C)$. Note that if $C_0, C_1, \ldots, C_n \in \mathcal{G}$, then

$$\begin{aligned} &\mu(\{x = (x_0, x_1, \ldots) : x_j \in C_j,\ 0 \leq j \leq n\}) \\ &\qquad\qquad = \nu(S^{-n}C_0 \cap S^{-(n-1)}C_1 \cap \ldots \cap C_n). \end{aligned}$$

The above shows that we have a pre-measure on the collection of all cylinders so by the Carathéodory Extension Theorem the measure μ is a well defined probability measure on $\mathcal{F}$. On the probability space $(X, \mathcal{F}, \mu)$ we consider the transformation $T : X \to X$ defined by

$$Tx = T(x_0, x_1, x_2, \ldots) = (Sx_0, Sx_1, Sx_2, \ldots) = (Sx_0, x_0, x_1, \ldots).$$

Since $T^{-1}A_{0,C} = A_{0,S^{-1}C}$ and $T^{-1}A_{i,C} = A_{i-1,C}$ for $i > 0$, we see that T is measurable and measure preserving. Furthermore, T is invertible with inverse defined by $T^{-1}(x_0, x_1, \ldots) = (x_1, x_2, \ldots)$. Since $TA_{i,C} = A_{i+1,C}$ for $i \geq 0$, we see that T^{-1} is invertible and measure preserving. Using the factor map $\psi : X \to Y$ defined by $\psi(x_0, x_1, \ldots) = x_0$, one can easily check that Definition 5.3.1 is satisfied, so that the invertible dynamical system $(X, \mathcal{F}, \mu, T)$ is a natural extension of $(Y, \mathcal{G}, \nu, S)$.

Theorem 5.3.1 (Rohlin). *Let $(Y, \mathcal{G}, \nu)$ be a probability space and $S : Y \to Y$ a measure preserving and non-invertible transformation.*

(i) *There exists a natural extension $(X, \mathcal{F}, \mu, T)$ of $(Y, \mathcal{G}, \nu, S)$.*

(ii) *If $(X, \mathcal{F}, \mu, T)$ and $(X', \mathcal{F}', \mu', T')$ are two natural extensions of $(Y, \mathcal{G}, \nu, S)$, then $(X, \mathcal{F}, \mu, T)$ and $(X', \mathcal{F}', \mu', T')$ are isomorphic.*

(iii) *If S is ergodic/weakly mixing/strongly mixing and $(X, \mathcal{F}, \mu, T)$ is a natural extension, then T is ergodic/weakly mixing/strongly mixing.*

From Theorem 5.3.1(ii) we see that a natural extension of a dynamical system is unique up to isomorphism. With this in mind we can therefore refer to *the* natural extension of a system.

Exercise 5.3.1. Prove that if $(X, \mathcal{F}, \mu, T)$ is a natural extension of $(Y, \mathcal{G}, \nu, S)$ and S is strongly mixing, then T is strongly mixing.

Exercise 5.3.2. Let $T : [0, 1) \to [0, 1)$ be the Lüroth map from Example 1.3.5 defined by

$$Tx = \begin{cases} n(n+1)x - n, & \text{if } x \in \left[\frac{1}{n+1}, \frac{1}{n}\right),\ n \geq 1, \\ 0, & \text{if } x = 0. \end{cases}$$

Recall from Exercise 1.3.2 that Lebesgue measure λ is a T-invariant measure and that for any point $x \in [0, 1)$ with $T^k x \neq 0$ for all k the sequence (a_k) defined by $a_k = a_k(x) = n$ if $T^{k-1}x \in \left[\frac{1}{n}, \frac{1}{n-1}\right)$, $n \geq 2$, gives

$$x = \frac{1}{a_1} + \frac{1}{a_1(a_1 - 1)a_2} + \cdots + \frac{1}{a_1(a_1 - 1) \cdots a_{k-1}(a_{k-1} - 1)a_k} + \cdots.$$

(a) Consider the Bernoulli shift $(X, \mathcal{C}, \mu, S)$, where $X = \{2, 3, \cdots\}^{\mathbb{N}}$ and μ is the product measure with $\mu(\{x : x_1 = j\}) = \frac{1}{j(j-1)}$. Show that $([0, 1), \mathcal{B}, \lambda, T)$ and $(X, \mathcal{C}, \mu, S)$ are isomorphic.

(b) Show that T is strongly mixing.

(c) Consider the product space $([0,1)^2, \mathcal{B}^2, \lambda^2)$. Define the transformation $\mathcal{T} : [0,1)^2 \to [0,1)^2$ by

$$\mathcal{T}(x,y) = \begin{cases} \left(Tx, \frac{y+n}{n(n+1)}\right), & \text{if } x \in \left[\frac{1}{n+1}, \frac{1}{n}\right), \\ (0,0), & \text{if } x = 0. \end{cases}$$

Show that $([0,1)^2, \mathcal{B}^2, \lambda^2, \mathcal{T})$ is the natural extension of $([0,1), \mathcal{B}, \lambda, T)$.

Exercise 5.3.3. Let $\beta > 1$ be the real number satisfying $\beta^3 = \beta^2 + \beta + 1$ (sometimes called the *tribonacci number*, see also (5.1)) and consider the β-transformation $T_\beta : [0,1) \to [0,1)$ given by $T_\beta x = \beta x \pmod 1$. Define a measure ν on the Lebesgue σ-algebra $\mathcal{B}$ by

$$\nu(A) = \int_A h \, d\lambda,$$

where

$$h(x) = \begin{cases} \frac{1}{\frac{1}{\beta} + \frac{2}{\beta^2} + \frac{3}{\beta^3}} \left(1 + \frac{1}{\beta} + \frac{1}{\beta^2}\right), & \text{if } x \in \left[0, \frac{1}{\beta}\right), \\ \frac{1}{\frac{1}{\beta} + \frac{2}{\beta^2} + \frac{3}{\beta^3}} \left(1 + \frac{1}{\beta}\right), & \text{if } x \in \left[\frac{1}{\beta}, \frac{1}{\beta} + \frac{1}{\beta^2}\right), \\ \frac{1}{\frac{1}{\beta} + \frac{2}{\beta^2} + \frac{3}{\beta^3}}, & \text{if } x \in \left[\frac{1}{\beta} + \frac{1}{\beta^2}, 1\right). \end{cases}$$

(a) Show that T_β is measure preserving with respect to ν.

Let

$$X = \left(\left[0, \frac{1}{\beta}\right) \times [0,1)\right) \cup \left(\left[\frac{1}{\beta}, \frac{1}{\beta} + \frac{1}{\beta^2}\right) \times \left[0, \frac{1}{\beta} + \frac{1}{\beta^2}\right)\right)$$
$$\cup \left(\left[\frac{1}{\beta} + \frac{1}{\beta^2}, 1\right) \times \left[0, \frac{1}{\beta}\right)\right).$$

Let $\mathcal{D}$ be the restriction of the two-dimensional Lebesgue σ-algebra on X, and λ^2 the normalized (two-dimensional) Lebesgue measure on $(X, \mathcal{D})$. Define on X the transformation $\mathcal{T}_\beta$ by

$$\mathcal{T}_\beta(x,y) = \left(T_\beta x, \frac{1}{\beta}(\lfloor \beta x \rfloor + y)\right).$$

(b) Show that $\mathcal{T}_\beta$ is measurable and measure preserving with respect to λ^2. Prove also that $\mathcal{T}_\beta$ is one-to-one and onto λ^2-a.e.

(c) Show that $\mathcal{T}_\beta$ is the natural extension of T_β.

CHAPTER 6

The Perron-Frobenius Operator

As a starting point in much of the theory developed so far we took a dynamical system $(X, \mathcal{F}, \mu, T)$ with a transformation $T : X \to X$ that is measure preserving with respect to μ. In this chapter we consider the existence of invariant measures.

6.1 ABSOLUTELY CONTINUOUS INVARIANT MEASURES

Typically a dynamical system has many different invariant measures.

Exercise 6.1.1. **(a)** Let X be a set and $T : X \to X$ a transformation. Suppose that $x \in X$ is a *periodic point* for T, i.e., it satisfies $T^n x = x$ for some $n \geq 1$ and $T^j x \neq x$ for all $j < n$. Prove that the measure μ defined by

$$\mu(A) = \frac{1}{n} \sum_{j=0}^{n-1} \delta_{T^j x}(A),$$

for any subset $A \subseteq X$, where δ_y denotes the Dirac measure at y, is invariant for T.

(b) Use part (a) to find infinitely many invariant measures for Arnold's cat map from Example 1.3.8.

So any periodic point of a transformation T can be associated to an invariant measure. The next example shows that there can still be many others.

Example 6.1.1. Consider the doubling map $Tx = 2x \pmod 1$ on the unit interval $[0,1]$. As we have seen in Example 1.3.3 Lebesgue measure is invariant for T. From the previous exercise we know that for example the measure $\frac{1}{2}\delta_{\frac{1}{3}} + \frac{1}{2}\delta_{\frac{2}{3}}$ is also invariant for T. Now, let $0 < p < 1$ and $(\{0,1\}^{\mathbb{N}}, \mathcal{C}, m_p, S)$ be the Bernoulli shift on two symbols, where $\mathcal{C}$ is the σ-algebra generated by the cylinders, m_p is the $(p, 1-p)$-Bernoulli measure and S is the left shift as in Example 1.3.9. It was shown in Example 2.3.2 that this system is ergodic for any choice of p. Let $\mathcal{B}$ be the Lebesgue σ-algebra on $[0,1]$, fix a $p \neq \frac{1}{2}$ and define the measure μ_p on $([0,1], \mathcal{B})$ by $\mu_p = m_p \circ \psi$, where

$$\psi(x) = \psi\Big(\sum_{n\geq 1} \frac{b_n}{2^n}\Big) = (b_n)_{n\geq 1}$$

is the sequence of binary digits of x generated by T, so that $(b_n)_{n\geq 1}$ does not end in an infinite string of 1's. It follows from Example 5.1.2 that ψ is an isomorphism between the systems $([0,1], \mathcal{B}, \mu_p, T)$ and $(\{0,1\}^{\mathbb{N}}, \mathcal{C}, m_p, S)$. Since m_p is ergodic for S, so is μ_p for T. It then follows from Theorem 3.1.2(ii) that λ and μ_p are singular.

For systems defined on a subset of $\mathbb{R}^n$, invariant measures that are singular with respect to Lebesgue measure see very little of the "observable" dynamics of the system. One can wonder if and under which conditions invariant measures that are absolutely continuous with respect to some reference measure on the space X exist. To this end we introduce the Perron-Frobenius operator. Let $(X, \mathcal{F}, m, T)$ be a dynamical system. To any non-negative function $f \in L^1(X, \mathcal{F}, m)$ we can associate a measure on X by setting

$$\mu(A) = \int_A f\, dm \qquad \text{for all } A \in \mathcal{F},$$

so that $\mu \ll m$. What does the measure $\mu \circ T^{-1}$ look like? Is it also absolutely continuous with respect to m? Suppose this is the case, then the corresponding density, let's call it $P_T f$, would satisfy

$$\int_A P_T f\, dm = \mu(T^{-1}A) = \int_{T^{-1}A} f\, dm.$$

To see that $P_T f$ exists, define the measure (check) ν on $(X, \mathcal{F})$ by setting

$$\nu(A) = \int_{T^{-1}A} f\, dm, \qquad \text{for all } A \in \mathcal{F}.$$

Note that by the integrability of f the measure ν is a finite measure. The non-singularity of T then implies that $\nu \ll m$, so that by the Radon-Nikodym Theorem a unique element $P_T f \in L^1(X, \mathcal{F}, m)$ exists such that

$$\nu(A) = \int_A P_T f \, dm, \qquad \text{for all } A \in \mathcal{F}.$$

This motivates the following definition.

Definition 6.1.1. Let $(X, \mathcal{F}, m)$ be a measure space and $T : X \to X$ a transformation. Then for each $f \in L^1(X, \mathcal{F}, m)$, denote by $P_T f$ the unique element in $L^1(X, \mathcal{F}, m)$ that satisfies

$$\int_A P_T f \, dm = \int_{T^{-1}A} f \, dm \qquad \text{for all } A \in \mathcal{F}.$$

The operator $P_T : L^1(X, \mathcal{F}, m) \to L^1(X, \mathcal{F}, m)$ is called the *Perron-Frobenius operator* for T.

Note that the definition of $P_T f$ is up to m-a.e. equivalence. The Perron-Frobenius operator has some nice properties listed below.

Proposition 6.1.1. *Let $(X, \mathcal{F}, m, T)$ be a dynamical system.*

(i) *P_T is linear.*

(ii) *P_T is positive.*

(iii) *P_T preserves the integral, i.e., for any $f \in L^1(X, \mathcal{F}, m)$ we have*

$$\int_X P_T f \, dm = \int_X f \, dm.$$

(iv) *$P_T : L^1(X, \mathcal{F}, m) \to L^1(X, \mathcal{F}, m)$ is a* contraction, *i.e., $\|P_T f\|_1 \leq \|f\|_1$ for any $f \in L^1(X, \mathcal{F}, m)$.*

Exercise 6.1.2. Prove Proposition 6.1.1.

The Perron-Frobenius operator also has the following composition property.

Proposition 6.1.2. *Let $T : X \to X$ and $S : X \to X$ be two transformations on the same measure space $(X, \mathcal{F}, m)$. Then $P_{T \circ S} = P_T \circ P_S$. In particular, $P_{T^n} = P_T^n$.*

Proof. Let $f \in L^1(X, \mathcal{F}, m)$. The non-singularity of T and S implies that also the transformation $T \circ S : X \to X$ is non-singular, so that the operator $P_{T\circ S}$ exists. For each $A \in \mathcal{F}$ we have

$$\begin{aligned}\int_A P_{T\circ S} f \, dm &= \int_{S^{-1}(T^{-1}A)} f \, dm = \int_{T^{-1}A} P_S f \, dm \\ &= \int_A P_T(P_S f) \, dm = \int_A P_T \circ P_S f \, dm.\end{aligned}$$

Hence, $P_{T\circ S} = P_T \circ P_S$ m-a.e. □

The following proposition shows why we are interested in the Perron-Frobenius operator.

Proposition 6.1.3. *Let $(X, \mathcal{F}, m, T)$ be a dynamical system. For a non-negative function $f \in L^1(X, \mathcal{F}, m)$ it holds that $P_T f = f$ if and only if T is measure preserving with respect to the measure μ on $(X, \mathcal{F})$ given by*

$$\mu(A) = \int_A f \, dm \qquad \text{for all } A \in \mathcal{F}.$$

Proof. Let $A \in \mathcal{F}$ be given. Then the statement follows from

$$\mu(T^{-1}A) = \int_{T^{-1}A} f \, dm = \int_A P_T f \, dm.$$ □

A function $f \in L^1(X, \mathcal{F}, m)$ that satisfies $P_T f = f$ is called a *fixed point* of the operator P_T. Hence, the Perron-Frobenius operator can be used to verify whether a measure that is absolutely continuous with respect to m is an invariant measure by checking whether the corresponding density is a fixed point of the operator. The next proposition shows that fixed points that are not necessarily non-negative give invariant measures through their positive and negative parts.

Proposition 6.1.4. *Let $(X, \mathcal{F}, m, T)$ be a dynamical system and let $f \in L^1(X, \mathcal{F}, m)$. Then $P_T f = f$ if and only if $P_T f^+ = f^+$ and $P_T f^- = f^-$, where $f^+ = \max\{f, 0\}$ and $f^- = \max\{-f, 0\}$.*

Proof. One direction follows directly from the linearity of P_T. For the other direction, assume that $P_T f = f$. Then from the linearity and positivity of P_T it follows that

$$f^+ = (P_T f)^+ = \max\{P_T f^+ - P_T f^-, 0\} \leq P_T f^+$$

and hence,

$$\int_X P_T f^+ - f^+ \, dm \geq 0.$$

Similarly $f^- \leq P_T f^-$ and $\int_X P_T f^- - f^- \, dm \geq 0$. So,

$$\begin{aligned} 0 &\leq \int_X P_T f^+ - f^+ \, dm + \int_X P_T f^- - f^- \, dm \\ &= \int_X P_T (f^+ + f^-) \, dm - \int_X f^+ + f^- \, dm \\ &= \|P_T |f|\|_1 - \|f\|_1 \leq 0, \end{aligned}$$

where the last step follows from Proposition 6.1.1(iv). Thus, $P_T f^+ = f^+$ and $P_T f^- = f^-$. □

The next theorem says that if T is ergodic with respect to m, then there can be at most one T-invariant probability measure that is absolutely continuous with respect to m.

Theorem 6.1.1. *Let $(X, \mathcal{F}, m, T)$ be a dynamical system with $T : X \to X$ ergodic with respect to m. Then there is at most one $h \in L^1(X, \mathcal{F}, m)$ that satisfies $h \geq 0$, $\int_X h \, dm = 1$ and $P_T h = h$.*

Proof. Suppose that $g, h \in L^1(X, \mathcal{F}, m)$ are two densities that satisfy the properties from the theorem. Set $f = g - h$. Then $\int_X f \, dm = 1 - 1 = 0$ and by linearity, $P_T f = f$. If either $f(x) \geq 0$ for m-a.e. $x \in X$ (i.e., $g \geq h$ m-a.e.) or $f(x) \leq 0$ for m-a.e. $x \in X$ (i.e., $h \geq g$ m-a.e.), then from $\int_X f \, dm = 0$ one gets $f = 0$ m-a.e., so $g = h$ m-a.e. and we are done.

So, assume that this is not the case. Write $f = f^+ - f^-$ and let $E_1 = \{x \in X : f^+(x) > 0\}$ and $E_2 = \{x \in X : f^-(x) > 0\}$. Then by the assumption $m(E_1) > 0$ and $m(E_2) > 0$. The sets E_1 and E_2 are the *supports* of the functions f^+ and f^-, respectively. We now use Proposition 6.1.4 to deduce a contradiction.

Since $P_T f^+ = f^+$ it holds that $P_T f^+ = 0$ on $X \setminus E_1$. Therefore,

$$0 = \int_{X \setminus E_1} P_T f^+ dm = \int_{X \setminus T^{-1} E_1} f^+ \, dm,$$

so $f^+(x) = 0$ for m-a.e. $x \in X \setminus T^{-1}E_1$. Hence, by the definition of E_1 we have $E_1 \subseteq T^{-1}E_1 \cup N$ for some set N of m-measure zero, so $m(E_1 \setminus T^{-1}E_1) = 0$. By the non-singularity of T it follows that for each $k, n \geq 0$

with $k \leq n$ there is an m-null set $N_{k,n}$ such that $T^{-k}E_1 \subseteq T^{-n}E_1 \cup N_{k,n}$ and similarly for E_2 there is a m-null set $M_{k,n}$ with $T^{-k}E_2 \subseteq T^{-n}E_2 \cup M_{k,n}$. This leads to

$$0 \leq m(T^{-k}E_1 \cap T^{-n}E_2) \leq m(T^{-\min\{k,n\}}E_1 \cap T^{-\min\{k,n\}}E_2) = 0 \quad (6.1)$$

for each $k, n \geq 0$, where the last step follows from the non-singularity of T combined with the fact that $E_1 \cap E_2 = \emptyset$. Both sets $\bigcup_{n\geq 0} T^{-n}E_i$, $i = 1, 2$, have positive measure, so that for $i = 1, 2$,

$$\begin{aligned}
0 &\leq m\Big(\bigcup_{n\geq 0} T^{-n}E_i \,\Delta\, T^{-1}\bigcup_{n\geq 0} T^{-n}E_i\Big) \\
&= m\Big(\Big(E_i \cup \bigcup_{n\geq 1} T^{-n}E_i\Big) \Delta \bigcup_{n\geq 1} T^{-n}E_i\Big) \\
&= m\Big(E_i \setminus \bigcup_{n\geq 1} T^{-n}E_i\Big) \\
&\leq m(E_i \setminus T^{-1}E_i) = 0.
\end{aligned}$$

Hence by ergodicity $m\left(\left(\bigcup_{n\geq 0} T^{-n}E_i\right)^c\right) = 0$. Since $m(E_1 \cap E_2) = 0$, it holds that $m(T^{-n}E_1 \cap T^{-n}E_2) = 0$ for all $n \geq 0$, so also

$$m\Big(\bigcup_{n\geq 0} T^{-n}E_1 \cap \bigcup_{n\geq 0} T^{-n}E_2\Big) = 0.$$

By (6.1),

$$\begin{aligned}
0 &< m\Big(\bigcup_{n\geq 0} T^{-n}E_1\Big) \\
&= m\Big(\bigcup_{n\geq 0} T^{-n}E_1 \cap \bigcup_{n\geq 0} T^{-n}E_2\Big) + m\Big(\bigcup_{n\geq 0} T^{-n}E_1 \cap \Big(\bigcup_{n\geq 0} T^{-n}E_2\Big)^c\Big) \\
&= 0,
\end{aligned}$$

a contradiction. Hence, $g = h$ m-a.e. □

6.2 EXACTNESS

Recall the definition of the Koopman operator U_T from (1.1) by $U_T(g) = g \circ T$. Here we restrict U_T to the space $L^\infty(X, \mathcal{F}, m)$ of essentially bounded measurable functions. The reason for this is that $L^\infty(X, \mathcal{F}, m)$

is the adjoint space to $L^1(X, \mathcal{F}, m)$, which we took as the domain for P_T, so that if $f \in L^1(X, \mathcal{F}, m)$ and $g \in L^\infty(X, \mathcal{F}, m)$, then $f \cdot g$ is integrable. The next proposition states that U_T is adjoint to P_T.

Proposition 6.2.1. *Let $(X, \mathcal{F}, m, T)$ be a dynamical system. Let $f \in L^1(X, \mathcal{F}, m)$ and $g \in L^\infty(X, \mathcal{F}, m)$. Then,*

$$\int_X (P_T f) \cdot g \, dm = \int_X f \cdot U_T g \, dm.$$

Proof. Let $f \in L^1(X, \mathcal{F}, m)$ and $A \in \mathcal{F}$ be given and set $g = 1_A$. Then

$$\begin{aligned}\int_X (P_T f) \cdot 1_A \, dm &= \int_A P_T f \, dm = \int_{T^{-1}A} f \, dm = \int_X f \cdot 1_{T^{-1}A} \, dm \\ &= \int_X f \cdot (1_A \circ T) \, dm = \int_X f \cdot U_T g \, dm.\end{aligned}$$

Hence, the statement is true for indicator functions and thus for simple functions. Since the simple functions are dense in $L^\infty(X, \mathcal{F}, m)$ (see Theorem 12.4.1) the statement holds for all $f \in L^1(X, \mathcal{F}, m)$, $g \in L^\infty(X, \mathcal{F}, m)$. □

This proposition allows us to give an alternative characterization of weak and strong mixing.

Theorem 6.2.1. *Let $(X, \mathcal{F}, \mu)$ be a probability space and $T : X \to X$ a measure preserving transformation.*

(i) *T is ergodic if and only if for all $f \in L^1(X, \mathcal{F}, \mu)$ and $g \in L^\infty(X, \mathcal{F}, \mu)$,*

$$\lim_{n\to\infty} \frac{1}{n} \sum_{i=0}^{n-1} \int_X (P_T^i f) \cdot g \, d\mu = \int_X f \, d\mu \int_X g \, d\mu.$$

(ii) *T is weakly mixing if and only if for all $f \in L^1(X, \mathcal{F}, \mu)$ and $g \in L^\infty(X, \mathcal{F}, \mu)$,*

$$\lim_{n\to\infty} \frac{1}{n} \sum_{i=0}^{n-1} \left| \int_X (P_T^i f) \cdot g \, d\mu - \int_X f \, d\mu \int_X g \, d\mu \right| = 0.$$

(iii) *T is strongly mixing if and only if for all $f \in L^1(X, \mathcal{F}, \mu)$ and $g \in L^\infty(X, \mathcal{F}, \mu)$,*

$$\lim_{n\to\infty} \int_X (P_T^n f) \cdot g \, d\mu = \int_X f \, d\mu \int_X g \, d\mu.$$

Proof. One direction follows immediately from Corollary 3.1, Definition 3.4.1 and Proposition 6.2.1 by considering indicator functions $f = 1_A$ and $g = 1_B$ for $A, B \in \mathcal{F}$. For the other direction by the same results the property holds for indicator functions 1_A and 1_B and thus also for simple functions. The result then follows since any $f \in L^1(X, \mathcal{F}, \mu)$ is the L^1-limit of a sequence of simple functions and any $g \in L^\infty(X, \mathcal{F}, \mu)$ is the uniform limit of a sequence of simple functions. □

We add another property to the list.

Definition 6.2.1. Let $(X, \mathcal{F}, \mu)$ be a probability space and $T : X \to X$ a measure preserving transformation. Then T is called *exact* if any set A in the *tail σ-algebra*

$$\bigcap_{n\geq 0} T^{-n}\mathcal{F}$$

either has $\mu(A) = 0$ or $\mu(A) = 1$.

Note that a set $A \in \bigcap_{n\geq 0} T^{-n}\mathcal{F}$ has the property that for each $n \geq 0$ there is a set $B_n \in \mathcal{F}$, such that $A = T^{-n}B_n$. From this we immediately see that any exact transformation is ergodic: if $E \in \mathcal{F}$ is a T-invariant set, then $E = T^{-n}E$ for all $n \geq 0$, so $E \in \bigcap_{n\geq 0} T^{-n}\mathcal{F}$ and thus $\mu(E) = 0$ or $\mu(E) = 1$.

Exercise 6.2.1. Use the Kolmogorov 0-1 Law (see Theorem 12.2.3) to prove that any one-sided Bernoulli shift is exact.

Example 6.2.1. Consider the binary odometer from Example 1.3.11. Our aim is to show that T is ergodic with respect to the product measure μ_p on $\{0,1\}^{\mathbb{N}}$ for all $0 < p < 1$. For any $(x_1, x_2, \ldots) \in \{0,1\}^{\mathbb{N}}$ and any $n \geq 1$, let $k(n,x) = \sum_{j=1}^{n} 2^{j-1}x_j$. Then,

$$T^{-k(n,x)}(x_1, x_2, \ldots) = (0, \ldots, 0, x_{n+1}, x_{n+2}, \ldots).$$

Any $\ell \in \{0, 1, \ldots, 2^n - 1\}$ can be written in its binary expansion as $\ell = \sum_{j=1}^{n} 2^{j-1}y_j$ for some $y_1, \ldots, y_n \in \{0,1\}$. From this we see that $T^{\ell-k(n,x)}(x_1, x_2, \ldots) = (y_1, \ldots, y_n, x_{n+1}, x_{n+2}, \ldots)$. Thus if f is any measurable function on $\{0,1\}^{\mathbb{N}}$, then

$$\begin{aligned} &\sum_{\ell=0}^{2^n-1} f(T^{\ell-k(n,x)}(x_1, x_2, \ldots,)) \\ &\qquad = \sum_{y_1, y_2, \ldots, y_n \in \{0,1\}} f(y_1, y_2, \ldots y_n, x_{n+1}, x_{n+2}, \ldots). \end{aligned} \tag{6.2}$$

The above shows that the left-hand side is independent of the first n coordinates. So if we denote by S the left shift on $\{0,1\}^{\mathbb{N}}$, then (6.2) can be written as

$$\sum_{\ell=0}^{2^n-1} f(T^{\ell-k(n,x)}(x_1, x_2, \ldots,)) = g(S^n(x_1, x_2, \ldots)),$$

for some measurable function g. Suppose now that f is a T-invariant function. Then the above leads to the statement $2^n f(x_1, x_2, \ldots) = g(S^n(x_1, x_2, \ldots))$ for any $n \geq 1$. Hence f is measurable with respect to the tail σ-algebra $\bigcap_{n=1}^{\infty} S^{-n}\mathcal{C}$, where $\mathcal{C}$ is the product σ-algebra on $\{0,1\}^{\mathbb{N}}$. By Exercise 6.2.1, this is the trivial σ-algebra and hence f is a constant μ_p-a.e. This proves that T is ergodic.

The following proposition gives an alternative characterization of exactness in case the transformation T is *forward measurable*, i.e., if for any $A \in \mathcal{F}$ it holds that also $TA \in \mathcal{F}$. Note that this condition holds for all examples in the book. The proposition says that if T is exact, then under iterations of T any positive measure set will eventually cover almost all of the space.

Proposition 6.2.2. *Let $(X, \mathcal{F}, \mu)$ be a probability space and $T : X \to X$ a forward measurable and measure preserving transformation. Then T is exact if and only if for every $A \in \mathcal{F}$ with $\mu(A) > 0$ it holds that*

$$\lim_{n\to\infty} \mu(T^n A) = 1. \tag{6.3}$$

Proof. First assume that (6.3) holds for each positive measure set. Let $A \in \bigcap_{n\geq 0} T^{-n}\mathcal{F}$ and assume for a contradiction that $0 < \mu(A) < 1$. Then for each $n \geq 0$ there is a set $B_n \in \mathcal{F}$ such that $A = T^{-n}B_n$. Fix an $n \geq 0$. T is measure preserving, so $0 < \mu(B_n) = \mu(A) < 1$ and from $T^n A = T^n(T^{-n}B_n) \subseteq B_n$ we obtain

$$\mu(T^n A) \leq \mu(A) < 1.$$

Since n was arbitrary, this contradicts (6.3). Hence, T is exact. For the other direction, assume T is exact and let $A \in \mathcal{F}$ be a set with $\mu(A) > 0$. Note that for any $n \geq 0$,

$$T^{-n}(T^n A) \subseteq T^{-(n+1)}(T^{n+1}A),$$

so that the limit set

$$B = \lim_{n\to\infty} T^{-n}(T^n A) = \bigcup_{n\geq 0} T^{-n}(T^n A) \in \bigcap_{n\geq 0} T^{-n}\mathcal{F}.$$

From $A \subseteq B$, it follows that $\mu(B) = 1$ and since T is measure preserving and the sequence $(T^{-n}(T^n A))$ is increasing, we obtain

$$\lim_{n\to\infty} \mu(T^n A) = \lim_{n\to\infty} \mu(T^{-n}(T^n A)) = \mu(B) = 1.$$

This gives the result. □

From this proposition it follows that an invertible measure preserving transformation cannot be exact. To see this, let $A \in \mathcal{F}$ be a measurable set with $0 < \mu(A) < 1$. Recall that invertibility of a transformation T implies that all sets $T^n A$, $n \geq 1$, are in $\mathcal{F}$, so T is forward measurable. Then for any $n \geq 1$,

$$\mu(T^n A) = \mu(T^{-n}(T^n A)) = \mu(A) < 1.$$

The following result on exactness is in the spirit of Theorem 6.2.1.

Theorem 6.2.2. *Let $(X, \mathcal{F}, \mu)$ be a probability space and $T : X \to X$ forward measurable and measure preserving. Then T is exact if and only if for all $f \in L^1(X, \mathcal{F}, \mu)$,*

$$\lim_{n\to\infty} \left\| P_T^n f - \int_X f \, d\mu \right\|_1 = 0.$$

Proof. Assume T is exact and let $f \in L^1(X, \mathcal{F}, \mu)$ be given. Note that for each $n \geq 1$, $T^{-n}\mathcal{F} \subseteq T^{-n+1}\mathcal{F}$, so the sequence $(T^{-n}\mathcal{F})$ is decreasing. Also note that by the measure preservingness of T for each $n \geq 1$,

$$\begin{aligned} \int_X \left| P_T^n f - \int_X f \, d\mu \right| d\mu &= \int_X \left| P_T^n f - \int_X f \, d\mu \right| \circ T^n \, d\mu \\ &= \int_X \left| P_T^n f \circ T^n - \int_X f \, d\mu \right| d\mu \end{aligned}$$

and that by the exactness of T

$$\int_X f \, d\mu = \mathbb{E}_\mu\Big(f \mid \bigcap_{n\geq 0} T^{-n}\mathcal{F} \Big).$$

Let $n \geq 1$. The function $P_T^n f \circ T^n$ is $T^{-n}\mathcal{F}$-measurable. Moreover, if we take $B \in \mathcal{F}$ we see with Proposition 6.1.2 that for $A = T^{-n}B$ we get

$$\begin{aligned} \int_A P_T^n f \circ T^n \, d\mu &= \int_{T^{-n}B} P_T^n f \circ T^n \, d\mu \\ &= \int_B P_{T^n} f \, d\mu = \int_{T^{-n}B} f \, d\mu = \int_A f \, d\mu. \end{aligned}$$

In other words,

$$P_T^n f \circ T^n = \mathbb{E}_\mu(f \,|\, T^{-n}\mathcal{F}).$$

Hence, since the sequence $(T^{-n}\mathcal{F})$ is decreasing,

$$\begin{aligned}\lim_{n\to\infty} \int_X \Big| P_T^n f - \int_X f\, d\mu \Big| d\mu \\ &= \lim_{n\to\infty} \int_X \Big| \mathbb{E}_\mu(f \,|\, T^{-n}\mathcal{F}) - \mathbb{E}_\mu\Big(f \,|\, \bigcap_{n\geq 0} T^{-n}\mathcal{F}\Big)\Big| d\mu = 0.\end{aligned}$$

For the other direction, let $A \in \mathcal{F}$ with $\mu(A) > 0$ be given. For each $n \geq 0$ we have

$$\begin{aligned}\mu(T^n A) &= \int_{T^n A} 1\, d\mu \\ &= \frac{1}{\mu(A)} \int_{T^n A} P_T^n 1_A \, d\mu - \frac{1}{\mu(A)} \int_{T^n A} P_T^n 1_A - \mu(A)\, d\mu \\ &\geq \frac{1}{\mu(A)} \int_{T^{-n}(T^n A)} 1_A \, d\mu - \frac{1}{\mu(A)} \int_X |P_T^n 1_A - \mu(A)|\, d\mu \\ &= 1 - \frac{1}{\mu(A)} \int_X |P_T^n 1_A - \mu(A)|\, d\mu.\end{aligned}$$

By assumption it holds that $\lim_{n\to\infty} \int_X |P_T^n 1_A - \mu(A)|\, d\mu = 0$, so the exactness of T follows. □

Note that the assumption of forward measurability in the previous theorem is only used in one direction.

Exercise 6.2.2. Let $T : X \to X$ be a measure preserving transformation on a probability space $(X, \mathcal{F}, \mu)$. Prove that if T is exact, then T is strongly mixing.

The next results, which is due to V. A. Rohlin and can be found in [55], gives another way to verify whether a transformation is exact.

Proposition 6.2.3. *Let $(X, \mathcal{F}, \mu)$ be a probability space and $T : X \to X$ a forward measurable and measure preserving transformation. Let $\mathcal{S} \subseteq \mathcal{F}$ be a countable collection of sets of positive μ-measure with the following properties.*

(i) *For every $E \in \mathcal{F}$ with $\mu(E) > 0$ there are pairwise disjoint sets $A_k \in \mathcal{S}$, $k \geq 1$, with $\mu(E \Delta \bigcup_{k\geq 1} A_k) = 0$.*

(ii) *There is a function* $n : \mathcal{S} \to \mathbb{N}$, *such that* $\mu(T^{n(A)}A) = 1$ *for each* $A \in \mathcal{S}$.

(iii) *There is a constant* $\gamma \geq 1$, *such that for each* $A \in \mathcal{S}$ *and for all measurable sets* $E \subseteq A$ *we have*

$$\mu(T^{n(A)}E) \leq \gamma \cdot \frac{\mu(E)}{\mu(A)}. \tag{6.4}$$

Then T *is exact.*

Proof. Let $E \in \mathcal{F}$ be arbitrary with $\mu(E) > 0$. We use Proposition 6.2.2 to prove the exactness of T. First note that by the measure preservingness of T we have $\mu(E) \leq \mu(T^{-1}(TE)) = \mu(TE)$, so the sequence $\big(\mu(T^n E)\big)_{n\geq 0}$ is increasing. Let $\delta > 0$. We claim that there must be a set $B \in \mathcal{S}$ with the property that

$$\mu(E \cap B) > \left(1 - \frac{\delta}{\gamma}\right)\mu(B). \tag{6.5}$$

To see that this is true, suppose that the opposite holds. Then by (i) there are pairwise disjoint sets $A_k \in \mathcal{S}$, $k \geq 1$, such that

$$\begin{aligned}\mu(E) = \mu\Big(E \cap \bigcup_{k\geq 1} A_n\Big) &\leq \sum_{k\geq 1}\left(1 - \frac{\delta}{\gamma}\right)\mu(A_k)\\ &= \left(1 - \frac{\delta}{\gamma}\right)\mu\Big(\bigcup_{k\geq 1} A_k\Big) = \left(1 - \frac{\delta}{\gamma}\right)\mu(E),\end{aligned}$$

a contradiction. So, let $B \in \mathcal{S}$ be a set that satisfies (6.5). Then by (6.4)

$$\begin{aligned}\mu(T^{n(B)}(B \setminus E \cap B)) &\leq \gamma\,\frac{\mu(B \setminus (E \cap B))}{\mu(B)}\\ &= \gamma\,\frac{\mu(B) - \mu(E \cap B)}{\mu(B)}\\ &< \gamma\left(1 - \left(1 - \frac{\delta}{\gamma}\right)\right) = \delta.\end{aligned}$$

Since

$$\begin{aligned}\mu(T^{n(B)}(B \setminus E \cap B)) &\geq \mu(T^{n(B)}B \setminus T^{n(B)}(E \cap B))\\ &\geq \mu(T^{n(B)}B) - \mu(T^{n(B)}(E \cap B)),\end{aligned}$$

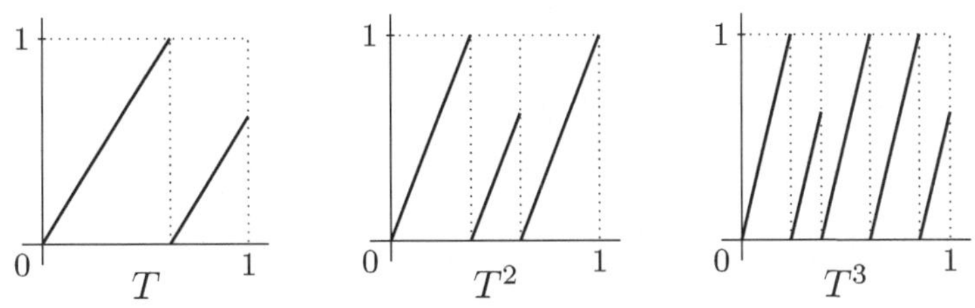

Figure 6.1 The first three iterations of the β-transformation from Example 6.2.2.

it follows using (ii) that

$$\mu(T^{n(B)}E) \geq \mu(T^{n(B)}(E\cap B)) \geq \mu(T^{n(B)}B) - \mu(T^{n(B)}(B\backslash E\cap B)) > 1-\delta.$$

The increasingness of the sequence $(\mu(T^n E))$ then implies that $\lim_{n\to\infty}\mu(T^n E) = 1$. □

Example 6.2.2. Consider the β-transformation $Tx = \beta x \pmod 1$ defined on the interval $[0,1)$, where $\beta = \frac{1+\sqrt{5}}{2}$ is the golden mean. Recall the density of the invariant measure μ for T that is absolutely continuous with respect to the Lebesgue measure given in (1.4). We use Proposition 6.2.3 to show that this map is exact. Note that $T[0,\frac{1}{\beta}) = [0,1)$ and $T[\frac{1}{\beta},1) = [0,\frac{1}{\beta})$. This implies that for each $n \geq 1$ we can partition the interval $[0,1)$ into half open intervals J_n, such that for each of these intervals J_n either $T^n J_n = [0,1)$ or $T^n J_n = [0,\frac{1}{\beta})$ and $T^{n+1}J_n = [0,1)$. See Figure 6.1 for the first three iterates of T. Write $\mathcal{P}_n = \{J_n\}$ for the collection of these intervals and let $\mathcal{S} = \bigcup_{n\geq 0}\mathcal{P}_n$. If you recall the notion of fundamental intervals from Example 2.3.4 for the Lüroth map, the elements from the collection $\mathcal{S}$ are the fundamental intervals of T in this case. We check the three conditions from Proposition 6.2.3.

- For each n, $\mu(J_n) \leq \frac{5+3\sqrt{5}}{10}\frac{1}{\beta^n}$ and since this decreases to 0 as $n\to\infty$, (i) of Proposition 6.2.3 is satisfied.
- For $J_n \in \mathcal{P}_n$ either $\mu(T^n J_n) = 1$ and we put $n(J_n) = n$ or $\mu(T^{n+1}J) = 1$ and we set $n(J_n) = n+1$. This gives (ii).
- Finally, let $J_n \in \mathcal{P}_n$ and a measurable set $E \subseteq J_n$ be given. Then

$$\mu(T^{n(J_n)}E) \leq \frac{5+3\sqrt{5}}{10}\,\beta^{n(J_n)}\,\lambda(E),$$

where λ denotes the one-dimensional Lebesgue measure. Since $\frac{1}{\beta} \notin J_n$, we have

$$\frac{\mu(E)}{\mu(J_n)} = \beta^{n(J_n)} \lambda(E),$$

so we obtain (6.4) with $\gamma = \frac{5+3\sqrt{5}}{10}$ and the exactness of T follows.

Using a change of variables formula we can in some cases get a nicer description of the Perron-Frobenius operator. Let $(X, \mathcal{F}, m, T)$ be a dynamical system. By non-singularity the measure $m \circ T^{-1}$ is absolutely continuous with respect to m. Let $\phi : X \to [0, \infty)$ be the Radon-Nikodym derivative $\phi := \frac{dm \circ T^{-1}}{dm}$. Then for any function $f \in L^0(X, \mathcal{F}, m)$ that satisfies $f \circ T \in L^1(X, \mathcal{F}, m)$ it holds that

$$\int_{T^{-1}A} f \circ T \, dm = \int_A f \cdot \phi \, dm \quad \text{for all } A \in \mathcal{F}. \tag{6.6}$$

This formula can be proved by checking the statement for indicator functions and then using the fact any f can be approximated by simple functions. Note that in case T is measure preserving, we get $\phi = 1$ and we obtain the result from Proposition 1.2.2.

Now assume that $T : X \to X$ is an invertible transformation on a measure space $(X, \mathcal{F}, m)$, such that the inverse T^{-1} is also non-singular. Then by (6.6) we have for any $A \in \mathcal{F}$ that

$$\int_A P_T f \, dm = \int_{T^{-1}A} f \, dm = \int_A (f \circ T^{-1}) \cdot \phi \, dm.$$

Hence,

$$P_T f = (f \circ T^{-1}) \cdot \phi \quad m\text{-a.e.} \tag{6.7}$$

In case $X \subseteq \mathbb{R}^d$, $\mathcal{F}$ is the Lebesgue σ-algebra and m is the d-dimensional Lebesgue measure, ϕ is just the determinant of the Jacobian matrix of T.

Example 6.2.3. The baker's transformation T from Exercise 1.3.1 is invertible. The Jacobian of T is given by

$$\begin{pmatrix} 2 & 0 \\ 0 & \frac{1}{2} \end{pmatrix}.$$

Therefore $\phi \equiv 1$ and by (6.7) the Perron-Frobenius operator becomes

$$P_T f(x, y) = f\Big(\frac{x}{2}, 2y\Big) 1_{[0, \frac{1}{2})}(y) + f\Big(\frac{x+1}{2}, 2y - 1\Big) 1_{[\frac{1}{2}, 1]}(y).$$

From this we immediately see that $P_T 1 = 1$, so that Lebesgue measure is invariant for T.

Exercise 6.2.3. Prove the change of variables formula from (6.6).

Exercise 6.2.4. Let $(X, \mathcal{F}, \mu)$ be a probability space and $T : X \to X$ a transformation.

(a) Let ν be a probability measure on $(X, \mathcal{F})$ that is equivalent to μ and let f be the Radon-Nikodym derivative $f = \frac{d\mu}{d\nu}$. Let $P_{T,\mu}$ and $P_{T,\nu}$ denote the Perron-Frobenius operators of T with respect to the measures μ and ν, respectively. Prove that for any $g \in L^1(X, \mathcal{F}, \mu)$,

$$P_{T,\mu} g = \frac{P_{T,\nu}(fg)}{f}.$$

(b) Let ν be a probability measure on $(X, \mathcal{F})$ that is absolutely continuous with respect to μ. Assume that μ is T-invariant and that T is strongly mixing with respect to μ. Prove that

$$\lim_{n \to \infty} \nu(T^{-n} A) = \mu(A)$$

holds for any $A \in \mathcal{F}$.

6.3 PIECEWISE MONOTONE INTERVAL MAPS

Several of our examples concern maps on the interval that have an underlying partition, such that on each partition element the transformation is monotone and smooth. For this type of maps the Perron-Frobenius operator can be written in a more manageable way and used to prove the existence of invariant measures that are absolutely continuous with respect to the Lebesgue measure.

In this paragraph the state space will be $([0,1], \mathcal{B}, \lambda)$ with $\mathcal{B}$ the Lebesgue σ-algebra and λ the Lebesgue measure. A collection $\{I_j\}_{j \in \Omega}$ of subintervals of $[0,1]$, with $\Omega \subseteq \mathbb{N}$, is called an *interval partition* of $[0,1]$ if for each $j, k \in \Omega$ it holds that $\lambda(I_j) > 0$, $\lambda(I_j \cap I_k) = 0$ if $j \neq k$ and $\lambda(\bigcup_{j \in \Omega} I_j) = 1$. We consider transformations in the following family.

Definition 6.3.1. A transformation $T : [0,1] \to [0,1]$ is called a *piecewise C^2 monotone interval map* if there is an interval partition $\{I_j\}_{j \in \Omega}$ of $[0,1]$ with Ω a finite set, such that

(1) $T_j := T|_{I_j}$ is C^2 and has a C^2 extension to the closure of I_j and

(2) $T' \neq 0$ on the interior of I_j for any $j \in \Omega$.

Such a T is non-singular with respect to λ, so the corresponding Perron-Frobenius operator is well defined.

Exercise 6.3.1. Prove that any piecewise C^2 monotone interval map is non-singular with respect to λ.

A piecewise C^2 monotone interval map T has a local inverse on any of the intervals I_j, so the change of variables formula (6.6) yields for any $f \in L^1([0,1], \mathcal{B}, \lambda)$ and $A \in \mathcal{B}$ that

$$\begin{aligned}\int_A P_T f \, d\lambda &= \sum_{j\in\Omega} \int_{T^{-1}A\cap I_j} f \, d\lambda \\ &= \sum_{j\in\Omega} \int_{A\cap TI_j} \frac{f(T_j^{-1}x)}{|T'(T_j^{-1}x)|} \, d\lambda(x) \\ &= \int_A \sum_{j\in\Omega} \frac{f(T_j^{-1}x)}{|T'(T_j^{-1}x)|} 1_{TI_j}(x) \, d\lambda(x).\end{aligned}$$

So, $P_T f$ is given by

$$P_T f(x) = \sum_{j\in\Omega} \frac{f(T_j^{-1}x)}{|T'(T_j^{-1}x)|} 1_{TI_j}(x). \tag{6.8}$$

Example 6.3.1. Consider the *logistic map* $Tx = 4x(1-x)$ on $[0,1]$, see Figure 6.2(a) for the graph. This is a piecewise C^2 monotone interval map with interval partition $\{[0, \frac{1}{2}), [\frac{1}{2}, 1]\}$. Note that it does not matter whether we include the endpoints in these intervals or not. By (6.8) for any $f \in L^1([0,1], \mathcal{B}, \lambda)$ and $A \in \mathcal{B}$ we can write

$$P_T f(x) = \frac{1}{4\sqrt{1-x}} \left(f\left(\frac{1+\sqrt{1-x}}{2}\right) + f\left(\frac{1-\sqrt{1-x}}{2}\right)\right).$$

Now take for f the function given by $f(x) = \frac{1}{\sqrt{x(1-x)}}$. From the fact that $1 - \frac{1\pm\sqrt{1-x}}{2} = \frac{1\mp\sqrt{1-x}}{2}$ we obtain

$$\begin{aligned}P_T\left(\frac{1}{\sqrt{x(1-x)}}\right) &= \frac{1}{4\sqrt{1-x}} \frac{2}{\sqrt{(\frac{1}{2} - \frac{1}{2}\sqrt{1-x})(\frac{1}{2} + \frac{1}{2}\sqrt{1-x})}} \\ &= \frac{1}{2\sqrt{1-x}} \frac{1}{\sqrt{\frac{1}{4} - \frac{1}{4}(1-x)}} = \frac{1}{\sqrt{x(1-x)}}.\end{aligned}$$

Since $\int_{[0,1]} \frac{1}{\sqrt{x(1-x)}} d\lambda(x) = 2 \arcsin \sqrt{1} = \pi$, it follows from Proposition 6.1.3 that the measure given by

$$\mu(A) = \int_A \frac{1}{\pi\sqrt{x(1-x)}} d\lambda(x) \quad \text{for all } A \in \mathcal{B}$$

is a T-invariant measure that is absolutely continuous with respect to λ.

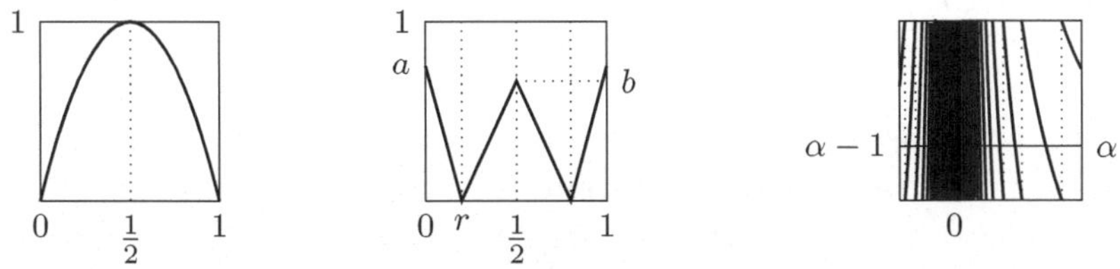

(a) Logistic map (b) W-shaped map (c) α-continued fraction map

Figure 6.2 The logistic map from Example 6.3.1 in (a), the W-shaped map from Example 6.3.3 with $a = \frac{3}{4}$, $b = \frac{2}{3}$ and $r = \frac{1}{5}$ in (b) and the α-continued fraction map with $\alpha = \frac{7}{10}$ from Exercise 6.3.3 in (c).

Example 6.3.2. For the doubling map $Tx = 2x \pmod 1$ from Example 1.3.3 we get for any $f \in L^1([0,1], \mathcal{B}, \lambda)$ that

$$P_T f(x) = \frac{1}{2} f\left(\frac{x}{2}\right) + \frac{1}{2} f\left(\frac{x+1}{2}\right).$$

One immediately sees that the constant function 1 is a fixed point of P_T, so that λ is T-invariant. Note that for each $n \geq 1$,

$$P_T^n f(x) = \frac{1}{2^n} \sum_{k=0}^{2^n-1} f\left(\frac{k}{2^n} + \frac{x}{2^n}\right),$$

i.e., $P_T^n f$ takes the average value of f over all the points $\{\frac{k}{2^n} + \frac{x}{2^n} : 0 \leq k \leq 2^n - 1\}$. Hence, $P_T^n f$ converges uniformly to the Riemann integral of f as $n \to \infty$. Then by Hölder's Inequality it also holds that

$$\lim_{n\to\infty} \left\| P_T^n f - \int_{[0,1]} f \, d\lambda \right\|_1 = 0,$$

which by Theorem 6.2.2 implies that T is exact.

The goal of this section is to prove the following theorem from 1973, which is due to A. Lasota and J. A. Yorke [34]. We call a piecewise C^2 monotone interval map $T : [0,1] \to [0,1]$ *expanding* if $\inf_{x\in[0,1]} |T'(x)| \geq c > 1$ for some constant c, where the infimum is taken over all points where T' is defined.

Theorem 6.3.1 (Lasota and Yorke). *Let $T : [0,1] \to [0,1]$ be a piecewise C^2 monotone expanding interval map. Then T admits an invariant probability measure that is absolutely continuous with respect to Lebesgue measure and has a density of bounded variation.*

The theorem asserts more than just the existence of an invariant probability measure μ that is absolutely continuous with respect to Lebesgue. It also states that the density $\frac{d\mu}{d\lambda}$ of this measure is of bounded variation. Let $a, b \in \mathbb{R}$ with $a < b$. The *total variation* of a function $f : [a,b] \to \mathbb{R}$ is defined by

$$Var_{[a,b]}(f) = \sup_{x_0=a<x_1<\cdots<x_n=b} \sum_{i=1}^{n} |f(x_i) - f(x_{i-1})|,$$

where the supremum is taken over all possible n and all possible collections of points $x_0 = a < x_1 < \cdots < x_n = b$. The function f is said to be of *bounded variation* if $Var_{[a,b]}(f) < \infty$ and we let $BV([a,b])$ denote the set of functions $f : [a,b] \to \mathbb{R}$ that are of bounded variation. The appendix contains a list of properties of the total variation and of functions of bounded variation that we will use in the proof of Theorem 6.3.1. In particular, any function $f : [a,b] \to \mathbb{R}$ of bounded variation is bounded by Proposition 12.7.1(i) and Lebesgue integrable by Theorem 12.7.1. The next lemma is an essential ingredient in the proof of Theorem 6.3.1.

Lemma 6.3.1. *Let $T : [0,1] \to [0,1]$ be a piecewise C^2 monotone expanding interval map and $f \in BV([0,1])$. Then*

$$Var_{[0,1]}(P_T f) \leq \frac{2}{\theta(T)} Var_{[0,1]}(f) + L(T)\|f\|_1, \tag{6.9}$$

where $\theta(T) = \inf_{x\in[0,1]} |T'(x)|$ and $L(T) > 0$ is a constant depending only on T. In particular, $P_T BV([0,1]) \subseteq BV([0,1])$.

The equation (6.9) is usually referred to the *Lasota-Yorke inequality.*

Proof. Let $(I_j)_{1\leq j\leq N}$ be the collection of fundamental intervals of T (so the maximal intervals on which T is monotone) and $f \in BV([0,1])$ a function of bounded variation. By (6.8) we know that

$$P_T f(x) = \sum_{j=1}^{N} \frac{f(T_j^{-1}x)}{|T'(T_j^{-1}x)|} 1_{TI_j}(x).$$

The total variation is subadditive, see Proposition 12.7.1(i), so

$$Var_{[0,1]}(P_T f) \leq \sum_{j=1}^{N} Var_{[0,1]}\Big(1_{TI_j}\Big(\frac{f}{|T'|} \circ T_j^{-1}\Big)\Big). \tag{6.10}$$

Note that TI_j is an interval. We apply Yorke's Inequality from Theorem 12.7.2 to $g = \frac{f}{|T'|} \circ T_j^{-1}$ to get

$$\begin{aligned} &Var_{[0,1]}\Big(1_{TI_j} \cdot \Big(\frac{f}{|T'|} \circ T_j^{-1}\Big)\Big) \\ &\qquad \leq 2Var_{TI_j}\Big(\frac{f}{|T'|} \circ T_j^{-1}\Big) + \frac{2}{\lambda(TI_j)} \int_{TI_j} \Big|\frac{f}{T'}\Big| \circ T_j^{-1} d\lambda. \end{aligned}$$

Since T_j^{-1} is monotone, we obtain from Proposition 12.7.2(i) and the change of variables formula (6.6) that

$$Var_{[0,1]}\Big(1_{TI_j} \cdot \Big(\frac{f}{|T'|} \circ T_j^{-1}\Big)\Big) \leq 2Var_{I_j}\Big(\frac{f}{|T'|}\Big) + \frac{2}{\lambda(TI_j)} \int_{I_j} |f| d\lambda. \tag{6.11}$$

T is expanding, so $\theta(T) = \inf_{x \in [0,1]} |T'(x)| > 1$. Moreover, T has a C^2 extension to the closure of each of the finitely many intervals I_j, so there is a $K > 0$ such that $|T''| \leq K$. Hence,

$$\Big|\frac{d}{dx}\frac{1}{T'(x)}\Big| = \frac{|T''(x)|}{(T'(x))^2} \leq \frac{K}{\theta(T)^2}$$

for all $x \in I$ where the derivatives are defined. From Proposition 12.7.2(ii) applied to $g = f$ and $h = \frac{1}{|T'|}$ we get

$$\begin{aligned} Var_{I_j}\Big(\frac{f}{|T'|}\Big) &\leq \frac{1}{\theta(T)} Var_{I_j}(f) + \int_{I_j} |f(x)| \Big|\frac{d}{dx}\frac{1}{T'(x)}\Big| d\lambda(x) \\ &\leq \frac{1}{\theta(T)} Var_{I_j}(f) + \frac{K}{\theta(T)^2} \int_{I_j} |f| d\lambda. \end{aligned} \tag{6.12}$$

Set $\eta = \min_{1 \leq j \leq N} \lambda(I_j)$. Then

$$\lambda(TI_j) > \theta(T)\lambda(I_j) \geq \theta(T)\eta. \tag{6.13}$$

Combining (6.11), (6.12) and (6.13) with (6.10) yields

$$Var_{[0,1]}(P_T f) \leq \frac{2}{\theta(T)} \sum_{j=1}^{N} Var_{I_j}(f) + \frac{2}{\theta(T)} \sum_{j=1}^{N} \left(\frac{K}{\theta(T)} + \frac{1}{\eta} \right) \int_{I_j} |f| d\lambda.$$

Since $(I_j)_{1 \leq j \leq N}$ is an interval partition the statement holds with $L(T) = \frac{2}{\theta(T)} \left(\frac{K}{\theta(T)} + \frac{1}{\eta} \right)$. □

Proof of Theorem 6.3.1. We first use Lemma 6.3.1 to obtain an appropriate bound on the total variation of any iterate T^n of T. If T is a piecewise C^2 monotone expanding interval map, then so is T^n for any n. Since $\theta(T^n) \geq \theta(T)^n$ we can fix a k such that $\theta(T)^k > 2$. Then from the previous lemma we obtain for T^k and any $f \in BV([0,1])$ that

$$Var_{[0,1]}(P_T^k f) \leq \rho Var_{[0,1]}(f) + L(T^k) \|f\|_1, \tag{6.14}$$

where $\rho = \frac{2}{\theta(T)^k} \in (0,1)$ and $L(T^k) > 0$. Let $n > 0$ be given and write $n = mk + \ell$ with $0 \leq \ell < k$. By Lemma 6.3.1 we know that $P_T^j f \in BV([0,1])$ for any j. Repeated application of (6.14) together with Proposition 6.1.1(iv) and Proposition 6.1.2 gives that

$$\begin{aligned} Var_{[0,1]}(P_T^n f) &= Var_{[0,1]}(P_T^{mk}(P_T^\ell f)) \\ &\leq \rho^m Var_{[0,1]}(P_T^\ell f) + L(T^k) \|f\|_1 \sum_{i=0}^{m-1} \rho^i. \end{aligned}$$

Applying Lemma 6.3.1 to $P_T^\ell f$ then leads to

$$\begin{aligned} Var_{[0,1]}(P_T^n f) &\leq \rho^m \left(\frac{2}{\theta(T^\ell)} Var_{[0,1]}(f) + L(T^\ell) \|f\|_1 \right) + L(T^k) \|f\|_1 \sum_{i=0}^{m-1} \rho^i \\ &\leq 2\rho^m Var_{[0,1]}(f) + L(T^\ell) \rho^m \|f\|_1 + L(T^k) \sum_{i=0}^{m-1} \rho^i \|f\|_1 \\ &\leq 2 Var_{[0,1]}(f) + C \|f\|_1, \end{aligned}$$

where we have used that $\rho \in (0,1)$ and have set $C = \dfrac{\max_{0 \leq \ell \leq k} L(T^\ell)}{1-\rho} > 0$, which is independent of n.

Now fix a function $f \in BV([0,1])$ with $f \geq 0$ and $\int_{[0,1]} f \, d\lambda = 1$. Define the sequence of functions (f_n) by

$$f_n = \frac{1}{n} \sum_{i=0}^{n-1} P_T^i f.$$

By the positivity of P_T we find $f_n \geq 0$ for each n and since P_T preserves the integral also $\|f_n\|_1 = \int_{[0,1]} f_n \, d\lambda = 1$ for any n, so each f_n is a probability density as well. Moreover,

$$Var_{[0,1]}(f_n) \leq \frac{1}{n} \sum_{i=0}^{n-1} Var_{[0,1]}(P_T^i f) \leq 2Var_{[0,1]}(f) + C$$

and by Proposition 12.7.2(iii),

$$\|f_n\|_\infty \leq Var_{[0,1]}(f_n) + \int_{[0,1]} |f_n| d\lambda \leq 2Var_{[0,1]}(f) + C + 1.$$

Hence $\|f_n\|_\infty$ and $Var_{[0,1]}(f_n)$ can be bounded uniformly in n. By Helly's First Theorem we find a subsequence $(f_{n_j}) \subseteq (f_n)$ that converges pointwise to some $h \in BV([0,1])$. By Proposition 6.1.1(iv) it holds that

$$\begin{aligned}
\|P_T h - h\|_1 &\leq \|P_T h - P_T f_{n_j}\|_1 + \|P_T f_{n_j} - f_{n_j}\|_1 + \|f_{n_j} - h\|_1 \\
&\leq 2\|f_{n_j} - h\|_1 + \left\| \frac{1}{n_j} \sum_{i=0}^{n_j-1} P_T^{i+1} f - \frac{1}{n_j} \sum_{i=0}^{n_j-1} P_T^i f \right\|_1 \\
&= 2\|f_{n_j} - h\|_1 + \frac{1}{n_j} \|P_T^{n_j} f - f\|_1 \\
&\leq 2\|f_{n_j} - h\|_1 + \frac{2}{n_j}.
\end{aligned}$$

Since $\sup_{j \geq 1} \|f_{n_j}\|_\infty < \infty$, we can conclude from the Dominated Convergence Theorem that $\int_{[0,1]} h \, d\lambda = 1$ and that $\lim_{j\to\infty} \|f_{n_j} - h\|_1 = 0$. So, $\|P_T h - h\|_1 = 0$ and hence $P_T h = h$ λ-a.e. Moreover, $h \geq 0$ and $\int_{[0,1]} h \, d\lambda = 1$. Therefore h is the density of an absolutely continuous invariant probability measure for T. □

Note that in the proof of Theorem 6.3.1, we started with an arbitrary probability density function $f \in BV([0,1])$ and then found that some subsequence of $(f_n) = \left(\frac{1}{n} \sum_{i=0}^{n-1} P_T^i f\right)$ converges to an invariant probability density function for T that is of bounded variation.

Example 6.3.3. Let $\frac{1}{2} \leq a, b \leq 1$ and $0 < r < \frac{1}{2}$ be three parameters and define the transformations $W_{a,b,r} : [0,1] \to [0,1]$ by

$$W_{a,b,r}(x) = \begin{cases} a(1 - \frac{x}{r}), & \text{if } 0 \leq x \leq r, \\ \frac{2b}{1-2r}(x - r), & \text{if } r < x \leq \frac{1}{2}, \\ W_{a,b,r}(1 - x), & \text{if } \frac{1}{2} < x \leq 1. \end{cases}$$

See Figure 6.2(b) for a graph. Then $W_{a,b,r}$ is a piecewise C^2 monotone interval map with partition $\{[0,r],(r,\frac{1}{2}],(\frac{1}{2},1-r),[1-r,1]\}$. Since $a,b \geq \frac{1}{2}$, the map $W_{a,b,r}$ is expanding, so by Theorem 6.3.1 an invariant probability measure that is absolutely continuous with respect to λ exists.

Exercise 6.3.2. **(a)** Find an invariant probability measure that is absolutely continuous with respect to λ for the map $W_{\frac{1}{2},\frac{1}{2},r}$ and prove that $W_{\frac{1}{2},\frac{1}{2},r}$ is ergodic for any $0 < r < \frac{1}{2}$.
(b) Verify that the probability density $h:[0,1]\to\mathbb{R}$ given by $h(x) = (2-2r)1_{[0,\frac{1}{2})}(x)+2r1_{[\frac{1}{2},1)}(x)$ is a fixed point of $P_{W_{1,\frac{1}{2},r}}$ for any $0<r<\frac{1}{2}$.

Example 6.3.4. Let $\beta \in (1,2)$ and let $T:[0,1]\to[0,1]$ be the β-transformation defined by $Tx = \beta x \pmod 1$ (compare Example 1.3.6). Then T is a piecewise C^2 monotone interval map and T is expanding. The Perron-Frobenius operator is given by

$$P_T f(x) = \frac{1}{\beta} f\Big(\frac{x}{\beta}\Big) + \frac{1}{\beta} f\Big(\frac{x+1}{\beta}\Big) 1_{[0,\beta-1)}(x)$$

for any $f \in L^1([0,1],\mathcal{B},\lambda)$. Consider the probability density $h:[0,1)\to[0,\infty)$ given by

$$h(x) = \frac{1}{C}\sum_{n\geq 0}\frac{1}{\beta^n} 1_{[0,T^n 1)}(x),$$

where $C = \int_{[0,1)}\sum_{n\geq 0}\frac{1}{\beta^n}1_{[0,T^n1)}\,d\lambda$ is a normalizing constant. First suppose that $x \in [0,\beta-1)$. If $T^n 1 < \frac{1}{\beta}$, then $\frac{x}{\beta}\in[0,T^n1)$ if and only if $x\in[0,T^{n+1}1)$, so that

$$1_{[0,T^n1)}\Big(\frac{x}{\beta}\Big) + 1_{[0,T^n1)}\Big(\frac{x+1}{\beta}\Big) = 1_{[0,T^n1)}\Big(\frac{x}{\beta}\Big) = 1_{[0,T^{n+1}1)}(x).$$

Similarly, if $T^n1 \geq \frac{1}{\beta}$, then $\frac{x+1}{\beta}\in[0,T^n1)$ if and only if $x\in[0,T^{n+1}1)$ and

$$1_{[0,T^n1)}\Big(\frac{x}{\beta}\Big) + 1_{[0,T^n1)}\Big(\frac{x+1}{\beta}\Big) = 1 + 1_{[0,T^n1)}\Big(\frac{x+1}{\beta}\Big) = 1 + 1_{[0,T^{n+1}1)}(x).$$

This gives

$$\begin{aligned}
P_T h(x) &= \frac{1}{\beta}\Big(h\Big(\frac{x}{\beta}\Big) + \frac{1}{\beta}h\Big(\frac{x+1}{\beta}\Big)\Big)\\
&= \frac{1}{C}\sum_{n\ge 0}\frac{1}{\beta^{n+1}}\Big(1_{[0,T^n 1)}\Big(\frac{x}{\beta}\Big) + 1_{[0,T^n 1)}\Big(\frac{x+1}{\beta}\Big)\Big)\\
&= \frac{1}{C}\sum_{n\ge 0}\frac{1}{\beta^{n+1}}\Big(1_{[0,T^{n+1}1)}(x) + 1_{[\frac{1}{\beta},1)}(T^n 1)\Big)\\
&= \frac{1}{C}\sum_{n\ge 0}\frac{1}{\beta^{n+1}}1_{[0,T^{n+1}1)}(x) + \frac{1}{C} = h(x),
\end{aligned}$$

where we have used that the β-expansion of 1 is given by the expression $1 = \sum_{n\ge 1}\frac{1}{\beta^n}1_{[\frac{1}{\beta},1)}(T^{n-1}1)$. Now assume that $x \in [\beta - 1, 1)$. Then $\frac{x+1}{\beta} \notin [0,1]$ and it still holds that if $T^n 1 < \frac{1}{\beta}$, then $\frac{x}{\beta} \in [0, T^n 1)$ if and only if $x \in [0, T^{n+1}1)$. We obtain

$$\begin{aligned}
P_T h(x) &= \frac{1}{\beta}h\Big(\frac{x}{\beta}\Big) = \frac{1}{C}\sum_{n\ge 0}\frac{1}{\beta^{n+1}}1_{[0,T^n 1)}\Big(\frac{x}{\beta}\Big)\\
&= \frac{1}{C}\sum_{n\ge 0}\frac{1}{\beta^{n+1}}\Big(1_{[0,T^{n+1}1)}(x) + 1_{[\frac{1}{\beta},1)}(T^n 1)\Big) = h(x).
\end{aligned}$$

So the measure μ on $([0,1),\mathcal{B})$ given by $\mu(A) = \int_A h\, d\lambda$ for each $A \in \mathcal{B}$ is an absolutely continuous invariant probability measure for T.

Over the years the result of A. Lasota and J. A. Yorke was extended in many different directions. The condition that the number of intervals of monotonicity of T is finite can be relaxed to include also countable partitions (see [57]) and also the condition that T needs to be expanding can be relaxed to include a countable number of points $x \in [0,1]$ where $T'(x) = 1$ as long as such points x are not attracting (see e.g. [62, 66]), even though it might then no longer be possible to find an invariant *probability* measure. Here we just mention without a proof (a simplified version of) the result by M. Rychlik on systems with countably many branches. The proof can be found in [57], see also [8, Theorem 5.4.1].

Theorem 6.3.2 (Rychlik). *Let $I_1, I_2, \ldots$ be a countable collection of closed intervals with disjoint interiors such that $\bigcup_{n=1}^{\infty} I_n = [0,1]$. Let $U \subseteq [0,1]$ be the open set consisting of the union of the interiors of the intervals $I_1, I_2, \ldots$ and $S = [0,1] \setminus U$. Let $T : [0,1] \to [0,1]$ be a transformation with the following properties:*

(1) *$T|_U$ is continuous;*

(2) *for any interval I_n the restriction of T to the interior of I_n admits an extension to a homeomorphism from I_n to a subinterval of $[0,1]$;*

(3) *the function g given by $g(x) = \frac{1}{|T'x|}$ for $x \in U$ and $g(x) = 0$ for $x \in S$ satisfies $\|g\|_\infty < 1$ and $Var_{[0,1]}(g) < \infty$.*

Then there exists an invariant probability measure for T that is absolutely continuous with respect to λ.

To give an example of how to apply this result, we turn to the Gauss map from Example 1.3.7. This map is one specific member of a family of transformations introduced first by H. Nakada in [45] for $\alpha \in [\frac{1}{2}, 1]$ in 1981 and extended to the whole parameter interval $[0,1]$ in 2008 in [39]. For any $\alpha \in [0,1]$ the *α-continued fraction map* $T_\alpha : [\alpha-1, \alpha] \to [\alpha-1, \alpha]$ is defined by

$$T_\alpha x = \begin{cases} \left|\frac{1}{x}\right| - \left\lfloor \left|\frac{1}{x}\right| + 1 - \alpha \right\rfloor, & \text{if } x \neq 0, \\ 0, & \text{if } x = 0. \end{cases} \tag{6.15}$$

We obtain the Gauss map for $\alpha = 1$. The maps are called continued fraction maps, because for each $\alpha \in [0,1]$ the map T_α can be used to generate continued fraction expansions of numbers in the interval $[\alpha - 1, \alpha]$ in a way very similar to what we described in Example 1.3.7, see Exercise 8.4.2.

Exercise 6.3.3. Use Theorem 6.3.2 to prove that for every $\alpha \in (0,1]$ the α-continued fraction map T_α has an invariant probability measure that is absolutely continuous with respect to the Lebesgue measure on $[\alpha - 1, \alpha]$.

There is much more that can be said about the Perron-Frobenius operator and the information it gives on the statistical properties of dynamical systems. We refer the interested reader to e.g. [33, 8, 4].

CHAPTER 7

Invariant Measures for Continuous Transformations

7.1 EXISTENCE

Suppose (X, d) is a compact metric space. The topology induced by the metric can be used to put a measure structure on X. Let $\mathcal{B}$ be the Borel σ-algebra, i.e., the smallest σ-algebra containing all the open sets. Let $M(X)$ denote the collection of all Borel probability measures on X. There is a natural embedding of the space X in $M(X)$ via the map $x \mapsto \delta_x$ assigning to each x the Dirac measure δ_x concentrated at x. This implies that $M(X)$ is non-empty. We summarize some of the properties of $M(X)$ here. More details can be found in Section 12.6 from the Appendix.

- $M(X)$ is a *convex* set, i.e., $p\mu + (1-p)\nu \in M(X)$ whenever $\mu, \nu \in M(X)$ and $0 \leq p \leq 1$.
- Each element $\mu \in M(X)$ is *regular*: for each $A \in \mathcal{B}$ and each $\varepsilon > 0$ there exist an open set O_ε and a closed set C_ε with $C_\varepsilon \subseteq A \subseteq O_\varepsilon$ and $\mu(O_\varepsilon \setminus C_\varepsilon) < \varepsilon$ (Theorem 12.6.1).
- A metric structure on $M(X)$ is given by weak convergence: a sequence $(\mu_n) \subseteq M(X)$ *converges weakly* to a measure $\mu \in M(X)$ if

$$\lim_{n\to\infty} \int_X f \, d\mu_n = \int_X f d\mu$$

for any continuous $f : X \to \mathbb{C}$. The space $M(X)$ is compact under this notion of convergence (Theorem 12.6.4).

On X we consider the dynamics given by a continuous map $T : X \to X$. Since $\mathcal{B}$ is generated by the open sets, T is measurable with respect to $\mathcal{B}$. We have already seen some examples that fit this setup.

Example 7.1.1. Let $k \geq 1$, $X = \{0, 1, \dots, k-1\}^{\mathbb{N}}$ or $X = \{0, 1, \dots, k-1\}^{\mathbb{Z}}$. On X we can consider the metric d defined by

$$d(x, y) = d((x_n), (y_n)) = 2^{-\min\{i \geq 1 \,:\, x_i \neq y_i\}} \tag{7.1}$$

for the one-sided sequences and

$$d(x, y) = d((x_n), (y_n)) = 2^{-\min\{|i| \geq 0 \,:\, x_i \neq y_i\}}$$

for the two-sided ones. This metric induces the product topology on X obtained from taking the discrete topology on the set $\{0, 1, \dots, k-1\}$. Open sets are unions of sets of the form

$$\{x \in X \,:\, x_i \in U_i, U_i \subseteq \{0, 1, \dots, k-1\} \text{ and } U_i \neq \{0, 1, \dots, k-1\} \text{ for only finitely many } i\}.$$

A basis for this topology is provided by the collection $\mathcal{S}$ of all sets that specify a finite number of coordinates. As the product of compact spaces X is compact.

Two examples of transformations defined on (X, d) are the left shift from Example 1.3.9 and the binary odometer in case $k = 2$ (see Example 1.3.11). Both maps are continuous since the pre-image of any set from $\mathcal{S}$ is a union of sets in $\mathcal{S}$. In case $X = \{0, 1\}^{\mathbb{N}}$ and T is the binary odometer, or $X = \{0, 1, \dots, k-1\}^{\mathbb{Z}}$ and T is the left shift, then these maps are homeomorphisms.

Example 7.1.2. Recall Arnold's cat map $T : \mathbb{T}^2 \to \mathbb{T}^2$ from Example 1.3.8 defined on the two-dimensional torus by an application of the matrix

$$A = \begin{pmatrix} 2 & 1 \\ 1 & 1 \end{pmatrix},$$

so

$$T(e^{2\pi i x}, e^{2\pi i y}) = (e^{2\pi i (2x+y)}, e^{2\pi i (x+y)}).$$

The metric on $\mathbb{T}^2$ is the Euclidean distance on arcs. We already noticed that T is invertible, and clearly it is continuous, since it is basically a linear transformation. Since the same holds for T^{-1}, T is a homeomorphism.

A continuous map $T : X \to X$ on a compact metric space (X, d) induces in a natural way an operator $\overline{T} : M(X) \to M(X)$ given by

$$(\overline{T}\mu)(A) = \mu(T^{-1}A), \quad A \in \mathcal{B}.$$

Then $\overline{T}^i \mu(A) = \mu(T^{-i}A)$. Using a standard argument, one can easily show that

$$\int_X f \, d(\overline{T}\mu) = \int_X f \circ T \, d\mu$$

for any integrable function f on X. Note that T is measure preserving with respect to $\mu \in M(X)$ if and only if $\overline{T}\mu = \mu$. Since X is a compact metric space, measures from $M(X)$ are uniquely determined by how they integrate continuous functions, see Theorem 12.6.2. Hence, $\mu \in M(X)$ is measure preserving if and only if

$$\int_X f \, d\mu = \int_X f \circ T \, d\mu$$

for all continuous functions f on X. Let

$$M(X,T) = \{\mu \in M(X) : \overline{T}\mu = \mu\}$$

be the collection of all T-invariant Borel probability measures. The first question we would like to answer is whether $M(X,T) \neq \emptyset$.

Theorem 7.1.1. *Let $T : X \to X$ be continuous, and (ν_n) a sequence in $M(X)$. Define a new sequence (μ_n) in $M(X)$ by*

$$\mu_n = \frac{1}{n} \sum_{i=0}^{n-1} \overline{T}^i \nu_n.$$

Then, any weak limit point μ of (μ_n) is an element of $M(X,T)$.

Proof. First note that by the compactness of $M(X)$, the sequence (μ_n) has a convergent subsequence (μ_{n_i}) with limit point $\mu \in M(X)$. For any continuous function f on X,

$$\begin{aligned}
\left|\int_X f\circ T\,d\mu - \int_X f\,d\mu\right| \\
&= \lim_{j\to\infty}\left|\int_X f\circ T\,d\mu_{n_j} - \int_X f\,d\mu_{n_j}\right| \\
&= \lim_{j\to\infty}\left|\frac{1}{n_j}\int_X \sum_{i=0}^{n_j-1}\Big(f(T^{i+1}x) - f(T^i x)\Big)\,d\nu_{n_j}(x)\right| \\
&= \lim_{j\to\infty}\left|\frac{1}{n_j}\int_X (f(T^{n_j}x) - f(x))\,d\nu_{n_j}(x)\right| \\
&\leq \lim_{j\to\infty}\frac{2\|f\|_\infty}{n_j} = 0.
\end{aligned}$$

Therefore $\int_X f\,d\mu = \int_X f\circ T\,d\mu$ and $\mu \in M(X,T)$. □

For example, to find a T-invariant measure, one can start with the Dirac measure δ_x concentrated on any point $x \in X$ and then look at the sequence (μ_n) given by $\mu_n = \frac{1}{n}\sum_{i=0}^{n-1}\delta_{T^i x}$. From the above theorem any limit point of a convergent subsequence of (μ_n) is an element of $M(X,T)$. The next result characterizes the ergodic members of $M(X,T)$.

Theorem 7.1.2. *Let T be a continuous transformation on a compact metric space X. Then the following hold.*

(i) *$M(X,T)$ is a compact convex subset of $M(X)$.*

(ii) *$\mu \in M(X,T)$ is an* extreme point *(i.e., μ cannot be written in a non-trivial way as a convex combination of elements of $M(X,T)$) if and only if T is ergodic with respect to μ.*

Proof. (i) Clearly $M(X,T)$ is convex. Now let (μ_n) be a sequence in $M(X,T)$ converging to $\mu \in M(X)$. We need to show that $\mu \in M(X,T)$. Since T is continuous, then for any continuous function f on X, the function $f\circ T$ is also continuous. Hence,

$$\begin{aligned}
\int_X f\circ T\,d\mu &= \lim_{n\to\infty}\int_X f\circ T\,d\mu_n \\
&= \lim_{n\to\infty}\int_X f\,d\mu_n \\
&= \int_X f\,d\mu.
\end{aligned}$$

Therefore, T is measure preserving with respect to μ, and $\mu \in M(X,T)$.

(ii) Suppose T is ergodic with respect to μ. If μ can be written as $\mu = p\nu + (1-p)\rho$ for some $\nu, \rho \in M(X,T)$ and some $0 < p \leq 1$, then ν is absolutely continuous with respect to μ. It then follows from Theorem 3.1.2 that $\nu = \mu$. Conversely, suppose that T is not ergodic with respect to μ. Then there exists a measurable set A such that $T^{-1}A = A$, and $0 < \mu(A) < 1$. Define measures μ_1, μ_2 on X by

$$\mu_1(B) = \frac{\mu(B \cap A)}{\mu(A)} \quad \text{and} \quad \mu_2(B) = \frac{\mu\left(B \cap (X \setminus A)\right)}{\mu(X \setminus A)}.$$

Since A and $X \setminus A$ are T-invariant sets, then $\mu_1, \mu_2 \in M(X,T)$, and $\mu_1 \neq \mu_2$ since $\mu_1(A) = 1$ while $\mu_2(A) = 0$. Furthermore, for any measurable set B,

$$\mu(B) = \mu(A)\mu_1(B) + (1 - \mu(A))\mu_2(B),$$

i.e., μ_1 is a non-trivial convex combination of elements of $M(X,T)$. Thus, μ is not an extreme point of $M(X,T)$. □

Recall that in the case of a general finite measure preserving system $(X, \mathcal{F}, \mu, T)$, the Pointwise Ergodic Theorem allows us to find for each $f \in L^1(X, \mathcal{F}, \mu)$ a measurable set X_f with $\mu(X_f) = 1$ such that

$$\lim_{n\to\infty} \frac{1}{n} \sum_{i=0}^{n-1} f(T^i x) = \int_X f \, d\mu$$

for each $x \in X_f$. The set X_f might depend on the function f (and does in most cases). When the underlying space X is a compact metric space and $T : X \to X$ is continuous, it is possible to find a single set Y of full measure that works for all continuous maps f. The proof uses the fact that the Banach space $C(X)$ of all complex valued continuous functions on X (under the supremum norm $\|f\|_\infty = \sup_{x\in X} |f(x)|$) is *separable*, i.e., $C(X)$ has a countable dense subset.

Theorem 7.1.3. *Let $T : X \to X$ be continuous and $\mu \in M(X,T)$ ergodic. Then there exists a measurable set Y with $\mu(Y) = 1$ and such that*

$$\lim_{n\to\infty} \frac{1}{n} \sum_{i=0}^{n-1} f(T^i x) = \int_X f \, d\mu$$

for all $x \in Y$ and $f \in C(X)$.

Proof. Since $C(X)$ is separable, we can choose a countable dense subset (g_m) in $C(X)$. By the Pointwise Ergodic Theorem, for each m there exists a subset Y_m with $\mu(Y_m) = 1$ and

$$\lim_{n\to\infty} \frac{1}{n} \sum_{i=0}^{n-1} g_m(T^i x) = \int_X g_m \, d\mu$$

for all $x \in Y_m$. Let $Y = \bigcap_{m=1}^{\infty} Y_m$, then $\mu(Y) = 1$, and

$$\lim_{n\to\infty} \frac{1}{n} \sum_{i=0}^{n-1} g_m(T^i x) = \int_X g_m \, d\mu$$

for all m and all $x \in Y$. Now, let $f \in C(X)$, $x \in Y$ and $\varepsilon > 0$ be given. Then there is an m, such that $\|f - g\|_\infty < \frac{\varepsilon}{3}$. Moreover, we can find an N such that for each $n \geq N$,

$$\left| \frac{1}{n} \sum_{i=0}^{n-1} g_m(T^i x) - \int_X g_m \, d\mu \right| < \frac{\varepsilon}{3}$$

and hence,

$$\begin{aligned} \left| \frac{1}{n} \sum_{i=0}^{n-1} f(T^i x) - \int_X f \, d\mu \right| &\leq \left| \frac{1}{n} \sum_{i=0}^{n-1} f(T^i x) - \frac{1}{n} \sum_{i=0}^{n-1} g_m(T^i x) \right| \\ &\quad + \left| \frac{1}{n} \sum_{i=0}^{n-1} g_m(T^i x) - \int_X g_m \, d\mu \right| \\ &\quad + \left| \int_X g_m \, d\mu - \int_X f \, d\mu \right| < \varepsilon. \end{aligned}$$

This gives the result. □

Theorem 7.1.4. *Let X be a compact metric space and $\mathcal{B}$ the Borel σ-algebra on X. Let $T : X \to X$ be continuous, and $\mu \in M(X,T)$. Then T is ergodic with respect to μ if and only if for μ-a.e. x one has*

$$\lim_{n\to\infty} \frac{1}{n} \sum_{i=0}^{n-1} \delta_{T^i x} = \mu.$$

Proof. Suppose T is ergodic with respect to μ. Notice that for any $f \in C(X)$,

$$\int_X f \, d(\delta_{T^i x}) = f(T^i x).$$

Hence by Theorem 7.1.3, there exists a measurable set Y with $\mu(Y) = 1$ such that

$$\lim_{n\to\infty} \frac{1}{n}\sum_{i=0}^{n-1} \int_X f\, d(\delta_{T^i x}) = \lim_{n\to\infty} \frac{1}{n}\sum_{i=0}^{n-1} f(T^i x) = \int_X f\, d\mu$$

for all $x \in Y$, and $f \in C(X)$. So, for all $x \in Y$ the sequence $\left(\frac{1}{n}\sum_{i=0}^{n-1} \delta_{T^i x}\right)$ converges weakly to μ.

Conversely, let $Y \in \mathcal{B}$ be a set with $\mu(Y) = 1$ and such that $\frac{1}{n}\sum_{i=0}^{n-1} \delta_{T^i x} \to \mu$ for all $x \in Y$. Then for any $f \in C(X)$ and any $h \in L^1(X, \mathcal{B}, \mu)$ one has for any $x \in Y$,

$$\lim_{n\to\infty} \frac{1}{n}\sum_{i=0}^{n-1} f(T^i x)h(x) = h(x) \int_X f\, d\mu.$$

By the Dominated Convergence Theorem,

$$\lim_{n\to\infty} \frac{1}{n}\sum_{i=0}^{n-1} \int_X f(T^i x)h(x)\, d\mu(x) = \int_X h\, d\mu \int_X f\, d\mu. \tag{7.2}$$

Now, let $F, H \in L^2(X, \mathcal{B}, \mu)$. Since μ is a probability measure we have $H \in L^1(X, \mathcal{B}, \mu)$ so that

$$\lim_{n\to\infty} \frac{1}{n}\sum_{i=0}^{n-1} \int_X f(T^i x)H(x)\, d\mu(x) = \int_X H\, d\mu \int_X f\, d\mu$$

for all $f \in C(X)$. Let $\varepsilon > 0$, since $C(X)$ is dense in $L^2(X, \mathcal{B}, \mu)$ (under the L^2-norm), there exists a $g \in C(X)$ such that $\|F - g\|_2 < \varepsilon$. This implies that $|\int F\, d\mu - \int g\, d\mu| \leq \|F - g\|_1 < \varepsilon$. Furthermore, by (7.2) there exists an N so that for $n \geq N$ one has

$$\left|\int_X \frac{1}{n}\sum_{i=0}^{n-1} g(T^i x)H(x)\, d\mu(x) - \int_X H\, d\mu \int_X g\, d\mu\right| < \varepsilon.$$

Thus, for $n \geq N$ one has using Hölder's Inequality and Proposition 1.2.3

that

$$
\begin{aligned}
\Big|\int_X \frac{1}{n}\sum_{i=0}^{n-1} F(T^i x)H(x)\,d\mu(x) - \int_X H\,d\mu \int_X F\,d\mu\Big| \\
&\leq \int_X \frac{1}{n}\sum_{i=0}^{n-1} \Big|F(T^i x) - g(T^i x)\Big|\,|H(x)|\,d\mu(x) \\
&\quad + \Big|\int_X \frac{1}{n}\sum_{i=0}^{n-1} g(T^i x)H(x)\,d\mu(x) - \int_X H\,d\mu \int_X g\,d\mu\Big| \\
&\quad + \Big|\int_X g\,d\mu \int_X H\,d\mu - \int_X F\,d\mu \int_X H\,d\mu\Big| \\
&< \varepsilon\|H\|_2 + \varepsilon + \varepsilon\|H\|_2.
\end{aligned}
$$

It follows that

$$
\lim_{n\to\infty} \frac{1}{n}\sum_{i=0}^{n-1} \int_X F(T^i x)H(x)\,d\mu(x) = \int_X H\,d\mu \int_X F\,d\mu
$$

for all $F, H \in L^2(X, \mathcal{B}, \mu)$. Taking F and H to be indicator functions, one gets that T is ergodic. □

Exercise 7.1.1. Let X be a compact metric space and $T : X \to X$ a homeomorphism. Let $x \in X$ be a periodic point of T of period n, i.e., $T^n x = x$ and $T^j x \neq x$ for $j < n$. Show that if $\mu \in M(X, T)$ is ergodic and $\mu(\{x\}) > 0$, then $\mu = \frac{1}{n}\sum_{i=0}^{n-1} \delta_{T^i x}$.

We end this section with a short discussion that the assumption of ergodicity is not very restrictive. In case (X, d) is a compact metric space, we have seen that $M(X, T)$ is a compact convex subset of the convex set $M(X)$ and that the extreme points of $M(X, T)$ are precisely the ergodic T-invariant Borel probability measures. Denote this set by $M_E(X, T)$. Any $f \in C(X)$ defines a continuous linear functional on $M(X)$ (with respect to the weak topology) by $f(\mu) = \int_X f\,d\mu$. Now recall Choquet's Theorem, see Theorem 12.6.6. If we simply take $Y = M(X, T)$ and $V = M(X)$ and let $\mu \in M(X, T)$, then Theorem 12.6.6 gives the existence of a probability measure ν on $M(X, T)$ with the property that for any $f \in C(X)$,

$$
\int_X f d\mu = \int_{M_E(X,T)} \int_X f\,d\mu_e\,d\nu.
$$

In other words, the integral of f with respect to μ can be decomposed into integrals with respect to the ergodic members of $M(X,T)$. This is a special case of a more general theorem, called the Ergodic Decomposition, which we state below. For the proof we refer to e.g. [17] or [63].

Theorem 7.1.5 (Ergodic Decomposition). *Let $T : X \to X$ be a measure preserving transformation on a Borel probability space $(X,\mathcal{F},\mu)$. Then there exists a Borel probability space $(Z,\mathcal{G},\nu)$ and an injective measurable map $z \mapsto \mu_z$ such that*

(i) μ_z is a T-invariant and ergodic probability measure on X for ν-a.e. $z \in Z$, and

(ii) $\mu = \int_Z \mu_z \, d\nu(z)$.

7.2 UNIQUE ERGODICITY AND UNIFORM DISTRIBUTION

Examining Theorem 7.1.3, one sees that for $\mu \in M(X,T)$ the set Y (associated with μ) can be taken to be T-invariant, since the limit of the ergodic averages is a T-invariant function. So it is feasible that Y^c (which is T-invariant as well) is the support of some other measure $\nu \in M(X,T)$ such that $\lim_{n\to\infty} \frac{1}{n}\sum_{i=0}^{n-1} f(T^i x) = \int_X f \, d\nu$ for all $f \in C(X)$ and all $x \in Y^c$. So what can we say in the case $X = Y$? This leads us to the notion of unique ergodicity which will be discussed below.

A continuous transformation $T : X \to X$ on a compact metric space (X,d) is called *uniquely ergodic* if there is only one T-invariant probability measure μ on X. In this case, $M(X,T) = \{\mu\}$, and μ is necessarily ergodic, since μ is an extreme point of $M(X,T)$. Recall that if $\nu \in M(X,T)$ is ergodic, then there exists a measurable subset Y such that $\nu(Y) = 1$ and

$$\lim_{n\to\infty} \frac{1}{n}\sum_{i=0}^{n-1} f(T^i x) = \int_X f \, d\nu$$

for all $x \in Y$ and all $f \in C(X)$. When T is uniquely ergodic we have the following much stronger result.

Theorem 7.2.1. *Let $T : X \to X$ be a continuous transformation on a compact metric space (X,d). Then the following are equivalent:*

(i) *For every $f \in C(X)$, the sequence $(\frac{1}{n}\sum_{j=0}^{n-1} f(T^j x))$ converges uniformly to a constant.*

(ii) *For every* $f \in C(X)$, *the sequence* $(\frac{1}{n}\sum_{j=0}^{n-1} f(T^j x))$ *converges pointwise to a constant.*

(iii) *There exists a* $\mu \in M(X,T)$ *such that for every* $f \in C(X)$ *and all* $x \in X$

$$\lim_{n\to\infty} \frac{1}{n}\sum_{i=0}^{n-1} f(T^i x) = \int_X f\, d\mu.$$

(iv) T *is uniquely ergodic.*

Proof. (i) $\Rightarrow$ (ii) is immediate.

(ii) $\Rightarrow$ (iii) Define $L : C(X) \to \mathbb{C}$ by

$$L(f)(x) = \lim_{n\to\infty} \frac{1}{n}\sum_{i=0}^{n-1} f(T^i x).$$

By assumption $L(f)$ is a constant function. It is easy to see that L is linear, continuous ($|L(f)| \leq \|f\|_\infty$), positive and satisfies $L(1) = 1$. Thus, by the Riesz Representation Theorem (Theorem 12.6.3) there exists a probability measure $\mu \in M(X)$ such that

$$L(f) = \int_X f\, d\mu$$

for all $f \in C(X)$. But

$$L(f \circ T) = \lim_{n\to\infty} \frac{1}{n}\sum_{i=0}^{n-1} f(T^{i+1} x) = L(f).$$

Hence,

$$\int_X f \circ T\, d\mu = \int_X f\, d\mu.$$

Thus, $\mu \in M(X,T)$, and for every $f \in C(X)$,

$$\lim_{n\to\infty} \frac{1}{n}\sum_{i=0}^{n-1} f(T^i x) = \int_X f\, d\mu$$

for all $x \in X$.

(iii) $\Rightarrow$ (iv) Suppose $\mu \in M(X,T)$ is such that for every $f \in C(X)$,

$$\lim_{n\to\infty} \frac{1}{n}\sum_{i=0}^{n-1} f(T^i x) = \int_X f\, d\mu$$

holds for all $x \in X$. Assume $\nu \in M(X,T)$, we will show that $\mu = \nu$. For any $f \in C(X)$, since the sequence $(\frac{1}{n}\sum_{j=0}^{n-1} f(T^j x))$ converges pointwise to the constant function $\int_X f\,d\mu$, and since each term of the sequence is bounded in absolute value by the constant $\|f\|_\infty = \sup_{x\in X}|f(x)|$, it follows by the Dominated Convergence Theorem that

$$\lim_{n\to\infty}\int_X \frac{1}{n}\sum_{i=0}^{n-1} f(T^i x)\,d\nu(x) = \int_X\int_X f\,d\mu\,d\nu = \int_X f\,d\mu.$$

But for each n,

$$\int_X \frac{1}{n}\sum_{i=0}^{n-1} f(T^i x)\,d\nu(x) = \frac{1}{n}\sum_{i=0}^{n-1}\int_X f(T^i x)\,d\nu(x) = \int_X f\,d\nu.$$

Thus, $\int_X f\,d\mu = \int_X f\,d\nu$, and $\mu = \nu$.

(iv) $\Rightarrow$ (i) The proof is done by contradiction. Assume $M(X,T) = \{\mu\}$ and suppose $h \in C(X)$ is such that the sequence $(\frac{1}{n}\sum_{j=0}^{n-1} h\circ T^j)$ does not converge uniformly to $\int_X h\,d\mu$. Then there exists an $\varepsilon > 0$ such that for each M there exist $n > M$ and $y_n \in X$ such that

$$\left|\frac{1}{n}\sum_{j=0}^{n-1} h(T^j y_n) - \int_X h\,d\mu\right| \geq \varepsilon.$$

Let

$$\mu_n = \frac{1}{n}\sum_{j=0}^{n-1}\delta_{T^j y_n} = \frac{1}{n}\sum_{j=0}^{n-1}\overline{T}^j\delta_{y_n}.$$

Then,

$$\left|\int_X h\,d\mu_n - \int_X h\,d\mu\right| \geq \varepsilon.$$

Since $M(X)$ is compact, there exists a subsequence (μ_{n_i}) converging to some $\nu \in M(X)$. Hence,

$$\left|\int_X h\,d\nu - \int_X h\,d\mu\right| \geq \varepsilon.$$

By Theorem 7.1.1, $\nu \in M(X,T)$ and by unique ergodicity $\mu = \nu$, which is a contradiction. □

Example 7.2.1. Let $\theta \in [0,1)\setminus\mathbb{Q}$ and $T_\theta(x) = x + \theta \pmod 1$ the corresponding irrational rotation. Then T_θ is uniquely ergodic. This is a consequence of the above theorem as well as *Weyl's Theorem* on uniform distribution: for any Riemann integrable function f on $[0,1)$, and any $x \in [0,1)$, one has

$$\lim_{n\to\infty} \frac{1}{n} \sum_{i=0}^{n-1} f(x + i\theta - \lfloor x + i\theta \rfloor) = \int_X f(y)\, dy.$$

We will prove this statement using Theorem 7.2.1(ii). In order to use the results of this section we need to look at the continuous version of T_θ. So consider the probability space $(\mathbb{S}^1, \mathcal{B}, m)$, where $\mathbb{S}^1$ is the unit circle with the Borel σ-algebra and m is the normalized Haar (or Lebesgue) measure. On it we consider the map R_θ which is the rotation by the vector $e^{2\pi i\theta}$, see also Example 1.3.2. Since R_θ is an isometry, it is a homeomorphism. To show R_θ is uniquely ergodic first note that the set of trigonometric polynomials $p(x) = \sum_{n=-N}^{N} c_n e^{2\pi i n x}$ are dense in $C([0,1))$ (under the supremum norm), so it is enough to consider continuous functions of the form $f(x) = e^{2\pi i k x}$ for some fixed k. Now for any x and any n,

$$\frac{1}{n}\sum_{j=0}^{n-1} f(R_\theta^j x) = \frac{1}{n}\sum_{j=0}^{n-1} e^{2\pi i k(x+j\theta)} = \frac{1}{n} e^{2\pi i k x} \sum_{j=0}^{n-1} \left(e^{2\pi i k\theta}\right)^j.$$

Thus,

$$\frac{1}{n}\sum_{j=0}^{n-1} f(R_\theta^j x) = \begin{cases} 1, & \text{if } k = 0, \\ \frac{e^{2\pi i k x}}{n} \cdot \frac{e^{2\pi i k n\theta}-1}{e^{2\pi i k\theta}-1}, & \text{if } k \neq 0. \end{cases}$$

Taking limits we have

$$\lim_{n\to\infty} \frac{1}{n}\sum_{j=0}^{n-1} f(R_\theta^j x) = \begin{cases} 1, & \text{if } k = 0, \\ 0, & \text{if } k \neq 0, \end{cases} = \int_{\mathbb{S}^1} f\, dm.$$

This shows that R_θ is uniquely ergodic, and the same holds for T_θ.

Example 7.2.2. Consider the sequence of *first digits*

$$S = (1, 2, 4, 8, 1, 3, 6, 1, \ldots)$$

obtained by writing the first decimal digit of each term in the sequence

$$\{2^n : n \geq 0\} = \{1, 2, 4, 8, 16, 32, 64, 128, \ldots\}.$$

The elements of S belong to the set $\{1, 2, \ldots, 9\}$. For each $k = 1, 2, \ldots, 9$, let $p_k(n)$ be the number of times the digit k appears in the first n terms of the *first digit* sequence S. The asymptotic relative frequency of the digit k is then $\lim_{n\to\infty} \frac{p_k(n)}{n}$. We want to identify this limit for each $k \in \{1, 2, \ldots, 9\}$. It looks like an ergodic average, but what is the underlying transformation? We analyze the sequence S further. The first digit of 2^i is k if and only if

$$k \cdot 10^r \leq 2^i < (k+1) \cdot 10^r$$

for some $r \geq 0$. In this case,

$$r + \log_{10} k \leq i \log_{10} 2 < r + \log_{10}(k+1).$$

This shows that $r = \lfloor i \log_{10} 2 \rfloor$, and

$$\log_{10} k \leq i \log_{10} 2 - \lfloor i \log_{10} 2 \rfloor < \log_{10}(k+1).$$

Let $J_k = [\log_{10} k, \log_{10}(k+1))$ and $\theta = \log_{10} 2$. Then, θ is irrational and $T_\theta^i 0 = i \log_{10} 2 - \lfloor i \log_{10} 2 \rfloor$. Using this, we see that the first digit of 2^i is k if and only if $T_\theta^i 0 \in J_k$, and so

$$\lim_{n\to\infty} \frac{p_k(n)}{n} = \lim_{n\to\infty} \frac{1}{n} \sum_{i=0}^{n-1} 1_{J_k}(T_\theta^i 0).$$

We cannot apply Theorem 7.2.1 directly because the function 1_{J_k} is not continuous, but this problem can be solved easily by squeezing 1_{J_k} between two continuous functions. In fact we need to use the continuous version R_θ of T_θ, but to keep the notation simple we keep using the space $([0,1), \mathcal{B}, \lambda, T_\theta)$. Let $\varepsilon > 0$, we can find two continuous functions f_1, f_2 such that $f_1 \leq 1_{J_k} \leq f_2$ and $\int (f_2 - f_1)\, d\lambda < \varepsilon$. Then,

$$\begin{aligned}
\limsup_{n\to\infty} \frac{1}{n} \sum_{i=0}^{n-1} 1_{J_k}(T_\theta^i 0) &\leq \lim_{n\to\infty} \frac{1}{n} \sum_{i=0}^{n-1} f_2(T_\theta^i 0) \\
&= \int f_2\, d\lambda \\
&< \lambda(J_k) + \varepsilon \\
&= \log_{10}\left(\frac{k+1}{k}\right) + \varepsilon,
\end{aligned}$$

and

$$\begin{aligned}
\liminf_{n\to\infty}\frac{1}{n}\sum_{i=0}^{n-1}1_{J_k}(T_\theta^i 0) &\geq \lim_{n\to\infty}\frac{1}{n}\sum_{i=0}^{n-1}f_1(T_\theta^i 0)\\
&= \int f_1\,d\lambda\\
&> \lambda(J_k)-\varepsilon\\
&= \log_{10}\left(\frac{k+1}{k}\right)-\varepsilon.
\end{aligned}$$

This shows that

$$\lim_{n\to\infty}\frac{p_k(n)}{n} = \lim_{n\to\infty}\frac{1}{n}\sum_{i=0}^{n-1}1_{J_k}(T_\theta^i 0) = \log_{10}\left(\frac{k+1}{k}\right).$$

Exercise 7.2.1. Let X be a compact metric space, and $\mathcal{B}$ the Borel σ-algebra on X. Let $T : X \to X$ be a continuous transformation. Let $N \geq 1$ and $x \in X$.

(a) Show that $T^N x = x$ if and only if $\frac{1}{N}\sum_{i=0}^{N-1}\delta_{T^i x} \in M(X,T)$.

(b) Suppose $X = \{1,2,\ldots,N\}$ and $Ti = i+1 \pmod N$. Show that T is uniquely ergodic. Determine the unique ergodic measure.

Exercise 7.2.2. Let X be a compact metric space, $\mathcal{B}$ the Borel σ-algebra on X and $T : X \to X$ a uniquely ergodic continuous transformation. Let μ be the unique ergodic measure and assume that $\mu(G) > 0$ for all non-empty open sets $G \subseteq X$.

(a) Show that for each non-empty open subset G of X there exists a continuous function $f \in C(X)$, and a closed subset F of G of positive μ measure such that $f(x) = 1$ for $x \in F$, $f(x) > 0$ for $x \in G$ and $f(x) = 0$ for $x \in X \setminus G$.

(b) Show that for each $x \in X$ and for every non-empty open set $G \subseteq X$, there exists an $n \geq 0$ such that $T^n x \in G$. Conclude that $\{T^n x : n \geq 0\}$ is dense in X.

Exercise 7.2.3. Consider the measure space $([0,1],\mathcal{B})$, where $\mathcal{B}$ is the Borel σ-algebra on $[0,1]$. Let $T : [0,1] \to [0,1]$ be given by $Tx = \frac{x}{2}$, and assume that T is measure preserving with respect to some probability measure μ on $\mathcal{B}$. Show that $\mu = \delta_0$, the Dirac measure concentrated at the point 0.

Exercise 7.2.4. In Example 1.3.11 the binary odometer $T : \{0,1\}^{\mathbb{N}} \to \{0,1\}^{\mathbb{N}}$ was defined by $T(1,1,1,\ldots) = (0,0,0,\ldots)$ and for $n \geq 1$,

$$T(\underbrace{1,\cdots,1}_{n-1},0,x_{n+1},\ldots) = (\underbrace{0,\cdots,0}_{n-1},1,x_{n+1},\ldots),$$

on the underlying space $\{0,1\}^{\mathbb{N}}$ equipped with the σ-algebra $\mathcal{C}$ generated by the cylinder sets, and uniform Bernoulli measure $\mu_{\frac{1}{2}}$. We already know from Example 1.3.11 that T is measure preserving with respect to $\mu_{\frac{1}{2}}$. Prove that T is uniquely ergodic.

7.3 SOME TOPOLOGICAL DYNAMICS

One can study the dynamics of a continuous transformation $T : X \to X$ on a compact metric space without referring to an invariant measure. In this section we introduce some concepts from topological dynamics that have clear analogues in the setting of ergodic theory. The lack of a measure implies that we cannot have a.e. results, but the statements will have to hold for all elements in the compact metric space (X, d).

Definition 7.3.1. A continuous transformation $T : X \to X$ is called *(one-sided) topologically transitive* if for some $x \in X$ the (forward) orbit $\{T^n x : n \geq 0\}$ is dense in X.
A homeomorphism $T : X \to X$ is called *topologically transitive* if for some $x \in X$ the (two-sided) orbit $\{T^n x : n \in \mathbb{Z}\}$ is dense in X.

Write $O_T(x)$ for the appropriate orbit of x under T, so $O_T(x) = \{T^n x : n \in \mathbb{Z}\}$ if T is a homeomorphism and $O_T(x) = \{T^n x : n \geq 0\}$ if T is continuous and not invertible. The following theorem summarizes some equivalent definitions of topological transitivity. The analogues for non-invertible continuous maps can be found in e.g. [65].

Theorem 7.3.1. *Let $T : X \to X$ be a homeomorphism of a compact metric space (X, d). Then the following are equivalent.*

(i) *T is topologically transitive.*

(ii) *If $A \subseteq X$ is a closed set satisfying $TA = A$, then either $A = X$ or A has empty interior (i.e., $X \setminus A$ is dense in X).*

(iii) *For any pair $U, V \subseteq X$ of non empty open sets there exists an $n \in \mathbb{Z}$ such that $T^n U \cap V \neq \emptyset$.*

Proof. Recall that $O_T(x)$ is dense in X if and only if for every non-empty open set U in X, $U \cap O_T(x) \neq \emptyset$.

(i) $\Rightarrow$ (ii) Let $x \in X$ be such that $\overline{O_T(x)} = X$, where $\overline{O_T(x)}$ denotes the closure of $O_T(x)$, and let $A \subseteq X$ be a closed set with $TA = A$. Suppose A does not have an empty interior. Then there is a non-empty open set $U \subseteq A$ and a $p \in \mathbb{Z}$ such that $T^p x \in U$. Since $T^n A = A$ for all $n \in \mathbb{Z}$ we get $O_T(x) \subseteq A$. The statement now follows since A is closed.

(ii) $\Rightarrow$ (iii) Let U, V be two non-empty open subsets of X. Write $W = \cup_{n\in\mathbb{Z}} T^n U$. Then W is open, non-empty and $TW = W$. Hence, $A = X \backslash W$ is closed, $TA = A$ and $A \neq X$. By (ii), A has empty interior, so W is dense in X. Thus $W \cap V \neq \emptyset$, which implies (iii).

(iii) $\Rightarrow$ (i) By Theorem 12.1.1(iv) there exists a countable basis $\mathcal{U} = \{U_k\}_{k\geq 1}$ for the topology of X. Then

$$\{x \in X : \overline{O_T(x)} = X\} = \bigcap_{k=1}^{\infty} \bigcup_{m\in\mathbb{Z}} T^m U_k. \tag{7.3}$$

One inclusion is clear. For the other inclusion note that for any $y \in X$ any neighborhood of y contains an element $U_n \in \mathcal{U}$ and thus also an element $T^m x \in O_T(x)$. Now fix a $k \geq 1$ and set $U = \cup_{m=-\infty}^{\infty} T^m U_k$. Then U is open, non-empty and $TU = U$. By (iii) for any open non-empty set V there is an $n \in \mathbb{Z}$ such that $U \cap V = T^n U \cap V \neq \emptyset$. So, for each $k \geq 1$ the set $\cup_{m=-\infty}^{\infty} T^m U_k$ is dense in X. From the Baire Category Theorem (see Theorem 12.1.1(vii)) we then see that also the intersection

$$\bigcap_{k=1}^{\infty} \bigcup_{m\in\mathbb{Z}} T^m U_k$$

is dense in X and thus non-empty. By (7.3) this gives a point $y \in X$ with a dense orbit. □

A direct consequence of the above theorem is the following corollary.

Corollary 7.3.1. *Let $T : X \to X$ be a homeomorphism of a compact metric space (X, d). Then T is topologically transitive if and only if there are no two disjoint non-empty T-invariant open sets.*

Note that Theorem 7.3.1(ii) resembles Proposition 2.2.1 and Theorem 7.3.1(iii) corresponds to Theorem 2.2.1(iii), so we could view topological transitivity as an analogue of ergodicity. This is also reflected in the following (partial) analogue of Theorem 2.2.2.

Proposition 7.3.1. *Let $T : X \to X$ be continuous and topologically transitive on a compact metric space (X, d). Then every continuous T-invariant function is constant.*

Exercise 7.3.1. Prove Proposition 7.3.1.

Transitivity requires the existence of one dense orbit. Minimality is a much stronger notion.

Definition 7.3.2. A continuous $T : X \to X$ on a compact metric space (X, d) is called *minimal* if the orbit of **every** $x \in X$ under T is dense in X.

Example 7.3.1. Let $X = \{1, 2, \ldots, 9\}$. The discrete metric given by

$$d_{disc}(x, y) = \begin{cases} 0, & \text{if } x = y, \\ 1, & \text{if } x \neq y, \end{cases}$$

induces the discrete topology on X and makes it into a compact metric space. If we now define $T : X \to X$ by the permutation $(12 \cdots 9)$, i.e., $T1 = 2$, $T2 = 3, \ldots, T9 = 1$, then it is easy to see that T is a homeomorphism. Moreover, $\{T^n x : n \in \mathbb{Z}\} = X$ for any $x \in X$, so T is minimal.

Example 7.3.2. Fix an integer $n \geq 2$. Let $X = \{0\} \cup \{\frac{1}{n^k} : k \in \mathbb{Z}_{\geq 0}\}$ equipped with the subspace topology inherited from $\mathbb{R}$. Then X is a compact metric space. Define $T : X \to X$ by $Tx = \frac{x}{n}$. Note that T is continuous, but not surjective. Observe that

$$\overline{\{T^n 1 : n \geq 0\}} = \overline{X \setminus \{0\}} = X.$$

So, 1 has a dense orbit and thus T is topologically transitive. On the other hand $\overline{\{T^n 0 : n \geq 0\}} = \{0\} \neq X$, so T is not minimal.

Exercise 7.3.2. Let (X, d) be a compact metric space and $T : X \to X$ a topologically transitive homeomorphism. Show that if T is an isometry (i.e., $d(Tx, Ty) = d(x, y)$ for all $x, y \in X$), then T is minimal.

Example 7.3.3. Consider the binary odometer T from Example 1.3.11. In Example 7.1.1 we have seen that T is a homeomorphism, and it is easily checked that T is an isometry. By Exercise 7.3.2, to prove that T is minimal it is enough to show that the point $(0, 0, \ldots)$ has a dense orbit.

So let $x = (x_1, x_2, \ldots) \in \{0,1\}^{\mathbb{N}}$. For $k \geq 1$, write $n(k) = \sum_{j=1}^{k} 2^{j-1} x_j$. Then,

$$T^{n(k)}(0,0,0,\ldots) = (x_1, \ldots, x_k, 0, 0, \ldots)$$

so that $d(x, T^{n(k)}(0,0,\ldots)) < 2^{-(k+1)}$. This proves that the orbit of $(0,0,0\ldots)$ is dense in X, and hence T is a minimal homeomorphism.

We have the following analogue of part of Theorem 7.3.1.

Proposition 7.3.2. *Let $T : X \to X$ be a homeomorphism on a compact metric space (X, d). Then T is minimal if and only if for any closed $A \subseteq X$ with $TA = A$ it holds that either $A = \emptyset$ or $A = X$.*

Proof. Suppose T is minimal and A is a closed set with $TA = A$. If $A \neq \emptyset$, then let $x \in A$. By the T-invariance of A, then $O_T(x) \subseteq A$. The minimality of T implies that

$$X = \overline{O_T(x)} \subseteq \overline{A} = A.$$

So $A = X$. For the other direction, let $x \in X$. By the continuity of T,

$$T\overline{O_T(x)} = \overline{TO_T(x)} = \overline{O_T(x)},$$

so by assumption $\overline{O_T(x)} = X$ or $\overline{O_T(x)} = \emptyset$ and since $x \in \overline{O_T(x)}$ we get $\overline{O_T(x)} = X$. □

Minimality implies that the dynamics of T can not be restricted to any subset of X and is in that sense an irreducibility condition. The following result shows that any homeomorphism on a compact metric space can be restricted to a subset on which the map is minimal. A closed T-invariant subset $A \subseteq X$ is called a *minimal set* for T if the restriction $T|_A : A \to A$ is minimal.

Theorem 7.3.2. *Any homeomorphism $T : X \to X$ on a compact metric space (X, d) has a minimal set.*

Proof. Let $\mathcal{A} = \{A \subseteq X\ :\ A$ is closed non-empty and $TA = A\}$. Then $X \in \mathcal{A}$ and moreover, for any $x \in X$ the set $\overline{O_T(x)} \in \mathcal{A}$. So $\mathcal{A} \neq \emptyset$. $\mathcal{A}$ is partially ordered by set inclusion and any totally ordered subset $\mathcal{K} \subseteq \mathcal{A}$ has the finite intersection property: any intersection of finitely many elements from $\mathcal{K}$ is non-empty. As a consequence of the compactness of X, any family of closed subsets of X that has the finite intersection property has a non-empty intersection. Hence, $\bigcap_{K \in \mathcal{K}} K \neq \emptyset$. Thus $\bigcap_{K \in \mathcal{K}} K$ is a lower bound for $\mathcal{K}$. Then by Zorn's Lemma there is a minimal element

$M \in \mathcal{A}$. This set M is closed, non-empty and $TM = M$. Furthermore, M must be a minimal set, since if there exists a point $x \in X$ with $\overline{O_T(x)} \neq M$, then $\overline{O_T(x)} \subseteq M$ and $\overline{O_T(x)} \in \mathcal{A}$, contradicting the minimality of M in $\mathcal{A}$. □

Example 7.3.4. Let $\mathbb{S}^1$ denote the unit circle in $\mathbb{C}$ and let $\theta \in (0,1)$. Use $R_\theta : \mathbb{S}^1 \to \mathbb{S}^1$ to denote the rotation on $\mathbb{S}^1$ with parameter θ given by $R_\theta z = e^{2\pi i\theta} z = e^{2\pi i(\theta+\omega)}$ for some $\omega \in [0, 2\pi)$ as in Example 1.3.2. We have already seen that in case θ is irrational the map R_θ is uniquely ergodic. Now we see that it is minimal as well.

Proposition 7.3.3. *Let $\theta \in (0,1)$. Then R_θ is minimal if and only if θ is irrational.*

Proof. Recall from Exercise 2.3.1 that if θ is rational, then the orbit of any point under R_θ is periodic. Hence, R_θ can not be minimal in this case. Let θ is irrational. Fix $\varepsilon > 0$ and $z \in \mathbb{S}^1$, so $z = e^{i\omega}$ for some $\omega \in [0, 2\pi)$. Then

$$\begin{aligned} R_\theta^n z = R_\theta^m z &\Leftrightarrow e^{i(\omega+n\theta)} = e^{i(\omega+m\theta)} \\ &\Leftrightarrow e^{i(n-m)\theta} = 1 \\ &\Leftrightarrow (n-m)\theta \in \mathbb{Z}. \end{aligned}$$

Thus, the points $R_\theta^n z$, $n \in \mathbb{Z}$, are all distinct and it follows by compactness that the sequence $(R_\theta^n z)_{n \geq 1}$ has a convergent subsequence. Therefore, we can find integers $n > m$ such that

$$d(R_\theta^n z, R_\theta^m z) < \varepsilon,$$

where d is the arc length distance function. Now, since R_θ is distance preserving with respect to this metric, we can set $\ell = n - m$ to obtain $d(R_\theta^\ell z, z) < \varepsilon$. By continuity R_θ^ℓ maps the connected, closed arc with endpoints z and $R_\theta^\ell z$ with length $< \varepsilon$ onto the connected closed arc from $R_\theta^\ell z$ to $R_\theta^{2\ell} z$ with length $< \varepsilon$ and this one onto the arc connecting $R_\theta^{2\ell} z$ to $R_\theta^{3\ell} z$ with length $< \varepsilon$, etc. Since these arcs have positive and equal length, they cover $\mathbb{S}^1$. The result now follows, since the arcs have length smaller than ε, and $\varepsilon > 0$ was arbitrary. □

We now give a generalization of Proposition 7.3.3 in higher dimensions. Recall that real numbers $x_1, x_2, \ldots, x_n$ are called *rationally independent* if $\sum_{i=1}^n k_i x_i = 0$ for some integers $k_1, \ldots, k_n$ implies that $k_1 = \cdots = k_n = 0$.

Proposition 7.3.4. *Let $\gamma = (e^{2\pi i\gamma_1}, \ldots, e^{2\pi i\gamma_n}) \in \mathbb{T}^n$, where $\mathbb{T}^n$ is the n-dimensional torus, and let $T_\gamma : \mathbb{T}^n \to \mathbb{T}^n$ be defined by*

$$T_\gamma(e^{2\pi i x_1}, \ldots, e^{2\pi i x_n}) = (e^{2\pi i(x_1+\gamma_1)}, \ldots, e^{2\pi i(x_n+\gamma_n)}).$$

Then T_γ is minimal if and only if $1, \gamma_1, \ldots, \gamma_n$ are rationally independent.

Proof. First note that T_γ is an isometry, so by Exercise 7.3.2 we can replace minimality by topological transitivity. Assume that $1, \gamma_1, \ldots, \gamma_n$ are rationally dependent, then there exist integers $k_1, \ldots, k_n$ not all zero and an integer m such that $\sum_{i=1}^n k_i\gamma_i = m$. The function $f : \mathbb{T}^n \to \mathbb{C}$ defined by

$$f(e^{2\pi i x_1}, \ldots, e^{2\pi i x_n}) = e^{2\pi i \sum_{j=1}^n k_j x_j}$$

is obviously non-constant and one can easily check that $f \circ T_\gamma = f$. Proposition 7.3.1 then implies that T_γ is not topologically transitive and hence is not minimal. To prove the converse, assume that $1, \gamma_1, \ldots, \gamma_n$ are rationally independent. We will show that T_γ is topologically transitive using Corollary 7.3.1. Let U be a non-empty T_γ-invariant open set. Then the indicator function 1_U is a T_γ-invariant function. Using the Fourier series expansion of 1_U, we have for Lebesgue almost all $(e^{2\pi i x_1}, \ldots, e^{2\pi i x_n}) \in \mathbb{T}^n$ that

$$1_U(e^{2\pi i x_1}, \ldots, e^{2\pi i x_n}) = \sum_{k_1,\ldots,k_n} c_{k_1,\ldots,k_n} e^{2\pi i \sum_{j=1}^n k_j x_j},$$

and hence

$$1_U(T_\gamma(e^{2\pi i x_1}, \ldots, e^{2\pi i x_n})) = \sum_{k_1,\ldots,k_n} c_{k_1,\ldots,k_n} e^{2\pi i \sum_{j=1}^n k_j\gamma_j} e^{2\pi i \sum_{j=1}^n k_j x_j}.$$

Since $1_U = 1_U \circ T_\gamma$, then by the uniqueness of the Fourier series we have $c_{k_1,\ldots,k_n}(1 - e^{2\pi i \sum_{j=1}^n k_j\gamma_j}) = 0$ for all $k_1, \ldots, k_n$. Hence either $c_{k_1,\ldots,k_n} = 0$ or $e^{2\pi i \sum_{j=1}^n k_j\gamma_j} = 1$. By rational independence, the latter can only happen if $k_1 = \cdots = k_n = 0$. This implies that $1_U = c_{0,0,\ldots,0}$ Lebesgue almost everywhere. Since U is non-empty, this shows that $1_U = 1$ Lebesgue almost everywhere. Any non-empty open set V has positive Lebesgue measure and thus $V \cap U \neq \emptyset$. Then Corollary 7.3.1 implies that T_γ is topologically transitive. Since T_γ is an isometry, it is minimal as well. □

Also mixing has a topological analogue.

Definition 7.3.3. A continuous map $T : X \to X$ on a compact metric space (X, d) is called *topologically mixing* if for any non-empty open sets $U, V \subseteq X$ there is an $N \geq 0$, such that $T^n U \cap V \neq \emptyset$ for all $n \geq N$.

Example 7.3.5. We revisit the Bernoulli shift from Example 7.1.1. We show that the one-sided version with $X = \{0, 1, \ldots, k-1\}^{\mathbb{N}}$ is topologically mixing, but with some minor changes the proof also works for the two-sided shift. Recall the metric on X from (7.1). An open ball under this metric is a set of the form

$$B(x, 2^{-\ell}) = \{y \in X \,:\, x_i = y_i \text{ for all } 1 \leq j \leq \ell\}.$$

If we consider two balls $B(x, 2^{-\ell})$ and $B(y, 2^{-m})$ for $x, y \in X$ with $x \neq y$ and $\ell, m \geq 1$, then the set $T^n B(x, 2^{-\ell}) \cap B(y, 2^{-m})$ contains all points $z = (z_i) \in X$ satisfying $z_i = y_i$ for all $1 \leq i \leq m$ and $z_i = x_{i-n}$ for all $n + 1 \leq i \leq n + \ell$. Hence, for any $n \geq m$ this set is non-empty. Since the collection of open balls forms a basis for the product topology, this implies that T is topologically mixing.

Example 7.3.6. Consider Arnold's cat map as in Example 7.1.2. The matrix A has eigenvalues $\ell = \frac{3+\sqrt{5}}{2} > 1$ and $0 < \frac{1}{\ell} < 1$. Corresponding eigenvectors are

$$v_\ell = \begin{pmatrix} \frac{1+\sqrt{5}}{2} \\ 1 \end{pmatrix} \quad \text{and} \quad v_{\frac{1}{\ell}} = \begin{pmatrix} \frac{1-\sqrt{5}}{2} \\ 1 \end{pmatrix}.$$

The eigenvectors are perpendicular, since A is symmetric, and T is expanding with a factor ℓ in the direction of v_ℓ. For $x \in \mathbb{T}^2$ the line in $\mathbb{T}^2$ through x in the direction of v_ℓ is therefore called the *unstable manifold* of x, denoted by $M_u(x)$. See Figure 7.1 for an illustration. The collection

$$\mathcal{M}_u = \{M_u(x) \,:\, x \in \mathbb{T}^2\}$$

is called the *unstable foliation* of T. Since $M_u(Tx) = TM_u(x)$, the unstable foliation is a T-invariant set. In the direction of the other eigenvector $v_{\frac{1}{\ell}}$, T is contracting with a factor ℓ and we can similarly construct the *stable manifold* $M_s(x)$ for any $x \in \mathbb{T}^2$ and the corresponding *stable foliation* $\mathcal{M}_s = \{M_s(x) \,:\, x \in \mathbb{T}^2\}$. This is a T-invariant set as well.

We show that T is topologically mixing. Let $x \in \mathbb{T}^2$ and $\varepsilon > 0$. Since 1 and $\frac{1+\sqrt{5}}{2}$ are rationally independent, by Proposition 7.3.4 the unstable manifold $M_u(x)$ is dense in $\mathbb{T}^2$. So, if we place open balls $B(y, \varepsilon)$ at all points $y \in M_u(x)$, we get an open cover of $\mathbb{T}^2$ and then by compactness a finite open subcover $\mathcal{U}$. The centers of the balls in $\mathcal{U}$ specify a finite number of points in $M_u(x)$. This implies that there exists a bounded segment $S \subseteq M_u(x)$ of which the ε-neighborhood

$$\{z \in \mathbb{T}^2 : \exists y \in S \text{ s.t. } z \in B(y, \varepsilon)\}$$

covers $\mathbb{T}^2$. If we take any translate of S, so $S + y$ for any $y \in \mathbb{T}^2$, this again gives a cover of $\mathbb{T}^2$ (since translations are isometries). Note that

$$S + y \subseteq M_u(x + y).$$

Let $D(\varepsilon)$ denote the length of the arc S. We can deduce that for any $\varepsilon > 0$ and any $y \in \mathbb{T}^2$ the ε-neighborhood of any arc of length $D(\varepsilon)$ in $M_u(y)$ covers $\mathbb{T}^2$.

Now let U, V be two non-empty open sets in $\mathbb{T}^2$. Let $B(x, \delta) \subseteq U$, $\delta > 0$, be a non-empty open ball inside U. Then U contains an arc of length 2δ in $M_u(x)$. Let $\overline{B(y, \varepsilon)} \subseteq V$ with $\varepsilon > 0$ be a closed ball inside V. Choose an N such that $\ell^N 2\delta > D(\varepsilon)$. For any $n \geq N$ the set $T^n U$ contains an arc of length at least $D(\varepsilon)$ from some unstable manifold $M_u(z)$, $z \in \mathbb{T}^2$. From the above we see that an ε-neighborhood of $T^n U$ covers $\mathbb{T}^2$. So it must intersect $\overline{B(y, \varepsilon)}$, and thus V as well. Hence, T is topologically mixing.

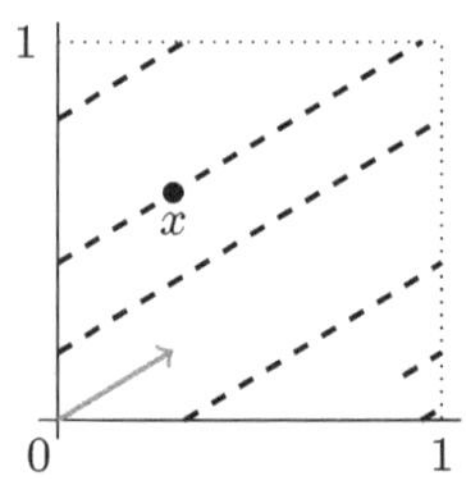

Figure 7.1 The torus $\mathbb{T}^2$ with part of the unstable manifold $M_u(x)$ indicated by the dashed lines. The arrow points in the direction of the eigenvector v_ℓ.

Finally, for a topological dynamical system there are notions analogous to factor maps and isomorphisms that allow one to carry dynamical properties from one system to another.

Definition 7.3.4. Let (X, d), (Y, ρ) be two compact metric spaces and let $T : X \to X$, $S : Y \to Y$ be continuous maps.

(i) A surjective, continuous map $\phi : X \to Y$ satisfying $\phi \circ T = S \circ \phi$ is called a *topological semi-conjugacy.* If such a map ϕ exists, then S is called a *topological factor* of T.

(ii) A homeomorphism $\phi : X \to Y$ satisfying $\phi \circ T = S \circ \phi$ is called a *(topological) conjugacy.* If such a map ϕ exists, then T and S are said to be *topologically conjugate.*

Proposition 7.3.5. *Let (X, d), (Y, ρ) be compact spaces and let $T : X \to X$, $S : Y \to Y$ be continuous maps with S a topological factor of T. The following hold.*

(i) *If T is topologically transitive, then S is topologically transitive.*

(ii) *If T is minimal, then S is minimal.*

(iii) *If T is mixing, then S is mixing.*

Exercise 7.3.3. Prove Proposition 7.3.5.

CHAPTER 8

Continued Fractions

Ergodic theory has been particularly useful in the area of continued fractions, where several proofs were simplified quite a bit by taking the dynamical approach. A *regular continued fraction expansion* of a number $x \in [0,1)$ is an expression of the form

$$x = \cfrac{1}{a_1 + \cfrac{1}{a_2 + \cfrac{1}{a_3 + \cfrac{1}{\ddots + \cfrac{1}{a_n}}}}} \qquad \text{or} \qquad x = \cfrac{1}{a_1 + \cfrac{1}{a_2 + \cfrac{1}{a_3 + \cfrac{1}{\ddots}}}},$$

with $a_i \in \mathbb{N}$ for all $i \geq 1$. The finite expansion occurs for $x \in \mathbb{Q}$ and the infinite one for $x \in [0,1) \setminus \mathbb{Q}$. Recall the definition of the Gauss map $T : [0,1) \to [0,1)$ from Example 1.3.7 by $T0 = 0$ and

$$Tx = \frac{1}{x} - \left\lfloor \frac{1}{x} \right\rfloor$$

for $x \neq 0$. The graph is shown in Figure 1.3(c). In Example 1.3.7 we saw that iterations of T produce for $x \in [0,1)$ expressions of the form

$$x = \cfrac{1}{a_1 + \cfrac{1}{a_2 + \ddots + \cfrac{1}{a_n + T^n x}}},$$

where the digits a_n, $n \geq 1$, are obtained by setting

$$a_n = a_n(x) = \begin{cases} 1, & \text{if } T^{n-1}x \in (\frac{1}{2}, 1), \\ k, & \text{if } T^{n-1}x \in (\frac{1}{k+1}, \frac{1}{k}],\ k \geq 2, \end{cases}$$

as long as $T^{n-1}x \neq 0$. It is a consequence of Euclid's algorithm that T produces a finite continued fraction expansion for each $x \in [0,1) \cap \mathbb{Q}$. In this chapter we first prove that also in case $x \in [0,1)$ is irrational this process converges and in the limit produces a regular continued fraction expansion of x. After that we discuss some of the approximation properties of the continued fractions and construct a natural extension for the map T.

8.1 REGULAR CONTINUED FRACTIONS

For numbers $a_1, a_2, a_3, \ldots$ (not necessarily integers), we use $[0; a_1, a_2, a_3, \ldots]$ to denote the continued fraction expansion

$$[0; a_1, a_2, a_3, \ldots] = \cfrac{1}{a_1 + \cfrac{1}{a_2 + \cfrac{1}{a_3 + \ddots}}}.$$

Fix some irrational $x \in [0,1) \setminus \mathbb{Q}$ and let $(a_n)_{n \geq 1} \in \mathbb{N}^{\mathbb{N}}$ be the infinite string of digits produced by the Gauss map, called the *partial quotients* of x or *continued fraction digits*. Set for each $n \geq 1$,

$$v_n = \frac{p_n}{q_n} := [0; a_1, a_2, \ldots, a_n].$$

These rational numbers are called the *convergents* of x. The goal of this section is to show that $x = \lim_{n\to\infty} \frac{p_n}{q_n}$. This is done by studying elementary properties of matrices associated with the continued fraction expansions.

Let $A \in \mathrm{SL}_2(\mathbb{Z})$, that is

$$A = \begin{pmatrix} r & p \\ s & q \end{pmatrix},$$

where $r, s, p, q \in \mathbb{Z}$ and *det* $A = rq - ps \in \{\pm 1\}$. Set $\mathbb{C}^* = \mathbb{C} \cup \{\infty\}$. Associated to A is the *Möbius* (or *fractional linear*) *transformation* $A : \mathbb{C}^* \to \mathbb{C}^*$ defined by

$$A(z) = \begin{pmatrix} r & p \\ s & q \end{pmatrix}(z) = \frac{rz + p}{sz + q},$$

where we let $\frac{1}{\infty} = 0$ and $\frac{1}{0} = \infty$. If we set for each $n \geq 1$

$$A_n := \begin{pmatrix} 0 & 1 \\ 1 & a_n \end{pmatrix}, \quad \text{and} \quad M_n := A_1 A_2 \cdots A_n, \tag{8.1}$$

then $A_n(z) = \frac{1}{a_n+z}$ and multiplication of the matrices in M_n corresponds to the composition of the associated Möbius transformations. Note that $v_n = M_n(0)$. This allows us to find recurrence relations for p_n and q_n. Writing

$$M_n = \begin{pmatrix} r_n & p_n \\ s_n & q_n \end{pmatrix},$$

for some r_n, s_n, it follows that

$$\begin{pmatrix} r_n & p_n \\ s_n & q_n \end{pmatrix} = M_n = M_{n-1}A_n = \begin{pmatrix} r_{n-1} & p_{n-1} \\ s_{n-1} & q_{n-1} \end{pmatrix} \begin{pmatrix} 0 & 1 \\ 1 & a_n \end{pmatrix}.$$

So $r_n = p_{n-1}$ and $s_n = q_{n-1}$ and we obtain the recurrence relations

$$\begin{aligned} p_{-1} &:= 1; \quad p_0 := a_0 = 0; \quad p_n = a_n p_{n-1} + p_{n-2}, \; n \geq 1, \\ q_{-1} &:= 0; \quad q_0 := 1; \quad\quad\quad\;\; q_n = a_n q_{n-1} + q_{n-2}, \; n \geq 1. \end{aligned} \tag{8.2}$$

From this it immediately follows that $p_n(x) = q_{n-1}(Tx)$ for all $n \geq 0$ and $x \in (0,1)$, and that

$$p_{n-1}q_n - p_n q_{n-1} = det\, M_n = (-1)^n, \quad \text{for all } n \geq 1. \tag{8.3}$$

This in turn gives

$$\gcd(p_n, q_n) = 1, \quad \text{for all } n \geq 1.$$

Setting

$$A_n^* := \begin{pmatrix} 0 & 1 \\ 1 & a_n + T^n x \end{pmatrix},$$

it follows from

$$x = M_{n-1}A_n^*(0) = [\,0;\, a_1, \ldots, a_{n-1}, a_n + T^n x\,],$$

that

$$x = \frac{p_n + p_{n-1}T^n x}{q_n + q_{n-1}T^n x}, \tag{8.4}$$

i.e., $x = M_n(T^n x)$. Combining (8.3) and (8.4) gives

$$x - \frac{p_n}{q_n} = \frac{(-1)^n \, T^n x}{q_n(q_n + q_{n-1}T^n x)}. \tag{8.5}$$

In fact, (8.5) yields information about the quality of approximation of the rational number $v_n = \frac{p_n}{q_n}$ to the irrational number x. Since $T^n x \in [0,1)$, it at once follows that

$$\left| x - \frac{p_n}{q_n} \right| < \frac{1}{q_n^2}, \qquad \text{for all } n \geq 0. \tag{8.6}$$

From $\frac{1}{T^n x} = a_{n+1} + T^{n+1}x$ we get

$$\left| x - \frac{p_n}{q_n} \right| = \frac{1}{q_n(q_n a_{n+1} + q_n T^{n+1}x + q_{n-1})}.$$

As can be seen from the recurrence relations in (8.2) the sequence (q_n) is increasing, so one even has

$$\frac{1}{2q_n q_{n+1}} < \left| x - \frac{p_n}{q_n} \right| < \frac{1}{q_n q_{n+1}}, \qquad \text{for all } n \geq 1. \tag{8.7}$$

Notice that the recurrence relations (8.2) yield that

$$v_n - v_{n-1} = \frac{(-1)^{n-1}}{q_{n-1}q_n}, \qquad \text{for all } n \geq 1. \tag{8.8}$$

This together with (8.5) gives

$$0 = v_0 < v_2 < v_4 < \cdots < x < \cdots < v_3 < v_1 < 1. \tag{8.9}$$

A natural question to ask now is: "Given a sequence of positive integers $(a_n)_{n \geq 1}$, does the limit $\lim_{n\to\infty} v_n$ exist? And if so, and the limit equals x, do we have that $x = [0; a_1, \ldots, a_n, \ldots]$?" The answer to both questions is "yes" as the following proposition shows.

Proposition 8.1.1. *Let $(a_n)_{n\geq 1}$ be a sequence of positive integers, and let the sequence of rational numbers $(v_n)_{n\geq 1}$ be given by*

$$v_n := [0; a_1, \ldots, a_n], \qquad \textit{for all } n \geq 1.$$

Then there exists an irrational number $x \in [0,1)$ for which

$$\lim_{n\to\infty} v_n = x$$

and we moreover have that $x = [0; a_1, a_2, \ldots, a_n, \ldots]$.

Proof. Let the sequence $(a_n)_{n\geq 1}$ be given and use A_n to denote the matrices specified by this sequence as in (8.1). Write $v_0 := 0$ and for $n \geq 1$, $v_n = \frac{p_n}{q_n}$. We saw before that

$$\begin{pmatrix} p_{n-1} & p_n \\ q_{n-1} & q_n \end{pmatrix} = A_1 \cdots A_n.$$

From (8.8), (8.9) and $v_0 = 0$ one sees that

$$v_n = \sum_{k=1}^{n} \frac{(-1)^{k-1}}{q_{k-1}q_k}.$$

Hence the Leibniz Test for the convergence of alternating series yields that $x = \lim_{n\to\infty} v_n$ exists. What is left is to show that $a_n = a_n(x)$ for $n \geq 1$, i.e., that $(a_n)_{n\geq 1}$ is the sequence of partial quotients of x. Since

$$v_n = [0; a_1, a_2, \ldots, a_n] = \frac{1}{a_1 + [0; a_2, a_3, \ldots, a_n]} = \frac{1}{a_1 + v_n^*}, \tag{8.10}$$

where $v_n^* = [0; a_2, a_3, \ldots, a_n]$, it is sufficient to show that

$$\left\lfloor \frac{1}{x} \right\rfloor = a_1.$$

Taking limits as $n \to \infty$ in (8.10) yields

$$x = \frac{1}{a_1 + x^*},$$

where $x^* = \lim_{n\to\infty} v_n^*$. From $0 < v_2^* < x^* < v_3^* < 1$ it follows that $\frac{1}{x} = a_1 + x^* \in (a_1, a_1 + 1)$, so $\lfloor \frac{1}{x} \rfloor = a_1$. □

Exercise 8.1.1. Let

$$\Delta(a_1, \ldots, a_k) := \{x \in [0,1) : a_1(x) = a_1, \ldots, a_k(x) = a_k\}, \tag{8.11}$$

where $a_j \in \mathbb{N}$ for each $1 \leq j \leq k$. Show that $\Delta(a_1, a_2, \ldots, a_k)$ is an interval in $[0,1)$ with endpoints

$$\frac{p_k}{q_k} \quad \text{and} \quad \frac{p_k + p_{k-1}}{q_k + q_{k-1}},$$

where $\frac{p_k}{q_k} = [0; a_1, a_2, \ldots, a_k]$. Conclude that

$$\lambda(\Delta(a_1, a_2, \ldots, a_k)) = \frac{1}{q_k(q_k + q_{k-1})},$$

where λ is Lebesgue measure on $[0,1)$. These are the fundamental intervals for the Gauss map T.

Exercise 8.1.2. Let $a_1, \dots, a_n \in \mathbb{N}$ and $0 \le a < b \le 1$. Write $\Delta_n = \Delta(a_1, \cdots a_n)$ for the fundamental intervals as in (8.11). Show that

$$T^{-n}[a,b) \cap \Delta_n = \begin{cases} \left[\dfrac{p_{n-1}a + p_n}{q_{n-1}a + q_n}, \dfrac{p_{n-1}b + p_n}{q_{n-1}b + q_n}\right), & \text{if } n \text{ is even,} \\ \\ \left(\dfrac{p_{n-1}b + p_n}{q_{n-1}b + q_n}, \dfrac{p_{n-1}a + p_n}{q_{n-1}a + q_n}\right], & \text{if } n \text{ is odd.} \end{cases}$$

Conclude that

$$\lambda(T^{-n}[a,b) \cap \Delta_n) = \lambda([a,b))\lambda(\Delta_n)\frac{q_n(q_{n-1} + q_n)}{(q_{n-1}b + q_n)(q_{n-1}a + q_n)},$$

where λ is Lebesgue measure on $[0,1)$.

8.2 ERGODIC PROPERTIES OF THE GAUSS MAP

In Exercise 1.3.4 we saw that the Gauss map is measure preserving with respect to the Gauss measure μ given by

$$\mu(B) = \int_B \frac{1}{\log 2}\frac{1}{1+x}\, d\lambda(x) \tag{8.12}$$

for Borel sets $B \subseteq [0,1)$. (Here and in the rest of the chapter log refers to the natural logarithm.) Note that the density $x \mapsto \frac{1}{\log 2}\frac{1}{1+x}$ is bounded from above and bounded away from 0, so μ and λ are equivalent. The next theorem gives the ergodicity of T with respect to μ.

Theorem 8.2.1. *Let $T : [0,1) \to [0,1)$ be the Gauss map given by $Tx = \frac{1}{x} - \lfloor \frac{1}{x} \rfloor$ defined on the measure space $([0,1), \mathcal{B}, \mu)$, where $\mathcal{B}$ is the Borel σ-algebra on $[0,1)$ and μ the Gauss measure from (8.12). Then, T is ergodic with respect to μ.*

Proof. We prove the theorem by applying Knopp's Lemma, Lemma 2.3.1, to the collection of fundamental intervals. First we estimate the Gauss measure of inverse images of Borel sets. Let I be an interval in $[0,1)$ with endpoints $a < b$, and $\Delta_n = \Delta(a_1, \dots, a_n)$ a fundamental interval of order n. From Exercise 8.1.2, we know that

$$\lambda(T^{-n}I \cap \Delta_n) = \lambda(I)\lambda(\Delta_n)\frac{q_n(q_{n-1} + q_n)}{(q_{n-1}b + q_n)(q_{n-1}a + q_n)}.$$

Since $q_{n-1} < q_n$ and $0 \le a < b \le 1$, it follows that

$$\frac{1}{2} < \frac{q_n}{q_{n-1}+q_n} < \frac{q_n(q_{n-1}+q_n)}{(q_{n-1}b+q_n)(q_{n-1}a+q_n)} < \frac{q_n(q_{n-1}+q_n)}{q_n^2} < 2.$$

Therefore we find that

$$\frac{1}{2}\lambda(I)\lambda(\Delta_n) < \lambda(T^{-n}I \cap \Delta_n) < 2\lambda(I)\lambda(\Delta_n).$$

Let A be a finite disjoint union of intervals. Since Lebesgue measure is additive one has

$$\frac{1}{2}\lambda(A)\lambda(\Delta_n) \le \lambda(T^{-n}A \cap \Delta_n) \le 2\lambda(A)\lambda(\Delta_n). \tag{8.13}$$

The collection of finite disjoint unions of intervals generates the Borel σ-algebra. It follows that (8.13) holds for any Borel set A. From (8.12) it is clear that

$$\frac{1}{2\log 2}\lambda(A) \le \mu(A) \le \frac{1}{\log 2}\lambda(A). \tag{8.14}$$

Then by (8.13) and (8.14) one has

$$\mu(T^{-n}A \cap \Delta_n) \ge \frac{\log 2}{4}\mu(A)\mu(\Delta_n). \tag{8.15}$$

Now let $\mathcal{S}$ be the collection of all fundamental intervals Δ_n. Since the set of all endpoints of these fundamental intervals is the set of all rationals in $[0,1)$, condition (a) of Knopp's Lemma is satisfied. Let $B \in \mathcal{B}$ is a T-invariant set, i.e., $T^{-1}B = B$, and suppose that $\mu(B) > 0$. It then follows from (8.15) that for every fundamental interval Δ_n

$$\mu(B \cap \Delta_n) \ge \frac{\log 2}{4}\mu(B)\mu(\Delta_n).$$

So condition (b) from Knopp's Lemma is satisfied with $\gamma = \frac{\log 2}{4}\mu(B)$. Thus $\lambda(B) = 1$ and by the equivalence of μ and λ also $\mu(B) = 1$. Hence T is ergodic. □

Recall Exercise 3.2.2 where we computed the frequency of the even and odd digits in typical Lüroth expansions. We can do a similar computation for the continued fractions generated by the Gauss map.

Exercise 8.2.1. For $x \in [0,1]$, let $(a_n(x))_{n\geq 1}$ denote the finite or infinite sequence of regular continued fraction digits of x as produced by the Gauss map.

(a) Let $\mathcal{E}$ denote the set of $x \in [0,1]$ for which the sequence $(a_n(x))_{n\geq 1}$ contains only odd integers. Prove that $\mathcal{E}$ has zero Lebesgue measure.

(b) Prove that for Lebesgue almost every $x \in [0,1]$ we have

$$\lim_{n\to\infty} \frac{\#\{1 \leq k \leq n : a_k(x) \text{ is odd}\}}{n} = \frac{\log \pi}{\log 2} - 1.$$

Hint: Use the identity

$$\prod_{k\geq 1} \left(1 + \frac{1}{4k^2-1}\right) = \frac{\pi}{2}.$$

In fact, using Proposition 6.2.3 we can even show that the Gauss map is exact. To see this, first note that the collection $\mathcal{S}$ of all fundamental intervals Δ_n satisfies the conditions of the proposition. Consider a Borel set $A \subseteq \Delta_n$ for some fundamental interval Δ_n. The n-th iterate T^n is invertible as a map from Δ_n to $[0,1)$. From (8.4) we know the local inverse is given by

$$T^{-n}y = \frac{p_n + p_{n-1}y}{q_n + q_{n-1}y},$$

so by using (8.3) and the change of variables formula (6.6) we get

$$\begin{aligned}
\lambda(A) &= \int_A d\lambda = \int_{T^n A} \frac{1}{(q_n + q_{n-1}y)^2} d\lambda(y) \\
&\geq \frac{1}{4q_n^2}\lambda(T^n A) \\
&\geq \frac{1}{4}\frac{1}{q_n(q_n+q_{n+1})}\lambda(T^n A) = \frac{1}{4}\lambda(\Delta_n)\lambda(T^n A),
\end{aligned}$$

where the last equality follows from Exercise 8.1.1. Now using (8.14) we obtain

$$\mu(A) \geq \frac{1}{2\log 2}\lambda(A) \geq \frac{1}{8\log 2}\lambda(T^n A)\lambda(\Delta_n) \geq \frac{\log 2}{8}\mu(T^n A)\mu(\Delta_n).$$

So, condition (6.4) from Proposition 6.2.3 is satisfied with $\gamma = \frac{8}{\log 2}$.

Using the Pointwise Ergodic Theorem we can give simple proofs of old and famous results of P. Lévy from 1929; see [35].

Proposition 8.2.1 (Lévy). *For Lebesgue almost all $x \in [0,1)$ one has*

$$\lim_{n\to\infty} \frac{1}{n}\log q_n = \frac{\pi^2}{12\log 2}, \tag{8.16}$$

$$\lim_{n\to\infty} \frac{1}{n}\log \lambda(\Delta_n) = \frac{-\pi^2}{6\log 2}, \quad \textit{and} \tag{8.17}$$

$$\lim_{n\to\infty} \frac{1}{n}\log \left|x - \frac{p_n}{q_n}\right| = \frac{-\pi^2}{6\log 2}. \tag{8.18}$$

In the proof of the proposition we will use that for each $n \geq 1$ the inequality $q_n(x) \geq F_n$ holds, where $F_1, F_2, F_3, \ldots$ is the Fibonacci sequence $1, 1, 2, 3, 5, \ldots$. To see this, just notice that by the recurrence relations in (8.2) $q_n = F_n$ precisely when $a_i = 1$ for all $1 \leq i \leq n$, which gives the smallest possible value of q_n.

Proof of Proposition 8.2.1. By the recurrence relations (8.2), for any irrational $x \in [0,1)$ one has $p_n(x) = q_{n-1}(Tx)$, so that

$$\frac{1}{q_n(x)} = \frac{1}{q_n(x)}\frac{p_n(x)}{q_{n-1}(Tx)}\frac{p_{n-1}(Tx)}{q_{n-2}(T^2x)}\cdots\frac{p_2(T^{n-2}x)}{q_1(T^{n-1}x)}$$

$$= \frac{p_n(x)}{q_n(x)}\frac{p_{n-1}(Tx)}{q_{n-1}(Tx)}\cdots\frac{p_1(T^{n-1}x)}{q_1(T^{n-1}x)}.$$

Taking logarithms yields

$$-\log q_n(x) = \log\frac{p_n(x)}{q_n(x)} + \log\frac{p_{n-1}(Tx)}{q_{n-1}(Tx)} + \cdots + \log\frac{p_1(T^{n-1}x)}{q_1(T^{n-1}x)}. \tag{8.19}$$

For any $k \in \mathbb{N}$, and any irrational $x \in [0,1)$, $\frac{p_k(x)}{q_k(x)}$ is a rational number close to x. Therefore we compare the right-hand side of (8.19) with

$$\log x + \log Tx + \log T^2x + \cdots + \log T^{n-1}x.$$

We have

$$-\log q_n(x) = \log x + \log Tx + \log T^2x + \cdots + \log T^{n-1}x + R(n,x).$$

In order to estimate the error term $R(n,x)$, we recall from Exercise 8.1.1 that x lies in the interval Δ_n, which has endpoints $\frac{p_n}{q_n}$ and $\frac{p_n+p_{n-1}}{q_n+q_{n-1}}$. Therefore and by Exercise 8.1.1, in case n is even, one has

$$0 < \log x - \log\frac{p_n}{q_n} = \left(x - \frac{p_n}{q_n}\right)\frac{1}{\xi} \leq \frac{1}{q_n(q_{n-1}+q_n)}\frac{1}{p_n/q_n} < \frac{1}{q_n} \leq \frac{1}{F_n},$$

where $\xi \in (\frac{p_n}{q_n}, x)$ is given by the Mean Value Theorem. A similar argument shows that

$$0 < \log \frac{p_n}{q_n} - \log x < \frac{1}{q_n} \leq \frac{1}{F_n}$$

in case n is odd. Thus

$$|R(n,x)| \leq \frac{1}{F_n} + \frac{1}{F_{n-1}} + \cdots + \frac{1}{F_1}.$$

The n-th Fibonacci number can be expressed in terms of the golden mean $G = \frac{1+\sqrt{5}}{2}$ and the small golden mean $g = \frac{\sqrt{5}-1}{2} = \frac{1}{G}$ by

$$F_n = \frac{G^n + (-1)^{n+1} g^n}{\sqrt{5}}.$$

It follows that $F_n \sim \frac{1}{\sqrt{5}} G^n$ as $n \to \infty$. Thus $\frac{1}{F_n} + \frac{1}{F_{n-1}} + \cdots + \frac{1}{F_1}$ is the n-th partial sum of a convergent series, and therefore

$$|R(n,x)| \leq \frac{1}{F_n} + \cdots + \frac{1}{F_1} \leq \sum_{n \geq 1} \frac{1}{F_n} < \infty.$$

Hence for each x for which

$$\lim_{n\to\infty} \frac{1}{n}(\log x + \log Tx + \log T^2 x + \cdots + \log T^{n-1} x)$$

exists, the limit

$$-\lim_{n\to\infty} \frac{1}{n} \log q_n(x)$$

exists as well, and both limits have the same value.

Now $\lim_{n\to\infty} \frac{1}{n}(\log x + \log Tx + \log T^2 x + \cdots + \log T^{n-1} x)$ is ideally suited for the Pointwise Ergodic Theorem; we only need to check that the conditions of the Pointwise Ergodic Theorem are satisfied and to calculate the integral. This is left as an exercise for the reader. This proves (8.16).

From Exercise 8.1.1 we have

$$\lambda(\Delta_n(a_1, \ldots, a_n)) = \frac{1}{q_n(q_n + q_{n-1})}.$$

Since $q_n^2 < q_n(q_n + q_{n-1}) < 2q_n^2$, this gives

$$-\log 2 - 2\log q_n < \log \lambda(\Delta_n) < -2\log q_n.$$

Then (8.17) is obtained from dividing by n and applying (8.16). Finally (8.18) follows from (8.7) and (8.16). □

Exercise 8.2.2. **(a)** Show that the arithmetic mean of the partial quotients $a_n = a_n(x)$ given by the Gauss map satisfies

$$\lim_{n\to\infty}(a_1 a_2 \dots a_n)^{1/n} = \prod_{k=1}^{\infty}\left(1+\frac{1}{k(k+2)}\right)^{\frac{\log k}{\log 2}} \qquad \lambda\text{-a.e.}$$

(b) Show that the geometric mean of the partial quotients satisfies $\lim_{n\to\infty}\frac{a_1+a_2+\cdots+a_n}{n} = \infty$ for λ a.e. $x \in [0,1)$.

8.3 THE DOEBLIN-LENSTRA CONJECTURE

A planar and a very useful version of a natural extension of the Gauss map was given by S. Ito, H. Nakada and S. Tanaka, see [46, 45].

Theorem 8.3.1 (Ito, Nakada and Tanaka). *Let $X = [0,1) \times [0,1]$ and $\mathcal{B}^2$ be the collection of Borel sets of X. Define the two-dimensional Gauss-measure ν on $(X, \mathcal{B}^2)$ by*

$$\nu(A) = \frac{1}{\log 2}\iint_A \frac{1}{(1+xy)^2}\, d\lambda(x)\, d\lambda(y), \qquad \textit{for all } A \in \mathcal{B}^2,$$

where λ denotes the one-dimensional Lebesgue measure. Finally, let the transformation $\mathcal{T} : X \to X$ be defined for $(x,y) \in X$ by

$$\mathcal{T}(0,y) = (0,y) \quad \textit{and} \quad \mathcal{T}(x,y) = \left(Tx, \frac{1}{\lfloor \frac{1}{x} \rfloor + y}\right), \quad \textit{for } x \neq 0. \tag{8.20}$$

Then $(X, \mathcal{B}^2, \nu, \mathcal{T})$ is the natural extension of $([0,1), \mathcal{B}, \mu, T)$. Furthermore, it is ergodic.

Figure 8.1 illustrates $\mathcal{T}$.

Proof. We show that $\mathcal{T}$ is measure preserving with respect to ν and leave the remainder of the proof that $\mathcal{T}$ is the natural extension of T as an exercise below. It then follows from the ergodicity of T that $\mathcal{T}$ is ergodic. The map $\mathcal{T}$ is λ^2-a.e. invertible on X. For $(x,y) \in (0,1)\times(\frac{1}{n+1}, \frac{1}{n})$, $n \geq 1$, let $(\xi, \eta) \in X$ be such, that $(\xi,\eta) = \mathcal{T}(x,y)$. Then

$$\xi = \frac{1}{x} - c \Leftrightarrow x = \frac{1}{c+\xi},$$

and

$$\eta = \frac{1}{c+y} \Leftrightarrow y = \frac{1}{\eta} - c.$$

Hence the above coordinate transformation has Jacobian J, which satisfies

$$J = \begin{vmatrix} \frac{\partial x}{\partial \xi} & \frac{\partial x}{\partial \eta} \\ \frac{\partial y}{\partial \xi} & \frac{\partial y}{\partial \eta} \end{vmatrix} = \begin{vmatrix} \frac{-1}{(c+\xi)^2} & 0 \\ 0 & \frac{-1}{\eta^2} \end{vmatrix} = \frac{1}{(c+\xi)^2}\frac{1}{\eta^2},$$

and therefore using the change of variables formula (6.6) we find

$$\begin{aligned}\nu(A) =& \frac{1}{\log 2}\iint_A \frac{1}{(1+xy)^2}\,\mathrm{d}\lambda(x)\,\mathrm{d}\lambda(y) \\ =& \frac{1}{\log 2}\iint_{\mathcal{T}A} \frac{1}{(1+\frac{1}{c+\xi}(\frac{1}{\eta}-c))^2}\frac{1}{(c+\xi)^2\eta^2}\,\mathrm{d}\lambda(\xi)\,\mathrm{d}\lambda(\eta) \\ =& \frac{1}{\log 2}\iint_{\mathcal{T}A} \frac{1}{(1+\xi\eta)^2}\,\mathrm{d}\lambda(\xi)\,\mathrm{d}\lambda(\eta) \\ =& \nu(\mathcal{T}A)\end{aligned}$$

for any $A \in \mathcal{B}^2$. This gives the result. □

Exercise 8.3.1. Prove the remainder of Theorem 8.3.1.

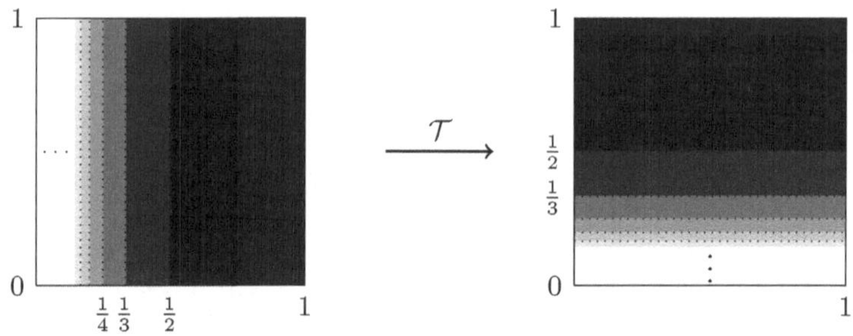

Figure 8.1 The natural extension map $\mathcal{T}$ from Theorem 8.3.1 maps regions on the left to regions of the same shade of grey on the right.

We define for every real number $x \in [0, 1)$ and every $n \geq 0$ the so-called *approximation coefficients* $\Theta_n(x)$ by

$$\Theta_n(x) = q_n^2 \left| x - \frac{p_n}{q_n} \right|. \tag{8.21}$$

Not only do these coefficients give a measure for the quality of approximation of the continued fraction convergents $\frac{p_n}{q_n}$ to an irrational number x, but they also determine the rationals that appear as such convergents.

It follows from (8.6) that $\Theta_n(x) < 1$ for all irrational $x \in [0,1)$ and $n \geq 0$. In 1981, H. W. Lenstra conjectured (this conjecture was formulated by W. Doeblin in a slightly different way in [16]) that for each $0 \leq z \leq 1$ there is a set B_z of full Lebesgue measure, such that for each $x \in B_z$ one has

$$\lim_{n\to\infty} \frac{1}{n}\#\{1 \leq j \leq n : \Theta_j(x) \leq z\} = F(z),$$

where F is the distribution function defined by

$$F(z) = \begin{cases} \dfrac{z}{\log 2}, & \text{if } 0 \leq z \leq \frac{1}{2}, \\ \\ \dfrac{1}{\log 2}(1 - z + \log(2z)), & \text{if } \frac{1}{2} \leq z \leq 1. \end{cases} \tag{8.22}$$

In other words: for Lebesgue almost all x the sequence $(\Theta_n(x))_{n\geq 1}$ has limiting distribution F.

The proof of this conjecture uses the Pointwise Ergodic Theorem applied to the natural extension $(X, \mathcal{B}^2, \nu, \mathcal{T})$. To see how we can apply this theorem, we first use (8.5) to rewrite $\Theta_n(x)$ as

$$\Theta_n(x) = q_n^2 \left| x - \frac{p_n}{q_n} \right| = \frac{T^n x}{1 + \frac{q_{n-1}}{q_n} \cdot T^n x}.$$

By the recurrence relations from (8.2) we have

$$\frac{q_{n-1}}{q_n} = \frac{q_{n-1}}{a_n q_{n-1} + q_{n-2}} = \frac{1}{a_n + \frac{q_{n-2}}{q_{n-1}}},$$

so setting $V_n = V_n(x) = \frac{q_{n-1}}{q_n}$ yields for $n \geq 1$ that

$$V_n = \frac{1}{a_n + V_{n-1}} = \cdots = \cfrac{1}{a_n + \cfrac{1}{a_{n-1} + \ddots + \cfrac{1}{a_1}}} = [0; a_n, a_{n-1}, \ldots, a_1].$$

Exercise 8.3.2. **(a)** Show that for any $x \in [0,1)$ one has $\mathcal{T}^n(x,0) = (T^n x, V_n)$.

(b) Show that for any $x \in [0,1)$ one has

$$\lim_{n\to\infty} (\mathcal{T}^n(x,0) - \mathcal{T}^n(x,y)) = 0,$$

where the convergence is uniform in y.

From the above, we can interpret V_n as "the past of x at time n" (in the same way as $T^n x$ is the "future of x at time n"). An immediate consequence of this and (8.5) is that

$$\Theta_n = \Theta_n(x) = \frac{T^n x}{1 + V_n(x) T^n x}, \quad n \geq 0. \tag{8.23}$$

Exercise 8.3.3. Show that

$$\Theta_{n-1} = \Theta_{n-1}(x) = \frac{V_n(x)}{1 + V_n(x) T^n x}, \quad n \geq 1. \tag{8.24}$$

An important consequence of this observation and Theorem 8.3.1 is the following theorem by H. Jager from 1986 stating that for Lebesgue almost all $x \in [0,1)$ the two-dimensional sequence $(T^n x, V_n(x))_{n \geq 0}$ is distributed over X according to the measure ν. See [25].

Theorem 8.3.2 (Jager). *For Lebesgue almost all $x \in [0,1)$ and any Borel set $B \subseteq X$ it holds that*

$$\lim_{n\to\infty} \frac{1}{n} \#\{0 \leq j \leq n-1 \,:\, (T^j x, V_j(x)) \in B\} = \nu(B).$$

Proof. Denote by E the subset of numbers $x \in [0,1)$ for which the sequence $(T_n x, V_n(x))_{n \geq 1}$ is **not** distributed according to the measure ν. Since the sequence $(\mathcal{T}^n(x,0) - \mathcal{T}^n(x,y))_{n \geq 0}$ converges to 0 uniformly in y by Exercise 8.3.2(b), it follows from Exercise 8.3.2(a) that for every pair $(x,y) \in E \times [0,1]$ the sequence $\mathcal{T}^n(x,y)_{n \geq 0}$ is **not** distributed according to the measure ν, i.e., there is a Borel set $B \subseteq X$ such that

$$\lim_{n\to\infty} \frac{1}{n} \sum_{j=0}^{n-1} 1_B(\mathcal{T}^j(x,y)) \neq \nu(B).$$

Now if $\lambda(E) > 0$, then $\lambda^2(E \times [0,1]) > 0$ and by definition of ν also $\nu(E \times [0,1]) > 0$. This would be in conflict with the ergodicity of $\mathcal{T}$ with respect to ν, obtained in Theorem 8.3.1. Hence, $\lambda(E) = 0$. □

Lenstra's Conjecture now follows directly from this theorem.

Proof of Lenstra's Conjecture. Let $A_z = \{(x,y) \in X \,:\, \frac{x}{1+xy} \leq z\}$. Then, $\nu(A_z) = F(z)$, and $\Theta_j(x) \leq z \Leftrightarrow \mathcal{T}^j(x,0) \in A_z$. Furthermore,

$$\frac{1}{n} \#\{1 \leq j \leq n \,:\, \Theta_j(x) \leq z\} = \frac{1}{n} \sum_{j=1}^{n} 1_{A_z}(\mathcal{T}^j(x,0)).$$

Taking limits and using the above theorem we get the required result. □

A corollary of Lenstra's Conjecture is that for Lebesgue almost all $x \in [0,1)$,

$$\lim_{n\to\infty} \frac{1}{n} \sum_{j=0}^{n-1} \Theta_j(x) = \frac{1}{4\log 2} = 0.360673\ldots. \tag{8.25}$$

Exercise 8.3.4. Use Exercise 8.3.2 and the Pointwise Ergodic Theorem applied to the map $G(x,y) = \frac{x}{1+xy}$ to prove (8.25).

8.4 OTHER CONTINUED FRACTION TRANSFORMATIONS

Regular continued fraction expansions are specific examples of the *semi-regular* continued fraction expansions that were introduced in 1913 by O. Perron; see for example [49]. These are finite or infinite expressions for real numbers of the form

$$x = a_0 + \cfrac{\epsilon_0}{a_1 + \cfrac{\epsilon_1}{a_2 + \ddots + \cfrac{\epsilon_{n-1}}{a_n + \ddots}}},$$

where $a_0 \in \mathbb{Z}$ and for each $n \geq 1$, $\epsilon_{n-1} \in \{-1,1\}$, $a_n \in \mathbb{N}$ and $a_n + \epsilon_n \geq 1$. For convergence reasons one asks that in case the expression is infinite, then $a_n + \epsilon_{n+1} \geq 1$ infinitely often.

Besides the Gauss map there are many other transformations that produce semi-regular continued fraction expansions by iteration. Several of them are defined based on the properties of the digits involved. Recall from Exercise 8.2.1 that the set of $x \in [0,1]$ that have only odd regular continued fraction digits has Lebesgue measure 0. The even and odd continued fraction transformations produce for each $x \in [-1,1]$ a semi-regular continued fraction expansion with only even and odd digits, respectively. They are given by (see Figures 8.2(a) and (b)).

$$T_e : [-1,1] \to [-1,1], x \mapsto \begin{cases} |\frac{1}{x}| - 2n, & \text{if } |x| \in (\frac{1}{2n+1}, \frac{1}{2n-1}], \ n \geq 1, \\ 0, & \text{if } x = 0, \end{cases}$$

and

$$T_o : [-1,1] \to [-1,1], x \mapsto \begin{cases} |\frac{1}{x}| - 1, & \text{if } |x| \in (\frac{1}{2}, 1], \\ |\frac{1}{x}| - (2n+1), & \text{if } |x| \in (\frac{1}{2n+2}, \frac{1}{2n}], \ n \geq 1, \\ 0, & \text{if } x = 0. \end{cases}$$

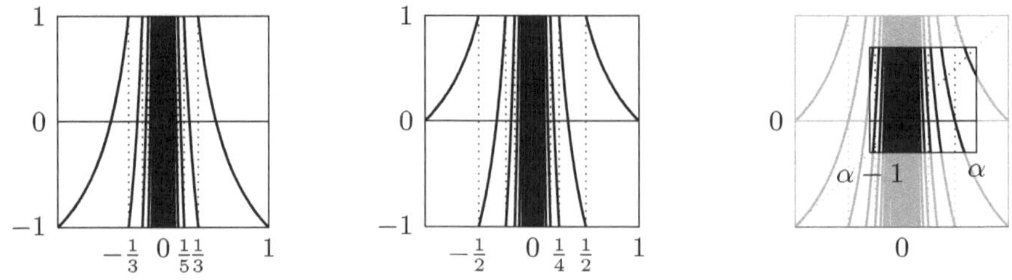

(a) T_e (b) T_o (c) T_α

Figure 8.2 The even (a) and odd (b) continued fraction maps and the α-continued fraction map (c) for $\alpha = \frac{7}{10}$.

A particularly useful continued fraction transformation is the *nearest integer* continued fraction map $T_{\frac{1}{2}}$, which for $x \in \left[-\frac{1}{2}, \frac{1}{2}\right)$ chooses as the digit a_1 the integer nearest to $\left|\frac{1}{x}\right|$. It is defined by

$$T_{\frac{1}{2}}(x) = \begin{cases} \left|\frac{1}{x}\right| - \left\lfloor \left|\frac{1}{x}\right| + \frac{1}{2} \right\rfloor, & \text{if } x \neq 0, \\ 0, & \text{if } x = 0. \end{cases}$$

The continued fraction convergents $\frac{\tilde{p}_n}{\tilde{q}_n}$ obtained from $T_{\frac{1}{2}}$ for irrational numbers $x \in \left(0, \frac{1}{2}\right)$ are a subsequence of the regular continued fraction convergents $\frac{p_n}{q_n}$ produced by the Gauss map for the same x (see [49]) and hence converge to x faster. Recall that the α-continued fraction transformations $\{T_\alpha : [\alpha-1, \alpha] \to [\alpha-1, \alpha]\}_{\alpha\in[0,1]}$ were given by

$$T_\alpha x = \begin{cases} \left|\frac{1}{x}\right| - \left\lfloor \left|\frac{1}{x}\right| + 1 - \alpha \right\rfloor, & \text{if } x \neq 0, \\ 0, & \text{if } x = 0, \end{cases}$$

see also (6.15). Figure 8.2(c) shows the graph of $T_{\frac{7}{10}}$. We can recognize $T_{\frac{1}{2}}$ for $\alpha = \frac{1}{2}$ and of course the Gauss map for $\alpha = 1$.

From all of these maps $\hat{T} = T_e, T_o, T_\alpha$ one can obtain continued fraction digits of real numbers in the domain of the map by iteration. In each case the procedure is essentially the same as for the Gauss map: if at time n the branch $x \mapsto \left|\frac{1}{x}\right| - k$ is applied, then the n-th digit a_n will be k and the value of ϵ_n is determined by the orientation of $x \mapsto \left|\frac{1}{x}\right| - k$. In this way, we can associate to each x that never gets mapped to 0 two sequences $(\epsilon_n)_{n\geq 0} \in \{0,1\}^{\mathbb{N}}$ and $(a_n)_{n\geq 1} \subseteq \mathbb{N}$ such that for each $n \geq 1$,

$$x = \cfrac{\epsilon_0}{a_1 + \cfrac{\epsilon_1}{a_2 + \ddots + \cfrac{\epsilon_{n-1}}{a_n + \hat{T}^n x}}}.$$

The next question is whether this process converges. An essential ingredient in the proof of Proposition 8.1.1 for the Gauss map is that the sequence (q_n) is increasing. This is not generally true for the convergents produced by the other maps. Instead of treating each case separately, we will provide one proof that shows the convergence for all the maps T_e, T_o, T_α simultaneously. This is done by placing the maps T_e, T_o, T_α in a larger framework. The following dynamical system was described in [27].

Let $S : \{0,1\}^{\mathbb{N}} \to \{0,1\}^{\mathbb{N}}$ be the left shift. We define the *random continued fraction map* $K : \{0,1\}^{\mathbb{N}} \times [-1,1] \to \{0,1\}^{\mathbb{N}} \times [-1,1]$ by setting for $x = (x_n)_{n\geq 1}$ that $K(x,0) = (Sx,0)$ and for $|y| \in \left(\frac{1}{k+1}, \frac{1}{k}\right]$,

$$K(x,y) = \left(Sx, \left|\frac{1}{y}\right| - (k + x_1)\right).$$

The transformation K can be interpreted as follows. For each $y \in [-1,1] \setminus \{0\}$ there is a unique $k \geq 1$ so that $|y| \in \left(\frac{1}{k+1}, \frac{1}{k}\right]$. Then both $\left|\frac{1}{y}\right| - k, \left|\frac{1}{y}\right| - k - 1 \in [-1,1]$. The first digit of x determines which of these two points becomes the image of y. Under a Bernoulli measure on $\{0,1\}^{\mathbb{N}}$ this resembles flipping a coin at each time step to determine what will be the next orbit point of y, which explains the name of the transformation.

For $n = 1$, set

$$d_1 = d_1(x,y) = \begin{cases} k + x_1, & \text{if } |y| \in \left(\frac{1}{k+1}, \frac{1}{k}\right], \\ \infty, & \text{if } y = 0, \end{cases}$$

and for $n \geq 2$, define $d_n(x,y) = d_1(K^{n-1}(x,y))$. Use $\pi : \{0,1\}^{\mathbb{N}} \times [-1,1] \to [-1,1]$ to denote the projection onto the second coordinate, so $\pi(x,y) = y$, and set $x_0 = 0$ if $y \geq 0$ and $x_0 = 1$ otherwise. Then the sequence (d_n) is such that we can write

$$\pi(K(x,y)) = (-1)^{x_0}\frac{1}{y} - d_1. \tag{8.26}$$

For each $n \geq 1$ with $\pi(K^n(x,y)) \neq 0$, we get

$$y = \cfrac{(-1)^{x_0}}{d_1 + \pi(K(x,y))} = \cfrac{(-1)^{x_0}}{d_1 + \cfrac{(-1)^{x_1}}{d_2 + \pi(K^2(x,y))}}$$
$$= \cdots = \cfrac{(-1)^{x_0}}{d_1 + \cfrac{(-1)^{x_1}}{d_2 + \ddots + \cfrac{(-1)^{x_{n-1}}}{d_n + \pi(K^n(x,y))}}}.$$

If there is a smallest integer n such that $d_n(x,y) = \infty$ (so $K^{n-1}(x,y) = (S^{n-1}x, 0)$), then

$$y = \cfrac{(-1)^{x_0}}{d_1 + \cfrac{(-1)^{x_1}}{d_2 + \ddots + \cfrac{(-1)^{x_{n-2}}}{d_{n-1}}}}.$$

Suppose that $d_n(x,y) < \infty$ for all $n \geq 1$. We would like to show that

$$y = \cfrac{(-1)^{x_0}}{d_1 + \cfrac{(-1)^{x_1}}{d_2 + \ddots + \cfrac{(-1)^{x_{n-1}}}{d_n + \ddots}}}.$$

For each $n \geq 1$, let $\frac{\hat{p}_n}{\hat{q}_n}$ denote the convergent for y given by the sequence $(d_n(x,y))$, i.e., write

$$\frac{\hat{p}_n}{\hat{q}_n} = \cfrac{(-1)^{x_0}}{d_1 + \cfrac{(-1)^{x_1}}{d_2 + \ddots + \cfrac{(-1)^{x_{n-1}}}{d_n}}}.$$

Following the approach from Section 8.1 we obtain the following relations:

$$\begin{aligned} \hat{p}_{-1} &= 1, \quad \hat{p}_0 = 0, \quad \hat{p}_n = d_n \hat{p}_{n-1} + (-1)^{x_{n-1}} \hat{p}_{n-2}, \\ \hat{q}_{-1} &= 0, \quad \hat{q}_0 = 1, \quad \hat{q}_n = d_n \hat{q}_{n-1} + (-1)^{x_{n-1}} \hat{q}_{n-2}. \end{aligned} \tag{8.27}$$

Using these recurrences, induction easily gives that

$$y = \frac{\hat{p}_n + \hat{p}_{n-1}\pi(K^n(x,y))}{\hat{q}_n + \hat{q}_{n-1}\pi(K^n(x,y))} \tag{8.28}$$

and that

$$\hat{p}_{n-1}\hat{q}_n - \hat{p}_n\hat{q}_{n-1} = (-1)^n(-1)^{x_0+\cdots+x_{n-1}}. \tag{8.29}$$

We will show that although the sequence $(\hat{q}_n)_{n\geq 1}$ is not necessarily increasing, we still have $\lim_{n\to\infty}\frac{1}{\hat{q}_n} = 0$. The following properties of the numbers $\hat{q}_n$ can easily be proved by induction.

Exercise 8.4.1. **(a)** Prove that $\hat{q}_n > 0$ for each $n \geq 2$ and that moreover, if $\hat{q}_n \leq \hat{q}_{n-1}$, then $x_{n-1} = 1$ and $d_n = 1$, so $x_n = 0$.
(b) Prove that if $\hat{q}_n \leq \hat{q}_{n-1}$, then $\hat{q}_{n-2} < \hat{q}_{n-1} < \hat{q}_{n+1}$ and in fact $\hat{q}_n \neq \hat{q}_{n-1}$.

Before we can prove that the process converges, we need a lower bound on the $\hat{q}_n$'s in case $\hat{q}_n < \hat{q}_{n-1}$. This is done in the next lemma.

Lemma 8.4.1. *Suppose $\hat{q}_n < \hat{q}_{n-1}$, so $d_{n-1} > 1$.*

(i) If $d_{n-1} > 2$, then $\hat{q}_n > \hat{q}_{n-2}$.

(ii) If $d_{n-1} = 2$ and $(x_k, d_k) = (1, 2)$ for all $1 \leq k \leq n-1$, then $\hat{q}_n = 1$ and $\hat{q}_{n-1} = n$.

(iii) Suppose $d_{n-1} = 2$ and there is a $1 \leq k < n-1$ such that $(x_k, d_k) \neq (1, 2)$. Let k be the largest such index. Then $\hat{q}_n > \hat{q}_{k-1}$.

Proof. From Exercise 8.4.1(a) we see that $\hat{q}_n < \hat{q}_{n-1}$ implies that $d_n = 1$, $x_n = 0$ and $x_{n-1} = 1$, so $d_{n-1} > 1$. Hence, by the recurrence relations from (8.27) $\hat{q}_n = \hat{q}_{n-1} - \hat{q}_{n-2}$.
(i) If $d_{n-1} > 2$, then

$$\begin{aligned}
\hat{q}_n &= d_{n-1}\hat{q}_{n-2} + (-1)^{x_{n-2}}\hat{q}_{n-3} - \hat{q}_{n-2} \\
&= (d_{n-1} - 1)\hat{q}_{n-2} + (-1)^{x_{n-2}}\hat{q}_{n-3} \\
&\geq 2\hat{q}_{n-2} + (-1)^{x_{n-2}}\hat{q}_{n-3}.
\end{aligned}$$

If $\hat{q}_{n-2} > \hat{q}_{n-3}$, then $\hat{q}_n \geq 2\hat{q}_{n-2} - \hat{q}_{n-3} > \hat{q}_{n-2}$. If $\hat{q}_{n-2} < \hat{q}_{n-3}$, then Exercise 8.4.1(a) gives that $x_{n-2} = 0$ and hence $\hat{q}_n \geq 2\hat{q}_{n-2} + \hat{q}_{n-3} > \hat{q}_{n-3} > \hat{q}_{n-2}$.

For both (ii) and (iii), note that if $d_k = 2$ and $x_k = 1$ for some $1 \le k \le n-1$, then $\hat{q}_k = 2\hat{q}_{k-1} - \hat{q}_{k-2}$, so

$$\hat{q}_k - \hat{q}_{k-1} = \hat{q}_{k-1} - \hat{q}_{k-2}. \tag{8.30}$$

(ii) From (8.30) it follows that

$$\hat{q}_n = \hat{q}_{n-1} - \hat{q}_{n-2} = \hat{q}_2 - \hat{q}_1 = 1.$$

Moreover, for each $1 \le k \le n-1$ it holds that

$$\hat{q}_k = 2\hat{q}_{k-1} - \hat{q}_{k-2} = \hat{q}_{k-1} + (\hat{q}_{k-1} - \hat{q}_{k-2}) = \hat{q}_{k-1} + 1.$$

Hence, $\hat{q}_{n-1} = n - 2 + \hat{q}_1 = n$.
(iii) Let k be as given in the lemma, so $(x_k, d_k) \ne (1,2)$ and $(x_j, d_j) = (1,2)$ for $k+1 \le j \le n-1$. Then,

$$\begin{aligned}\hat{q}_n &= \hat{q}_{n-1} - \hat{q}_{n-2} = \hat{q}_{k+1} - \hat{q}_k \\ &= d_{k+1}\hat{q}_k + (-1)^{x_k}\hat{q}_{k-1} - \hat{q}_k = \hat{q}_k + (-1)^{x_k}\hat{q}_{k-1}.\end{aligned}$$

If $x_k = 0$, then $\hat{q}_n = \hat{q}_k + \hat{q}_{k-1} > \hat{q}_{k-1}$. If $x_k = 1$, then $d_k \ge 3$ and Exercise 8.4.1(a) implies $\hat{q}_k > \hat{q}_{k-1}$. This gives

$$\hat{q}_n = \hat{q}_k - \hat{q}_{k-1} = (d_k - 1)\hat{q}_{k-1} + (-1)^{x_{k-1}}\hat{q}_{k-2} \ge 2\hat{q}_{k-1} + (-1)^{x_{k-1}}\hat{q}_{k-2}.$$

As in the proof of part (i) we now get that if $\hat{q}_{k-1} > \hat{q}_{k-2}$, then $\hat{q}_n > \hat{q}_{k-1}$ and if $\hat{q}_{k-1} < \hat{q}_{k-2}$, then $x_{k-1} = 0$ and we also get $\hat{q}_n > \hat{q}_{k-1}$. □

The following result was obtained in [27].

Proposition 8.4.1. *Let $y \in [-1,1]\backslash\mathbb{Q}$. For each $x \in \{0,1\}^{\mathbb{N}}$, the digits $d_n(x,y)$ give a continued fraction expansion of y.*

Proof. For all $n \ge 1$ we obtain from (8.28) and (8.29) that

$$\begin{aligned}\left|y - \frac{\hat{p}_n}{\hat{q}_n}\right| &= \left|\frac{\hat{q}_n(\hat{p}_n + \hat{p}_{n-1}\pi(K^n(x,y))) - \hat{p}_n(\hat{q}_n + \hat{q}_{n-1}\pi(K^n(x,y)))}{\hat{q}_n(\hat{q}_n + \hat{q}_{n-1}\pi(K^n(x,y)))}\right| \\ &= \left|\frac{\pi(K^n(x,y))(\hat{q}_n\hat{p}_{n-1} - \hat{p}_n\hat{q}_{n-1})}{\hat{q}_n(\hat{q}_n + \hat{q}_{n-1}\pi(K^n(x,y)))}\right| \\ &\le \frac{1}{|\hat{q}_n(\hat{q}_n + \hat{q}_{n-1}\pi(K^n(x,y)))|} \le \frac{1}{\hat{q}_n|\hat{q}_n - \hat{q}_{n-1}|} \le \frac{1}{\hat{q}_n},\end{aligned}$$

where in the last step we have also used Exercise 8.4.1(b). We now show that $\lim_{n\to\infty} \frac{1}{\hat{q}_n} = 0$. Let $\varepsilon > 0$. By Exercise 8.4.1(b) there exists a subsequence $(\hat{q}_{n_k})_{k\geq 0}$ such that $\lim_{k\to\infty} \frac{1}{\hat{q}_{n_k}} = 0$ and hence there exists an N_1 such that $\frac{1}{\hat{q}_{N_1}} < \varepsilon$. If there is no $n > N_1$, such that $\hat{q}_n < \hat{q}_{n-1}$, then we are done. If there is, let k be the smallest such index. If $(x_{N_1+1}, d_{N_1+1}) \neq (1,2)$, by Lemma 8.4.1 $\hat{q}_k \geq \hat{q}_{N_1}$ and then same holds for all other $n > N_1$. If $(x_{N_1+1}, d_{N_1+1}) = (1,2)$, then set $N = k+1$. We then have $\frac{1}{\hat{q}_N} < \frac{1}{\hat{q}_{k-1}} < \frac{1}{\hat{q}_{N_1}} < \varepsilon$. Moreover, for all $n > N$ we have $\hat{q}_n > \hat{q}_{k-1}$, since $(x_k, d_k) = (0,1)$. This shows that the limit exists and is equal to 0. Hence we get

$$y = \cfrac{(-1)^{x_0}}{d_1 + \cfrac{(-1)^{x_1}}{d_2 + \cfrac{(-1)^{x_2}}{d_3 + \ddots}}}$$

for each $y \in [-1,1]\backslash\mathbb{Q}$. □

The Gauss map is obtained from K by taking $x = (0,0,0,\ldots)$ and we get the 0-continued fraction map $T_0 : [-1,0] \to [-1,0]$ from $x = (1,1,1,\ldots)$. This map is a shifted version of what is sometimes called the Rényi map, see also Exercise 11.1.1. The odd continued fraction expansions can be found inductively by choosing for each $n \geq 1$ the coordinate x_n such that $d_n(x,y)$ is odd and similarly for the even continued fractions.

Exercise 8.4.2. Let $T_\alpha : [\alpha - 1, \alpha] \to [\alpha - 1, \alpha]$, $\alpha \in [0,1]$, be the α-continued fraction map. For each $y \in [\alpha-1,\alpha]$ that satisfies $T_\alpha^n y \neq 0$ for each $n \geq 1$, define a sequence of signs $(s_n) \subseteq \{0,1\}^{\mathbb{N}}$ and a sequence of digits $(d_n) \subseteq \mathbb{N}^{\mathbb{N}}$, such that for each $k \geq 1$,

$$y = \cfrac{(-1)^{s_0}}{d_1 + \cfrac{(-1)^{s_1}}{d_2 + \ddots + \cfrac{(-1)^{s_{k-1}}}{d_k + T_\alpha^k y}}}.$$

Prove that

$$y = \cfrac{(-1)^{s_0}}{d_1 + \cfrac{(-1)^{s_1}}{d_2 + \cfrac{(-1)^{s_2}}{d_3 + \ddots}}}.$$

CHAPTER 9

Entropy

9.1 RANDOMNESS AND INFORMATION

In some of the examples of dynamical systems we saw, we attached to each orbit a sequence of symbols. For example, orbits under the doubling map correspond to binary digit sequences. There is a more general setup for this.

Definition 9.1.1. Suppose $(X, \mathcal{F}, \mu, T)$ is a dynamical system with μ a probability measure. A collection of measurable sets $\alpha = \{A_i\} \subseteq \mathcal{F}$ is called a *partition* for T if

- $\mu(A_i) > 0$ for all i,
- $\mu(A_i \cap A_j) = 0$ for $i \neq j$, and,
- $\mu(\bigcup_i A_i) = 1$.

The elements of a partition are called its *atoms*. We will only consider finite or countable partitions.

As for the doubling map, given a partition $\alpha = \{A_i\}$ of X, we can assign to μ-almost all $x \in X$ a digit sequence $(a_n(x))$ by setting $a_n(x) = i$ if $T^{n-1}x \in A_i$, $n \geq 1$. Suppose we are in the situation where we cannot observe the action of the dynamical system $(X, \mathcal{F}, \mu, T)$ directly, but we can only see the consecutive symbols $a_1, a_2, a_3, \ldots$. The information value of this sequence obviously depends on the transformation T and on the chosen partition (the partition $\alpha = \{X\}$ gives no information). We would like to define a non-negative quantity $h_\mu(T)$, called the measure theoretic (or metric) entropy, which in some sense measures the asymptotic average information generated by each application of T and is independent of the chosen partition α. We want to define $h_\mu(T)$ in such a way that

(i) $h_\mu(T)$ reflects the average amount of information gained by an application of T, where information gained or uncertainty removed are seen as proportional quantities, and that

(ii) $h_\mu(T)$ is isomorphism invariant, so that isomorphic transformations have equal entropy.

The connection between entropy (that is randomness, uncertainty) and the transmission of information was first studied by C. Shannon in the famous paper [59] from 1948. He considered sequences of symbols emitted from some data source and fed into a transmission channel. Think for example of Morse code or compact discs. To study the effect of noise on these sequences the output from the transition channel was viewed as a stationary stochastic process. Entropy was devised to measure the uncertainty in the value of this outcome. Before we turn to dynamics, the first basic question we need to answer is how do we quantify the information gained by the occurrence of an event E? A possible choice would be $-\log \mu(E)$ (the base of the logarithm is chosen to be 2, but any base would work). To check whether this is a good choice we examine two natural situations. Firstly, if E and F are two independent events, then

$$-\log \mathbb{P}(E \cap F) = -\log \mathbb{P}(E) - \log \mathbb{P}(F),$$

which is exactly what we expect: the information transmitted by $E \cap F$ is the sum of the information transmitted by each one individually. Secondly, if E is a sure event, i.e., $\mu(E) = 1$, then $\log \mu(E) = 0$, which is also what we expect: we gain no information by the occurrence of E since we already know that almost surely any outcome is in E.

Now that we have a reasonable way to quantify the information gained by the occurrence of an event, how do we quantify the information gained by a stationary process or in general a probability measure preserving dynamical system? Let us first start simple, and consider an independent process, i.e., a Bernoulli shift. What we see are infinite sequences $x_1, x_2, x_3, \ldots$ from the set of symbols $\{0, 1, \ldots, k-1\}$ (or some other finite set of symbols). Suppose that the probability of receiving symbol i at any given time is p_i, and that each symbol is transmitted independently of what has been transmitted earlier. Of course each $p_i \geq 0$ and $\sum_i p_i = 1$. As we have seen in earlier chapters, we view this process as the dynamical system $(X, \mathcal{C}, \mu, T)$, where $X = \{0, 1, \ldots, k-1\}^{\mathbb{N}}$, $\mathcal{C}$ the σ-algebra generated by cylinder sets of the form

$$\{x \in X : x_1 = i_1, \ldots, x_n = i_n\},$$

μ the product measure assigning to each coordinate probability p_i of seeing the symbol i, and T the left shift. We define the entropy of this system by

$$H(p_0, \dots, p_{k-1}) = h(T) := -\sum_{i=0}^{k-1} p_i \log p_i. \tag{9.1}$$

If we define $-\log p_i$ as the amount of uncertainty in transmitting the symbol i, then H is the average amount of information gained (or uncertainty removed) per symbol (notice that H is in fact an expected value). To see why this is an appropriate definition, notice that if the source is degenerate, that is, $p_i = 1$ for some i (i.e., the source only transmits the symbol i), then $H = 0$. In this case we indeed have no randomness. Another reason to see why this definition is appropriate, is that H is maximal if $p_i = \frac{1}{k}$ for all i, and this agrees with the fact that the source is most random when all the symbols are equiprobable. To see this maximum, consider the function $\Phi : [0, \infty) \to \mathbb{R}$ defined by

$$\Phi(t) = \begin{cases} 0, & \text{if } t = 0, \\ -t \log t, & \text{if } 0 < t \leq 1. \end{cases}$$

Then Φ is continuous and concave, and Jensen's Inequality implies that for any $p_0, \dots, p_{k-1}$ with $p_i \geq 0$ and $p_0 + \cdots + p_{k-1} = 1$,

$$\frac{1}{k} H(p_0, \dots, p_{k-1}) = \frac{1}{k} \sum_{i=0}^{k-1} \Phi(p_i) \leq \Phi\Big(\frac{1}{k} \sum_{i=0}^{k-1} p_i\Big) = \Phi\Big(\frac{1}{k}\Big) = \frac{1}{k} \log k,$$

so $H(p_0, \dots, p_{k-1}) \leq \log k$ for all probability vectors $(p_0, \dots, p_{k-1})$. But

$$H\Big(\frac{1}{k}, \dots, \frac{1}{k}\Big) = \log k,$$

so the maximum value is attained at $(\frac{1}{k}, \dots, \frac{1}{k})$.

In the above example of a Bernoulli shift the symbols were transmitted independently. In general, the symbol generated might depend on what has been received before. In fact these dependencies are often "built-in" to be able to check the transmitted sequence of symbols on errors. Such dependencies must be taken into consideration in the calculation of the average information per symbol. This can be achieved if one replaces the symbols i by blocks of symbols of a particular length. More precisely, for every n, let $\mathcal{C}_n$ be the collection of all possible cylinder sets specifying n coordinates, and define

$$H_n := -\sum_{C\in\mathcal{C}_n} \mathbb{P}(C)\log\mathbb{P}(C).$$

Then $\frac{1}{n}H_n$ can be seen as the average information per symbol when a block of length n is transmitted. The entropy of the source is now defined by

$$h := \lim_{n\to\infty}\frac{H_n}{n}. \tag{9.2}$$

The existence of the limit in (9.2) is a consequence of Proposition 9.2.3 below and follows from the fact that H_n is a *subadditive sequence*, i.e., $H_{n+m} \le H_n + H_m$, and Proposition 9.2.2.

9.2 DEFINITIONS AND PROPERTIES

Consider now and for the remainder of this chapter a measure preserving system $(X, \mathcal{F}, \mu, T)$ with μ a probability measure. How can one define the entropy of this system similar to the case of a Bernoulli shift? We return to the setup from the beginning of the chapter. The symbols $\{0, 1, \ldots, k-1\}$ compare to a partition $\alpha = \{A_0, A_1, \ldots, A_{k-1}\}$ of X and with each point $x \in X$, we associate an infinite sequence $a_1, a_2, a_3, \ldots$, where a_i is j if and only if $T^{i-1}x \in A_j$. We define the *entropy of the partition* α by

$$H_\mu(\alpha) := -\sum_{A\in\alpha}\mu(A)\log\mu(A) \in [0,\infty].$$

It is finite if α is a finite partition. Our aim is to define the entropy of the transformation T independently of the partition we choose. First we need a few facts about partitions. We will use the following definition also for collections of sets that are not necessarily partitions.

Definition 9.2.1. Let α and β be two collections of subsets of a space X. We say that β is a *refinement* of α, and write $\alpha \le \beta$, if for every $B \in \beta$ there exists an $A \in \alpha$ such that $B \subseteq A$. In case the space X is a measure space, it is enough for the inclusion to hold up to sets of measure zero. The collection

$$\alpha \vee \beta := \{A \cap B \,:\, A \in \alpha,\ B \in \beta\}$$

is called the *common refinement* of α and β.

Exercise 9.2.1. Let α and β be two partitions of the same space X.

(a) Show that $\alpha \vee \beta$ and $T^{-1}\alpha := \{T^{-1}A : A \in \alpha\}$ are both partitions of X.

(b) Show that if β is finite and $\alpha \leq \beta$, then each atom of α is a finite (a.e. disjoint) union of atoms of β.

Given two partitions α and β of X, we define the *conditional entropy of α given β* by

$$H_\mu(\alpha|\beta) := -\sum_{A\in\alpha}\sum_{B\in\beta}\mu(A\cap B)\log\left(\frac{\mu(A\cap B)}{\mu(B)}\right)$$

(under the convention that $0\log 0 := 0$). The quantity $H_\mu(\alpha|\beta)$ is interpreted as the average uncertainty about which element of the partition α the point x will enter (under T) if we already know which element of β the point x will enter.

Proposition 9.2.1. *Let α, β and γ be partitions of X. Then,*

(i) $H_\mu(T^{-1}\alpha) = H_{\mu\circ T^{-1}}(\alpha)$, *so* $H_\mu(T^{-1}\alpha) = H_\mu(\alpha)$ *if T is measure preserving for μ;*

(ii) $H_\mu(\alpha \vee \beta) = H_\mu(\alpha) + H_\mu(\beta|\alpha)$;

(iii) $H_\mu(\beta|\alpha) \leq H_\mu(\beta)$;

(iv) $H_\mu(\alpha \vee \beta) \leq H_\mu(\alpha) + H_\mu(\beta)$;

(v) *If $\alpha \leq \beta$, then* $H_\mu(\alpha) \leq H_\mu(\beta)$;

(vi) $H_\mu(\alpha \vee \beta|\gamma) = H_\mu(\alpha|\gamma) + H_\mu(\beta|\alpha \vee \gamma)$;

(vii) *If $\alpha \leq \beta$, then* $H_\mu(\gamma|\beta) \leq H_\mu(\gamma|\alpha)$;

(viii) *If $\alpha \leq \beta$, then* $H_\mu(\alpha|\beta) = 0$.

(ix) *We call two partitions α and β* independent *if*

$$\mu(A\cap B) = \mu(A)\mu(B) \quad \textit{for all } A \in \alpha,\ B \in \beta.$$

If α and β are independent partitions, one has that

$$H_\mu(\alpha \vee \beta) = H_\mu(\alpha) + H_\mu(\beta).$$

Proof. For (ii),

$$\begin{aligned}
H_\mu(\alpha \vee \beta) &= -\sum_{A\in\alpha}\sum_{B\in\beta} \mu(A\cap B)\log\mu(A\cap B) \\
&= -\sum_{A\in\alpha}\sum_{B\in\beta} \mu(A\cap B)\log\frac{\mu(A\cap B)}{\mu(A)} \\
&\quad -\sum_{A\in\alpha}\sum_{B\in\beta} \mu(A\cap B)\log\mu(A) \\
&= H_\mu(\beta|\alpha) + H_\mu(\alpha).
\end{aligned}$$

We now show part (iii), that $H_\mu(\beta|\alpha) \leq H_\mu(\beta)$. Recall that the function $\Phi(t) = -t\log t$ for $t \geq 0$ is concave. Thus,

$$\begin{aligned}
H_\mu(\beta|\alpha) &= -\sum_{B\in\beta}\sum_{A\in\alpha} \mu(A\cap B)\log\frac{\mu(A\cap B)}{\mu(A)} \\
&= -\sum_{B\in\beta}\sum_{A\in\alpha} \mu(A)\frac{\mu(A\cap B)}{\mu(A)}\log\frac{\mu(A\cap B)}{\mu(A)} \\
&= \sum_{B\in\beta}\sum_{A\in\alpha} \mu(A)\Phi\left(\frac{\mu(A\cap B)}{\mu(A)}\right) \\
&\leq \sum_{B\in\beta}\Phi\left(\sum_{A\in\alpha}\mu(A)\frac{\mu(A\cap B)}{\mu(A)}\right) \\
&= \sum_{B\in\beta}\Phi(\mu(B)) = H_\mu(\beta).
\end{aligned}$$

The proofs of the other parts are left as an exercise. □

Exercise 9.2.2. Prove the rest of the properties of Proposition 9.2.1.

Now given a partition α of X consider the partition $\bigvee_{i=0}^{n-1} T^{-i}\alpha$, whose atoms are of the form

$$A_{i_0} \cap T^{-1}A_{i_1} \cap \cdots \cap T^{-(n-1)}A_{i_{n-1}},$$

consisting of all points $x \in X$ with the property that $x \in A_{i_0}$, $Tx \in A_{i_1}, \ldots, T^{n-1}x \in A_{i_{n-1}}$.

Exercise 9.2.3. Show that if α is a finite partition of $(X, \mathcal{F}, \mu, T)$, then

$$H_\mu\Big(\bigvee_{i=0}^{n-1} T^{-i}\alpha\Big) = H_\mu(\alpha) + \sum_{j=1}^{n-1} H_\mu\Big(\alpha|\bigvee_{i=1}^{j} T^{-i}\alpha\Big).$$

To define the notion of the entropy of a transformation with respect to a partition, we need the following two propositions.

Proposition 9.2.2. *If* (a_n) *is a subadditive sequence of real numbers, i.e.,* $a_{n+p} \le a_n + a_p$ *for all* n, p*, then*

$$\lim_{n\to\infty} \frac{a_n}{n}$$

exists.

Proof. Fix any $m > 0$. For any $n > 0$ one has $n = km + i$ for some i between $0 \le i \le m-1$. By subadditivity it follows that

$$\frac{a_n}{n} = \frac{a_{km+i}}{km+i} \le \frac{a_{km}}{km} + \frac{a_i}{km} \le k\frac{a_m}{km} + \frac{a_i}{km}.$$

Note that if $n \to \infty$, $k \to \infty$ and so $\limsup\limits_{n\to\infty} \dfrac{a_n}{n} \le \dfrac{a_m}{m}$. Since m is arbitrary one has

$$\limsup_{n\to\infty} \frac{a_n}{n} \le \inf \frac{a_m}{m} \le \liminf_{n\to\infty} \frac{a_n}{n}.$$

Therefore $\lim\limits_{n\to\infty} \dfrac{a_n}{n}$ exists, and equals $\inf\limits_{n} \dfrac{a_n}{n}$. □

Proposition 9.2.3. *Let* α *be a finite partition of* $(X, \mathcal{F}, \mu, T)$*, where* T *is a measure preserving transformation. Then,* $\lim_{n\to\infty} \frac{1}{n} H_\mu(\bigvee_{i=0}^{n-1} T^{-i}\alpha)$ *exists.*

Proof. Let $a_n = H_\mu(\bigvee_{i=0}^{n-1} T^{-i}\alpha) \ge 0$. Then, by Proposition 9.2.1(iv) and (i), we have

$$\begin{aligned} a_{n+p} &= H_\mu\Big(\bigvee_{i=0}^{n+p-1} T^{-i}\alpha\Big) \\ &\le H_\mu\Big(\bigvee_{i=0}^{n-1} T^{-i}\alpha\Big) + H_\mu\Big(\bigvee_{i=n}^{n+p-1} T^{-i}\alpha\Big) \\ &= a_n + H_\mu\Big(\bigvee_{i=0}^{p-1} T^{-i}\alpha\Big) \\ &= a_n + a_p. \end{aligned}$$

Hence, by Proposition 9.2.2,

$$\lim_{n\to\infty} \frac{a_n}{n} = \lim_{n\to\infty} \frac{1}{n} H_\mu\Big(\bigvee_{i=0}^{n-1} T^{-i}\alpha\Big)$$

exists. □

We are now in position to give the definition of the entropy of the transformation T.

Definition 9.2.2. Let $(X, \mathcal{F}, \mu)$ be a probability space and $T : X \to X$ measure preserving. The *entropy of* T *with respect to the finite partition* α is given by

$$h_\mu(\alpha, T) := \lim_{n\to\infty} \frac{1}{n} H_\mu\Big(\bigvee_{i=0}^{n-1} T^{-i}\alpha\Big),$$

where

$$H_\mu\Big(\bigvee_{i=0}^{n-1} T^{-i}\alpha\Big) = -\sum_{D\in\bigvee_{i=0}^{n-1} T^{-i}\alpha} \mu(D)\log(\mu(D)).$$

Finally, the *measure theoretic* or *metric entropy of the transformation* T is given by

$$h_\mu(T) := \sup_\alpha h_\mu(\alpha, T),$$

where the supremum is taken over all finite partitions of X.

The following theorem gives an equivalent definition of metric entropy.

Theorem 9.2.1. *Let* $(X, \mathcal{F}, \mu)$ *be a probability space and* $T : X \to X$ *measure preserving. Let* α *be a finite partition of* X*. The entropy of* T *with respect to* α *is also given by*

$$h_\mu(\alpha, T) = \lim_{n\to\infty} H_\mu\Big(\alpha | \bigvee_{i=1}^{n-1} T^{-i}\alpha\Big).$$

Proof. Notice that the sequence $\big(H_\mu(\alpha| \bigvee_{i=1}^{n} T^{-i}\alpha)\big)_{n\geq 1}$ is bounded from below, and is non-increasing by Proposition 9.2.1(vii). Hence $\lim_{n\to\infty} H_\mu(\alpha| \bigvee_{i=1}^{n} T^{-i}\alpha)$ exists. Furthermore,

$$\lim_{n\to\infty} H_\mu\Big(\alpha| \bigvee_{i=1}^{n} T^{-i}\alpha\Big) = \lim_{n\to\infty} \frac{1}{n}\sum_{j=1}^{n} H_\mu\Big(\alpha| \bigvee_{i=1}^{j} T^{-i}\alpha\Big).$$

From Exercise 9.2.3, we have

$$H_\mu\Big(\bigvee_{i=0}^{n-1} T^{-i}\alpha\Big) = H_\mu(\alpha) + \sum_{j=1}^{n-1} H_\mu\Big(\alpha| \bigvee_{i=1}^{j} T^{-i}\alpha\Big).$$

Now, dividing by n, and taking the limit as $n \to \infty$, one gets the desired result. □

The next theorem shows that entropy can be used to distinguish non-isomorphic dynamical systems.

Theorem 9.2.2. *Entropy is an isomorphism invariant.*

Proof. Let $(X, \mathcal{F}, \mu, T)$ and $(Y, \mathcal{G}, \nu, S)$ be two isomorphic measure preserving systems with $\psi : X \to Y$ the corresponding isomorphism. We need to show that $h_\mu(T) = h_\nu(S)$. Let $\beta = \{B_1, \dots, B_n\}$ be any finite partition of Y, then

$$\psi^{-1}\beta = \{\psi^{-1}B_1, \dots, \psi^{-1}B_n\}$$

is a partition of X. Set $A_i = \psi^{-1}B_i$, for $1 \le i \le n$. Since $\psi : X \to Y$ is an isomorphism, we have that $\nu = \mu \circ \psi^{-1}$ and $\psi \circ T = S \circ \psi$, so that for any $n \ge 0$ and $B_{i_0}, \dots, B_{i_{n-1}} \in \beta$,

$$\begin{aligned}
&\nu(B_{i_0} \cap S^{-1}B_{i_1} \cap \cdots \cap S^{-(n-1)}B_{i_{n-1}}) \\
&\qquad = \mu\left(\psi^{-1}B_{i_0} \cap \psi^{-1}S^{-1}B_{i_1} \cap \cdots \cap \psi^{-1}S^{-(n-1)}B_{i_{n-1}}\right) \\
&\qquad = \mu\left(\psi^{-1}B_{i_0} \cap T^{-1}\psi^{-1}B_{i_1} \cap \cdots \cap T^{-(n-1)}\psi^{-1}B_{i_{n-1}}\right) \\
&\qquad = \mu\left(A_{i_0} \cap T^{-1}A_{i_1} \cap \cdots \cap T^{-(n-1)}A_{i_{n-1}}\right).
\end{aligned}$$

Setting

$$A(n) = A_{i_0} \cap T^{-1}A_{i_1} \cap \cdots \cap T^{-(n-1)}A_{i_{n-1}}$$

and

$$B(n) = B_{i_0} \cap S^{-1}B_{i_1} \cap \cdots \cap S^{-(n-1)}B_{i_{n-1}},$$

we find that

$$\begin{aligned}
h_\nu(S) &= \sup_\beta h_\nu(\beta, S) = \sup_\beta \lim_{n\to\infty} \frac{1}{n} H_\nu\Big(\bigvee_{i=0}^{n-1} S^{-i}\beta\Big) \\
&= \sup_\beta \lim_{n\to\infty} -\frac{1}{n} \sum_{B(n)\in\bigvee_{i=0}^{n-1} S^{-i}\beta} \nu(B(n)) \log \nu(B(n)) \\
&= \sup_{\psi^{-1}\beta} \lim_{n\to\infty} -\frac{1}{n} \sum_{A(n)\in\bigvee_{i=0}^{n-1} T^{-i}\psi^{-1}\beta} \mu(A(n)) \log \mu(A(n)) \\
&= \sup_{\psi^{-1}\beta} h_\mu(\psi^{-1}\beta, T) \\
&\le \sup_\alpha h_\mu(\alpha, T) = h_\mu(T),
\end{aligned}$$

where in the last inequality the supremum is taken over all possible finite partitions α of X. Thus $h_\nu(S) \le h_\mu(T)$. By symmetry we find $h_\nu(S) = h_\mu(T)$, and the proof is complete. □

The previous result shows that one can check whether two systems are isomorphic by computing the metric entropy. The result only goes in one direction: if two systems have different metric entropies, then they are not isomorphic. To see that the other implication does not necessarily hold, recall the positive and negative β-transformations T_β and S_β, respectively, from Example 5.1.4. In Example 9.4.1 below we see that for any $\beta \in (1,2)$ the map T_β has metric entropy $h_{\mu_\beta}(T_\beta) = \log \beta$ for the invariant measure μ_β identified in Example 6.3.4. Exercise 9.4.6 below considers the negative β-transformation S_β for the specific value $\hat{\beta} > 1$ satisfying $\hat{\beta}^3 - \hat{\beta}^2 - 1 = 0$. It gives an invariant measure $\nu_{\hat{\beta}}$ for $S_{\hat{\beta}}$ that is absolutely continuous with respect to λ and has $h_{\nu_{\hat{\beta}}}(S_{\hat{\beta}}) = \log \hat{\beta}$. Since $\hat{\beta} < \frac{1+\sqrt{5}}{2}$, we know from Example 5.1.4 that the dynamical systems $([0,1], \mathcal{B}, \mu_{\hat{\beta}}, T_{\hat{\beta}})$ and $([0,1], \mathcal{B}, \nu_{\hat{\beta}}, S_{\hat{\beta}})$ are not isomorphic, even though they have the same metric entropies. In fact, using combined results by F. Hofbauer and G. Keller from [21, 22, 23] one can show that the same is true for any other value $\beta \in (1, \frac{1+\sqrt{5}}{2})$.

There is one important class of dynamical systems for which the metric entropy is a complete invariant, namely two-sided Bernoulli shifts. A famous result by D. Ornstein, see [47], states that two two-sided Bernoulli shifts are metrically isomorphic if and only if their metric entropies are equal.

9.3 CALCULATION OF ENTROPY AND EXAMPLES

Calculating the entropy of a transformation directly from the definition does not seem very feasible, since it requires taking the supremum over **all** finite partitions. However, the entropy of a partition is relatively easy to compute if one has full information about the partition under consideration. So the question is whether it is possible to find a partition α of X for which $h_\mu(\alpha, T) = h_\mu(T)$. Naturally, such a partition contains all the information "transmitted" by T. To answer this question we need some notations and definitions.

For $\alpha = \{A_1, \ldots, A_N\}$ and $m, n \in \mathbb{Z}$ with $n \leq m$, let

$$\sigma\left(\bigcup_{i=n}^{m} T^{-i}\alpha\right)$$

be the smallest σ-algebra containing all elements from the partition $\bigvee_{i=n}^{m} T^{-i}\alpha$. We only consider forward iterates of T in case T is invertible.

We call a partition α a *generator* with respect to an invertible transformation T if $\sigma\left(\bigcup_{i=-\infty}^{\infty} T^{-i}\alpha\right) = \mathcal{F}$, where $\mathcal{F}$ is the σ-algebra on X and $\sigma\left(\bigcup_{i=-\infty}^{\infty} T^{-i}\alpha\right)$ is the smallest σ-algebra containing all elements of partitions $\bigvee_{i=n}^{m} T^{-i}\alpha$ for all $m, n \in \mathbb{Z}$ with $n \leq m$. If T is non-invertible, then α is said to be a generator if $\sigma\left(\bigcup_{i=0}^{\infty} T^{-i}\alpha\right) = \mathcal{F}$. Naturally, this equality is modulo sets of measure zero.

We now state two famous theorems known as the *Kolmogorov-Sinai Theorem* and *Krieger's Generator Theorem*. A first version of the Kolmogorov-Sinai Theorem was provided by A. N. Kolmogorov in his lectures and the proof for the general case was given by Ya. G. Sinai in [61] from 1959. Not unexpectedly, Krieger's Generator Theorem was proven by W. Krieger, see [32] from 1970. Before we state the theorem, we would like to remark that although we have considered only finite partitions on X, however all the definitions and results hold if we were to consider countable partitions of finite entropy.

Theorem 9.3.1 (Kolmogorov-Sinai Theorem). *Let $T : X \to X$ be a measure preserving transformation on the probability space $(X, \mathcal{F}, \mu)$ and α a finite or countable partition with $H_\mu(\alpha) < \infty$. If α is a generator with respect to T, then $h_\mu(T) = h_\mu(\alpha, T)$.*

Theorem 9.3.2 (Krieger's Generator Theorem). *If T is an ergodic measure preserving transformation with $h_\mu(T) < \infty$, then T has a finite generator.*

The proof of Theorem 9.3.1 is given in Exercise 9.3.1 below. For the proof of Theorem 9.3.2, we refer the interested reader to the books [50, 65].

Exercise 9.3.1. Let $(X, \mathcal{F}, \mu)$ be a probability space and $T : X \to X$ a measure preserving transformation.

(a) Suppose α is a finite partition of X. Show that $h_\mu(\alpha, T) = h_\mu\left(\bigvee_{i=1}^{n} T^{-i}\alpha, T\right)$, for any $n \geq 1$.

(b) Let α and β be finite partitions. Show that $h_\mu(\beta, T) \leq h_\mu(\alpha, T) + H_\mu(\beta|\alpha)$.

(c) Suppose α is a finite generator, i.e., $\sigma\left(\bigcup_{i=0}^{\infty} T^{-i}\alpha\right) = \mathcal{F}$. Using parts (a) and (b), show that for any finite partition β of X one has $h_\mu(\beta, T) \leq h_\mu(\alpha, T)$. Conclude that $h_\mu(\alpha, T) = h_\mu(T)$.

We will use these two theorems to calculate the entropy of a Bernoulli shift.

Example 9.3.1 (Entropy of a Bernoulli Shift). Let T be the left shift on $X = \{0, 1, \ldots, k-1\}^{\mathbb{N}}$ endowed with the σ-algebra $\mathcal{C}$ generated by the cylinder sets, and product measure μ giving symbol i probability p_i, where $p_0 + p_1 + \cdots + p_{k-1} = 1$. Our aim is to calculate $h_\mu(T)$. To this end we need to find a partition α which generates the σ-algebra $\mathcal{C}$ under the action of T. The natural choice of α is what is known as the *time-zero partition* $\alpha = \{A_0, \ldots, A_{k-1}\}$, where

$$A_i := \{x \in X : x_1 = i\} \quad i = 0, \ldots, k-1.$$

Notice that for all $m \geq 0$,

$$T^{-m}A_i = \{x \in X : x_{m+1} = i\},$$

and

$$\bigcap_{j=0}^{m-1} T^{-j}A_{i_j} = \{x \in X : x_{j+1} = i_j,\ 0 \leq j \leq m-1\}.$$

In other words, $\bigvee_{i=0}^{m-1} T^{-i}\alpha$ is precisely the collection of cylinder sets of length m and these by definition generate $\mathcal{C}$. Hence, α is a generating partition, so that

$$h_\mu(T) = h_\mu(\alpha, T) = \lim_{m\to\infty} \frac{1}{m} H_\mu\left(\bigvee_{i=0}^{m-1} T^{-i}\alpha\right).$$

First notice that, since μ is a product measure, the partitions

$$\alpha,\ T^{-1}\alpha,\ \ldots\ ,\ T^{-(m-1)}\alpha$$

are all independent since each partition specifies a different coordinate. So

$$\begin{aligned} &H_\mu(\alpha \vee T^{-1}\alpha \vee \cdots \vee T^{-(m-1)}\alpha) \\ &\qquad = H_\mu(\alpha) + H_\mu(T^{-1}\alpha) + \cdots + H_\mu(T^{-(m-1)}\alpha) \\ &\qquad = mH_\mu(\alpha) \;=\; -m\sum_{i=0}^{k-1} p_i \log p_i. \end{aligned}$$

Thus,

$$h_\mu(T) = \lim_{m\to\infty} \frac{1}{m}(-m)\sum_{i=0}^{k-1} p_i \log p_i = -\sum_{i=0}^{k-1} p_i \log p_i.$$

Exercise 9.3.2. Let T be the left shift on $X = \{0, 1, \ldots, k-1\}^{\mathbb{Z}}$ endowed with the σ-algebra $\mathcal{C}$ generated by the cylinder sets, and the Markov measure μ given by the stochastic matrix $P = (p_{ij})$, and the probability vector $\pi = (\pi_0, \ldots, \pi_{k-1})$ with $\pi P = \pi$. Show that

$$h_\mu(T) = -\sum_{j=0}^{k-1}\sum_{i=0}^{k-1} \pi_i p_{ij} \log p_{ij}.$$

Exercise 9.3.3. Let $(X, \mathcal{F}, \mu)$ be a probability space and $T : X \to X$ a measure preserving transformation. Let $k > 0$.

(a) Show that for any finite partition α of X one has $h_\mu(\bigvee_{i=0}^{k-1} T^{-i}\alpha, T^k) = kh_\mu(\alpha, T)$.

(b) Prove that $kh_\mu(T) \leq h_\mu(T^k)$.

(c) Prove that $h_\mu(\alpha, T^k) \leq kh_\mu(\alpha, T)$.

(d) Prove that $h_\mu(T^k) = kh_\mu(T)$.

Recall from Theorem 12.3.2 that the fact that we are working on a Lebesgue space implies the existence of an increasing sequence of finite partitions $\alpha_1 \leq \alpha_2 \leq \ldots$ such that $\sigma(\cup_n \alpha_n) = \mathcal{F}$. This allows us to calculate entropy in the following way.

Proposition 9.3.1. *If $\alpha_1 \leq \alpha_2 \leq \ldots$ is an increasing sequence of finite partitions on $(X, \mathcal{F}, \mu, T)$ such $\sigma(\alpha_n) \nearrow \mathcal{F}$, then $h_\mu(T) = \lim_{n\to\infty} h_\mu(\alpha_n, T)$.*

Proof. It is enough to show that for any finite partition β, one has $h_\mu(\beta, T) \leq \lim_{n\to\infty} h_\mu(\alpha_n, T)$. From Exercise 9.3.1(b),

$$h_\mu(\beta, T) \leq h_\mu(\alpha_n, T) + H_\mu(\beta|\alpha_n).$$

Since $\sigma(\alpha_n) \nearrow \mathcal{F}$, then by the Martingale Convergence Theorem (see Theorem 12.4.7) together with the Dominated Convergence Theorem one has

$$\lim_{n\to\infty} H_\mu(\beta|\alpha_n) = \lim_{n\to\infty} H_\mu(\beta|\sigma(\alpha_n)) = H_\mu(\beta|\mathcal{F}) = 0.$$

Thus, $h_\mu(\beta, T) \leq \lim_{n\to\infty} h_\mu(\alpha_n, T)$, and hence

$$h_\mu(T) = \lim_{n\to\infty} h_\mu(\alpha_n, T). \qquad \square$$

Exercise 9.3.4. Prove that $h_{\mu_1\times\mu_2}(T_1 \times T_2) = h_{\mu_1}(T_1) + h_{\mu_2}(T_2)$.

9.4 THE SHANNON-MCMILLAN-BREIMAN THEOREM

In the previous sections we have considered only finite partitions on X, however as mentioned earlier, all the definitions and results hold if we were to consider countable partitions of finite entropy. Before we state and prove the Shannon-McMillan-Breiman Theorem, we need to introduce the information function associated with a partition.

Let $(X, \mathcal{F}, \mu)$ be a probability space, and $\alpha = \{A_1, A_2, \ldots\}$ a finite or a countable partition of X into measurable sets. For each $x \in X$, let $\alpha(x)$ be the element of α to which x belongs. Then, the *information function* associated to α is defined to be

$$I_\alpha(x) = -\log \mu(\alpha(x)) = -\sum_{A \in \alpha} 1_A(x) \log \mu(A).$$

Note that I_α is only defined up to μ-a.e. equivalence. Also note that

$$\int_X I_\alpha \, d\mu = H_\mu(\alpha).$$

For two finite or countable partitions α and β of X, we define the *conditional information function* of α given β by

$$I_{\alpha|\beta}(x) = -\sum_{B \in \beta} \sum_{A \in \alpha} 1_{(A \cap B)}(x) \log \left(\frac{\mu(A \cap B)}{\mu(B)} \right).$$

We claim that

$$I_{\alpha|\beta}(x) = -\log \mathbb{E}_\mu(1_{\alpha(x)}|\sigma(\beta))(x) = -\sum_{A \in \alpha} 1_A(x) \log \mathbb{E}_\mu(1_A|\sigma(\beta))(x), \tag{9.3}$$

where $\sigma(\beta)$ is the σ-algebra generated by the finite or countable partition β. This follows from the fact (which is easy to prove using the definition of conditional expectations) that if β is finite or countable, then for any $f \in L^1(X, \mathcal{F}, \mu)$, one has

$$\mathbb{E}_\mu(f|\sigma(\beta)) = \sum_{B \in \beta} 1_B \frac{1}{\mu(B)} \int_B f d\mu.$$

Clearly, $H_\mu(\alpha|\beta) = \int_X I_{\alpha|\beta} \, d\mu$.

Exercise 9.4.1. Let α and β be finite or countable partitions of X. Show that $I_{\alpha \vee \beta} = I_\alpha + I_{\beta|\alpha}$.

Now suppose $T : X \to X$ is a measure preserving transformation on $(X, \mathcal{F}, \mu)$, and let $\alpha = \{A_1, A_2, \ldots\}$ be any countable partition. Since $T^{-1}\alpha = \{T^{-1}A_1, T^{-1}A_2, \ldots\}$ is also a countable partition and T is measure preserving one has,

$$\begin{aligned} I_{T^{-1}\alpha}(x) &= -\sum_{A_i \in \alpha} 1_{T^{-1}A_i}(x) \log \mu(T^{-1}A_i) \\ &= -\sum_{A_i \in \alpha} 1_{A_i}(Tx) \log \mu(A_i) = I_\alpha(Tx). \end{aligned} \tag{9.4}$$

Furthermore,

$$\lim_{n\to\infty} \frac{1}{n+1} H\left(\bigvee_{i=0}^{n} T^{-i}\alpha\right) = \lim_{n\to\infty} \frac{1}{n+1} \int_X I_{\bigvee_{i=0}^{n} T^{-i}\alpha}\, d\mu = h_\mu(\alpha, T).$$

The Shannon-McMillan-Breiman Theorem, named after the contributions of C. Shannon ([59], 1948), B. McMillan ([41], 1953) and L. Breiman ([9], 1957 with a correction in 1960), says that if T is ergodic and if α has finite entropy, then in fact the integrand $\frac{1}{n+1} I_{\bigvee_{i=0}^{n} T^{-i}\alpha}$ converges a.e. to $h_\mu(\alpha, T)$. Before we proceed we need the following proposition.

Proposition 9.4.1. *Let $\alpha = \{A_1, A_2, \ldots\}$ be a countable partition with finite entropy. For each $n \geq 1$, let $f_n = I_{\alpha | \bigvee_{i=1}^{n} T^{-i}\alpha}$, and let $f^* = \sup_{n\geq 1} f_n$. Then, for each $t \geq 0$ and for each $A \in \alpha$,*

$$\mu\left(\{x \in A : f^*(x) > t\}\right) \leq 2^{-t}.$$

Furthermore, $f^* \in L^1(X, \mathcal{F}, \mu)$.

Proof. Let $t \geq 0$ and $A \in \alpha$. For $n \geq 1$, let

$$f_n^A(x) = -\log \mathbb{E}_\mu \left(1_A | \sigma\Big(\bigcup_{i=1}^{n} T^{-i}\alpha\Big)\right)(x),$$

and

$$B_n = \{x \in X : f_1^A(x) \leq t, \ldots, f_{n-1}^A(x) \leq t, f_n^A(x) > t\}.$$

Notice that for $x \in A$ one has $f_n(x) = f_n^A(x)$, and for $x \in B_n$ one has $\mathbb{E}_\mu\left(1_A | \sigma(\bigcup_{i=1}^{n} T^{-i}\alpha)\right)(x) < 2^{-t}$. Since $B_n \in \sigma\left(\bigcup_{i=1}^{n} T^{-i}\alpha\right)$, then

$$\begin{aligned} \mu(B_n \cap A) &= \int_{B_n} 1_A \, d\mu \\ &= \int_{B_n} \mathbb{E}_\mu \left(1_A | \sigma\Big(\bigcup_{i=1}^{n} T^{-i}\alpha\Big)\right) d\mu \\ &\leq \int_{B_n} 2^{-t} \, d\mu(x) = 2^{-t}\mu(B_n). \end{aligned}$$

Thus,

$$\begin{aligned}
\mu(\{x \in A : f^*(x) > t\}) &= \mu(\{x \in A : f_n(x) > t, \text{ for some } n\}) \\
&= \mu\left(\{x \in A : f_n^A(x) > t, \text{ for some } n\}\right) \\
&= \mu\left(\bigcup_{n=1}^{\infty} A \cap B_n\right) \\
&= \sum_{n \geq 1} \mu(A \cap B_n) \\
&\leq 2^{-t} \sum_{n \geq 1} \mu(B_n) \leq 2^{-t}.
\end{aligned}$$

We now show that $f^* \in L^1(X, \mathcal{F}, \mu)$. First notice that

$$\mu(\{x \in A : f^*(x) > t\}) \leq \mu(A),$$

hence,

$$\mu(\{x \in A : f^*(x) > t\}) \leq \min(\mu(A), 2^{-t}).$$

Using Fubini's Theorem, and the fact that $f^* \geq 0$ one has

$$\begin{aligned}
\int_X f^* \, d\mu &= \int_0^\infty \mu(\{x \in X : f^*(x) > t\}) \, dt \\
&= \int_0^\infty \sum_{A \in \alpha} \mu(\{x \in A : f^*(x) > t\}) \, dt \\
&= \sum_{A \in \alpha} \int_0^\infty \mu(\{x \in A : f^*(x) > t\}) \, dt \\
&\leq \sum_{A \in \alpha} \int_0^\infty \min(\mu(A), 2^{-t}) \, dt \\
&= \sum_{A \in \alpha} \int_0^{-\log \mu(A)} \mu(A) \, dt + \sum_{A \in \alpha} \int_{-\log \mu(A)}^\infty 2^{-t} \, dt \\
&= -\sum_{A \in \alpha} \mu(A) \log \mu(A) + \sum_{A \in \alpha} \frac{\mu(A)}{\ln 2} \\
&= H_\mu(\alpha) + \frac{1}{\ln 2} < \infty.
\end{aligned}$$

□

So far we have defined the notion of conditional information $I_{\alpha|\beta}$ when α and β are countable partitions. We can generalize the definition to the case α is a countable partition and $\mathcal{G}$ is a σ-algebra by setting (see (9.3)),

$$I_{\alpha|\mathcal{G}}(x) = -\log \mathbb{E}_\mu(1_{\alpha(x)}|\mathcal{G}).$$

Then

$$I_{\alpha|\sigma(\bigcup_{i=1}^{\infty} T^{-i}\alpha)}(x) = \lim_{n\to\infty} I_{\alpha|\sigma(\bigcup_{i=1}^{n} T^{-i}\alpha)}(x). \tag{9.5}$$

Exercise 9.4.2. Give a proof of (9.5) using the Martingale Convergence Theorem, see Theorem 12.4.7.

Theorem 9.4.1 (Shannon-McMillan-Breiman Theorem). *Suppose T is an ergodic measure preserving transformation on a probability space $(X, \mathcal{F}, \mu)$, and let α be a countable partition with $H_\mu(\alpha) < \infty$. Then,*

$$\lim_{n\to\infty} \frac{1}{n+1} I_{\bigvee_{i=0}^{n} T^{-i}\alpha}(x) = h_\mu(\alpha, T) \quad \mu - a.e.$$

Proof. For each $n \leq 1$, let $f_n(x) = I_{\alpha|\bigvee_{i=1}^{n} T^{-i}\alpha}(x)$. Using (9.4) and the fact that for any two partitions β and γ, $I_{\beta\vee\gamma}(x) = I_\beta(x) + I_{\gamma|\beta}(x)$, we obtain

$$\begin{aligned}
I_{\bigvee_{i=0}^{n} T^{-i}\alpha}(x) &= I_{\bigvee_{i=1}^{n} T^{-i}\alpha}(x) + I_{\alpha|\bigvee_{i=1}^{n} T^{-i}\alpha}(x)\\
&= I_{\bigvee_{i=0}^{n-1} T^{-i}\alpha}(Tx) + f_n(x)\\
&= I_{\bigvee_{i=1}^{n-1} T^{-i}\alpha}(Tx) + I_{\alpha|\bigvee_{i=1}^{n-1} T^{-i}\alpha}(Tx) + f_n(x)\\
&= I_{\bigvee_{i=0}^{n-2} T^{-i}\alpha}(T^2x) + f_{n-1}(Tx) + f_n(x)\\
&\ \ \vdots\\
&= I_\alpha(T^n x) + f_1(T^{n-1}x) + \cdots + f_{n-1}(Tx) + f_n(x).
\end{aligned}$$

Let $f(x) = I_{\alpha|\sigma(\bigcup_{i=1}^{\infty} T^{-i}\alpha)}(x) = \lim_{n\to\infty} f_n(x)$. Notice that $f \in L^1(X, \mathcal{F}, \mu)$ since $\int_X f\, d\mu = h_\mu(\alpha, T)$. Now letting $f_0 = I_\alpha$, we have

$$\begin{aligned}
\frac{1}{n+1} I_{\bigvee_{i=0}^{n} T^{-i}\alpha}(x) &= \frac{1}{n+1}\sum_{k=0}^{n} f_{n-k}(T^k x)\\
&= \frac{1}{n+1}\sum_{k=0}^{n} f(T^k x) + \frac{1}{n+1}\sum_{k=0}^{n}(f_{n-k} - f)(T^k x).
\end{aligned}$$

By the Pointwise Ergodic Theorem,

$$\lim_{n\to\infty} \frac{1}{n+1}\sum_{k=0}^{n} f(T^k x) = \int_X f\, d\mu = h_\mu(\alpha, T) \quad \mu - \text{a.e.}$$

We now study the sequence $\left(\frac{1}{n+1}\sum_{k=0}^{n}(f_{n-k}-f)(T^kx)\right)$. Let

$$F_N = \sup_{k\geq N}|f_k - f|, \text{ and } f^* = \sup_{n\geq 1} f_n.$$

Notice that $0 \leq F_N \leq f^* + f$, hence $F_N \in L^1(X,\mathcal{F},\mu)$ and $\lim_{N\to\infty} F_N(x) = 0$ μ-a.e. By the Lebesgue Dominated Convergence Theorem, one has $\lim_{N\to\infty}\int F_N\, d\mu = 0$. Also for any k, $|f_{n-k} - f| \leq f^* + f$, so that $|f_{n-k} - f| \in L^1(X,\mathcal{F},\mu)$ and $\lim_{n\to\infty}|f_{n-k} - f| = 0$ μ-a.e.

For any $N \geq 1$, and for all $n \geq N$ one has

$$\begin{aligned}\frac{1}{n+1}\sum_{k=0}^{n}|f_{n-k}-f|(T^kx) &= \frac{1}{n+1}\sum_{k=0}^{n-N}|f_{n-k}-f|(T^kx)\\ &\quad + \frac{1}{n+1}\sum_{k=n-N+1}^{n}|f_{n-k}-f|(T^kx)\\ &\leq \frac{1}{n+1}\sum_{k=0}^{n-N}F_N(T^kx)\\ &\quad + \frac{1}{n+1}\sum_{k=0}^{N-1}|f_k-f|(T^{n-k}x).\end{aligned}$$

If we take the limit as $n \to \infty$, then by Exercise 3.1.1, the second term tends to 0 μ-a.e. and by the Pointwise Ergodic Theorem the first term tends to $\int F_N\, d\mu$. Now taking the limit as $N \to \infty$, one sees that

$$\frac{1}{n+1}\sum_{k=0}^{n}|f_{n-k}-f|(T^kx) = 0 \quad \mu - \text{a.e.}$$

Hence,

$$\lim_{n\to\infty}\frac{1}{n+1}I_{\bigvee_{i=0}^{n}T^{-i}\alpha}(x) = h_\mu(\alpha,T) \quad \mu - \text{a.e.} \qquad \square$$

The above theorem can be interpreted as providing an estimate of the size of the atoms of $\bigvee_{i=0}^{n}T^{-i}\alpha$. For n sufficiently large, a typical element $A \in \bigvee_{i=0}^{n}T^{-i}\alpha$ satisfies

$$-\frac{1}{n+1}\log\mu(A) \approx h_\mu(\alpha,T)$$

or

$$\mu(A) \approx 2^{-(n+1)h_\mu(\alpha,T)}.$$

Furthermore, if α is a generating partition (i.e., $\sigma\big(\bigcup_{i=0}^{\infty} T^{-i}\alpha\big) = \mathcal{F}$), then in the conclusion of the Shannon-McMillan-Breiman Theorem one can replace $h_\mu(\alpha, T)$ by $h_\mu(T)$.

Exercise 9.4.3. Let $(X, \mathcal{F}, \mu, T)$ be a measure preserving and ergodic dynamical system on a probability space. Suppose α is a finite or countable partition of X with $H_\mu(\alpha) < \infty$. For $x \in X$, let $\alpha_n(x)$ be the element of the partition $\bigvee_{i=0}^{n-1} T^{-i}\alpha$ that contains x. Suppose λ is another probability measure on $(X, \mathcal{F})$ for which there are constants $0 < C_1 < C_2$ such that $C_1\lambda(A) < \mu(A) < C_2\lambda(A)$ for all $A \in \mathcal{F}$. Show that the conclusion of the Shannon-McMillan-Breiman Theorem holds if we replace μ by λ, i.e.,

$$\lim_{n\to\infty} -\frac{\log \lambda(\alpha_n(x))}{n} = h_\mu(\alpha, T) \quad \mu- \text{ a.e. with respect to } \lambda.$$

Exercise 9.4.4. let $T : [0, 1) \to [0, 1)$ be the β-transformation given by $Tx = \beta x \pmod 1$, where $\beta = \frac{1+\sqrt{5}}{2}$ (see also Example 1.3.6). Use the Shannon-McMillan-Breiman Theorem and Exercise 9.4.3 to calculate the entropy $h_\mu(T)$ of T with respect to the invariant measure μ given by

$$\mu(A) = \int_A g\, d\lambda,$$

with

$$g(x) = \begin{cases} \dfrac{5+3\sqrt{5}}{10}, & \text{if } 0 \le x < \dfrac{\sqrt{5}-1}{2}, \\[2ex] \dfrac{5+\sqrt{5}}{10}, & \text{if } \dfrac{\sqrt{5}-1}{2} \le x < 1. \end{cases}$$

Exercise 9.4.5. Use the Shannon-McMillan-Breiman Theorem, the Kolmogorov-Sinai Theorem and Exercise 9.4.3 to show that if T is the Gauss map and μ is Gauss measure (see Example 1.3.7 and Chapter 8), then $h_\mu(T) = \frac{\pi^2}{6\log 2}$.

As a tool for computing metric entropy we give a version of Rohlin's Formula. The result is valid for a large class of dynamical systems. Here we limit ourselves to a version for piecewise monotone interval maps that we state and prove below. The transformations we have in mind are the following.

Definition 9.4.1. A transformation $T : [0, 1) \to [0, 1)$ is called a *number theoretic fibered map* (NTFM) if it satisfies the following two conditions.

(a) There exists a finite or countable interval partition $\alpha = (A_j)_{j\in D}$ such that T restricted to each atom of α (fundamental interval of T) is strictly monotone and continuous. Furthermore, α generates $\mathcal{B}$ and $H_\mu(\alpha) < \infty$.

(b) There exists a T-invariant probability measure μ equivalent to λ and for which two constants $c_1, c_2 > 0$ exist such that $c_1 \le \frac{d\mu}{d\lambda} \le c_2$.

The name comes from the fact that iterations of T assign to each point $x \in [0,1)$ a digit sequence $(a_n(x))$ with digits in D. Since α generates $\mathcal{B}$ these sequences are uniquely determined λ-a.e. We refer to the resulting sequence as the *T-expansion of x*. Almost all known number expansions on $[0,1)$ are generated by an NTFM . Among them are the base N expansions ($Tx = Nx \pmod 1$, where N is a positive integer), β-expansions ($Tx = \beta x \pmod 1$, where $\beta > 1$ is a real number), regular continued fraction expansions ($Tx = \frac{1}{x} \pmod 1$), Lüroth expansions (see Example 1.3.5) and many others, see also [14].

Theorem 9.4.2 (Rohlin's Formula). *Let $T : [0,1) \to [0,1)$ be an NTFM for a partition α. Assume furthermore that for each $j \in D$ the restriction $T_j := T|_{A_j} : A_j \to TA_j$ is a diffeommorphism. Then*

$$h_\mu(T) = \int_{[0,1)} \log|T'|\, d\mu.$$

Proof. Since α is a generating partition, by Theorem 9.2.1 in combination with the Kolmogorov-Sinai Theorem,

$$h_\mu(T) = \lim_{n\to\infty} H_\mu\Big(\alpha|\bigvee_{i=1}^{n} T^{-i}\alpha\Big) = \lim_{n\to\infty} \int_{[0,1)} I_{\alpha|\bigvee_{i=1}^{n} T^{-i}\alpha}\, d\mu.$$

From Proposition 9.4.1 and by setting, as in the proof of the Shannon-McMillan-Breiman Theorem, $f = I_{\alpha|\sigma(\bigcup_{i=1}^{\infty} T^{-i}\alpha)} = I_{\alpha|T^{-1}\mathcal{B}}$, we obtain using the Dominated Convergence Theorem that

$$h_\mu(T) = \int_{[0,1)} f\, d\mu.$$

Note that

$$f = -\sum_{k\in D} 1_{A_k} \log \mathbb{E}_\mu(1_{A_k}|T^{-1}\mathcal{B}). \tag{9.6}$$

From the proof of Theorem 6.2.2 it follows that for any $n \ge 1$ and any $g \in L^1([0,1), \mathcal{B}, \mu)$,

$$P_{T,\mu}^n g \circ T^n = \mathbb{E}_\mu(g|T^{-n}\mathcal{B}), \tag{9.7}$$

where $P_{T,\mu}$ is the Perron-Frobenius operator of T with respect to μ. Even though an NTFM is not necessarily C^2 on the sets A_j and the partition α might be countable, the conditions of the theorem guarantee that the expression for $P_{T,\lambda}$ from (6.8) is still valid for any non-negative integrable function. So we can combine (6.8) with Exercise 6.2.4(a) to obtain for each $j,k \in D$ that

$$\begin{aligned}\mathbb{E}_\mu(1_{A_k}|T^{-1}\mathcal{B}) &= \frac{P_{T,\lambda}(\frac{d\mu}{d\lambda}1_{A_k})}{\frac{d\mu}{d\lambda}} \circ T \\ &= \frac{1}{\frac{d\mu}{d\lambda}\circ T}\sum_{j\in D}\frac{(\frac{d\mu}{d\lambda}\circ T_j^{-1}\circ T)(1_{A_k}\circ T_j^{-1}\circ T)}{|T'\circ T_j^{-1}\circ T|}1_{T^{-1}(TA_k)} \\ &= \frac{1}{\frac{d\mu}{d\lambda}\circ T}\frac{(\frac{d\mu}{d\lambda}\circ T_k^{-1}\circ T)(1_{A_k}\circ T_k^{-1}\circ T)}{|T'\circ T_k^{-1}\circ T|},\end{aligned}$$

where T_j denotes the local inverse of T on A_j. From (9.6) we then see that

$$f = \log\left(\frac{d\mu}{d\lambda}\circ T\right) - \log\left(\frac{d\mu}{d\lambda}\right) + \log|T'|$$

and since μ is T-invariant, Proposition 1.2.2 implies the result. □

Remark 9.4.1. Note that in the proof of Rohlin's Formula we only used that $\frac{d\mu}{d\lambda} \in L^1([0,1),\mathcal{B},\mu)$ and not that $c_1 \le \frac{d\mu}{d\lambda} \le c$ as in condition (b) of the definition of NTFM. This condition is only necessary for Lochs' Theorem in the next section.

Example 9.4.1. Let $\beta \in (1,2)$ and consider the β-transformation $Tx = \beta x \pmod 1$. Set $\alpha = \{[0,\frac{1}{\beta}),[\frac{1}{\beta},1]\}$. From $(T^n)'x = \beta^n$ for any $x \in [0,1]$, it follows that $\lambda(I) \le \frac{1}{\beta^n}$ for any $I \in \bigvee_{i=0}^{n-1} T^{-i}\alpha$. This implies that property (a) from the definition of NTFM holds. Example 6.3.4 contains a formula for the invariant density of an invariant measure μ for T that is equivalent to Lebesgue measure. In fact this density is bounded from above and bounded away from zero. Hence, T has property (b). By Theorem 9.4.2 then

$$h_\mu(T) = \int_{[0,1]} \log|T'|\,d\mu = \log\beta.$$

Exercise 9.4.6. Let $\beta > 1$ satisfy $\beta^3 - \beta^2 - 1 = 0$ and consider the negative β-transformation $T:[0,1]\to[0,1]$ as in Example 5.1.4, so

$$Tx = \begin{cases} 1-\beta x, & \text{if } 0 \le x < \frac{1}{\beta}, \\ 2-\beta x, & \text{if } \frac{1}{\beta} \le x \le 1. \end{cases}$$

(a) Prove that T is measure preserving with respect to the measure μ on $[0,1]$ given by

$$\mu(A) = \frac{1}{C}\int_A (\beta-1)1_{[0,\frac{1}{\beta}-\frac{1}{\beta^2})} + \frac{1}{\beta}1_{[2-\beta,\frac{1}{\beta})} + 1_{[\frac{1}{\beta},1]}\,d\lambda$$

for each $A \in \mathcal{B}$, where C is the normalizing constant.

(b) Prove that $h_\mu(T) = \log\beta$.

9.5 LOCHS' THEOREM

In 1964 in [37], G. Lochs compared decimal and continued fraction expansions of real numbers in $[0,1)$. Let $x \in [0,1)$ be an irrational number, and suppose $x = .d_1d_2\ldots$ is the decimal expansion of x (which is generated by iterating the $\times 10$ map $Sx = 10x \pmod 1$). Suppose further that

$$x = \cfrac{1}{a_1 + \cfrac{1}{a_2 + \cfrac{1}{a_3 + \cfrac{1}{\ddots}}}} = [0; a_1, a_2, \ldots] \tag{9.8}$$

is its regular continued fraction expansion (generated by the Gauss map $Tx = \frac{1}{x} - \pmod 1$)). Let $y = .d_1d_2\cdots d_n$ be the rational number determined by the first n decimal digits of x, and let $z = y + 10^{-n}$. Then, $[y,z)$ is the decimal cylinder of order n containing x, which we also denote by $B_n(x)$. Now let

$$y = \cfrac{1}{b_1 + \cfrac{1}{b_2 + \ddots + \cfrac{1}{b_\ell}}}$$

and

$$z = \cfrac{1}{c_1 + \cfrac{1}{c_2 + \ddots + \cfrac{1}{c_k}}}$$

be the continued fraction expansions of y and z. Let

$$m(n,x) = \max\left\{i \le \min\{\ell,k\} \,:\, b_j = c_j \text{ for all } j \le i\right\}. \tag{9.9}$$

In other words, if $B_n(x)$ denotes the decimal cylinder consisting of all points y in $[0, 1)$ such that the first n decimal digits of y agree with those of x, and if $C_j(x)$ denotes the continued fraction cylinder of order j containing x, i.e., $C_j(x)$ is the set of all points in $[0, 1)$ such that the first j digits in their continued fraction expansion are the same as those of x, then $m(n, x)$ is the largest integer such that $B_n(x) \subset C_{m(n,x)}(x)$. So, $m(n, x)$ is the number of regular continued fraction digits of x that can be determined from knowing the first n decimal digits of x. Lochs proved the following theorem. As before, let λ denote the Lebesgue measure on $[0, 1)$.

Theorem 9.5.1 (Lochs' Theorem). *For λ-a.e. $x \in [0, 1)$,*

$$\lim_{n\to\infty} \frac{m(n,x)}{n} = \frac{6 \log 2 \log 10}{\pi^2}.$$

In this section, we will prove a generalization of Lochs' Theorem that allows one to compare any two known expansions of numbers. We show that Lochs' Theorem is true for any two sequences of interval partitions on $[0, 1)$ satisfying the conclusion of the Shannon-McMillan-Breiman Theorem. The content of this section as well as the proofs can also be found in [13]. We begin with a few definitions that will be used in the arguments to follow.

Let P be an interval partition of $[0, 1)$ (see also Section 6.3). For $x \in [0, 1)$, we let $P(x)$ denote the interval of P containing x.

Definition 9.5.1. Let $\mathcal{P} = (P_n)_{n\geq 1}$ be a sequence of interval partitions of $[0, 1)$ and let $c \geq 0$. We say that $\mathcal{P}$ *has entropy c a.e. with respect to* λ if

$$\lim_{n\to\infty} -\frac{\log \lambda(P_n(x))}{n} = c \quad \lambda - \text{a.e.}$$

Note that we do not assume that each P_n is refined by P_{n+1}. Suppose that $\mathcal{P} = (P_n)_{n\geq 1}$ and $\mathcal{Q} = (Q_n)_{n\geq 1}$ are two sequences of interval partitions of $[0, 1)$. For each $n \in \mathbb{N}$ and $x \in [0, 1)$, define

$$m_{\mathcal{P},\mathcal{Q}}(n, x) = \sup \{m : P_n(x) \subseteq Q_m(x)\}.$$

The following result is the main ingredient of the proof of the generalization of Lochs' Theorem that we present below.

Theorem 9.5.2. *Let $\mathcal{P} = (P_n)_{n\geq 1}$ and $\mathcal{Q} = (Q_n)_{n\geq 1}$ be two sequences of interval partitions of $[0, 1)$. Suppose that for some constants $c > 0$*

and $d > 0$, $\mathcal{P}$ has entropy c a.e. with respect to λ and $\mathcal{Q}$ has entropy d a.e. with respect to λ. Then

$$\lim_{n\to\infty} \frac{m_{\mathcal{P},\mathcal{Q}}(n,x)}{n} = \frac{c}{d} \quad \lambda \text{ a.e.}$$

Proof. First we show that

$$\limsup_{n\to\infty} \frac{m_{\mathcal{P},\mathcal{Q}}(n,x)}{n} \leq \frac{c}{d} \quad \lambda-\text{ a.e.}$$

Fix $\varepsilon > 0$. Let $x \in [0,1)$ be a point at which the convergence conditions of the hypotheses are met. Fix $\eta > 0$ so that $\dfrac{c+\eta}{c - \frac{c}{d}\eta} < 1 + \varepsilon$. Choose N so that for all $n \geq N$

$$\lambda(P_n(x)) > 2^{-n(c+\eta)}$$

and

$$\lambda(Q_n(x)) < 2^{-n(d-\eta)}.$$

Fix n so that $\min\left\{n, \frac{c}{d}n\right\} \geq N$, and let m' denote any integer greater than $(1+\varepsilon)\,\frac{c}{d}n$. By the choice of η,

$$\lambda(P_n(x)) > \lambda(Q_{m'}(x))$$

so that $P_n(x)$ is not contained in $Q_{m'}(x)$. Therefore

$$m_{\mathcal{P},\mathcal{Q}}(n,x) \leq (1+\varepsilon)\frac{c}{d}n$$

and so

$$\limsup_{n\to\infty} \frac{m_{\mathcal{P},\mathcal{Q}}(n,x)}{n} \leq (1+\varepsilon)\frac{c}{d} \quad \lambda-\text{a.e.}$$

Since $\varepsilon > 0$ was arbitrary, we have the desired result.

Now we show that

$$\liminf_{n\to\infty} \frac{m_{\mathcal{P},\mathcal{Q}}(n,x)}{n} \geq \frac{c}{d} \quad \lambda-\text{ a.e.}$$

Fix $\varepsilon \in (0,1)$. Choose $\eta > 0$ so that $\zeta := \varepsilon c - \eta\left(1 + (1-\varepsilon)\frac{c}{d}\right) > 0$. For each $n \in \mathbb{N}$ let $\bar{m}(n) = \left\lfloor (1-\varepsilon)\frac{c}{d}n \right\rfloor$. For brevity, for each $n \in \mathbb{N}$ we call an element of P_n (respectively Q_n) (n,η)-good if

$$\lambda(P_n(x)) < 2^{-n(c-\eta)}$$

(respectively

$$\lambda(Q_n(x)) > 2^{-n(d+\eta)}).$$

For each $n \in \mathbb{N}$, let

$$D_n(\eta) = \left\{ x : \begin{array}{c} P_n(x) \text{ is } (n,\eta)\text{-good and } Q_{\bar{m}(n)}(x) \text{ is } (\bar{m}(n),\eta)\text{-good} \\ \text{and } P_n(x) \not\subseteq Q_{\bar{m}(n)}(x) \end{array} \right\}.$$

If $x \in D_n(\eta)$, then $P_n(x)$ contains an endpoint of the $(\bar{m}(n),\eta)$-good interval $Q_{\bar{m}(n)}(x)$. By the definition of $D_n(\eta)$ and $\bar{m}(n)$,

$$\frac{\lambda(P_n(x))}{\lambda(Q_{\bar{m}(n)}(x))} < 2^{-n\zeta}.$$

Since no more than one atom of P_n can contain a particular endpoint of an atom of $Q_{\bar{m}(n)}$, we see that

$$\lambda(D_n(\eta)) \leq \sum_{x \in D_n(\eta)} P_n(x) \leq 2 \sum_{x \in D_n(\eta)} 2^{-n\zeta} Q_{\bar{m}(n)}(x) < 2 \cdot 2^{-n\zeta}$$

and so

$$\sum_{n=1}^{\infty} \lambda(D_n(\eta)) < \infty.$$

By the Borel-Cantelli Lemma, this implies that

$$\lambda(\{x \in [0,1) \, : \, x \in D_n(\eta) \text{ i.o.}\}) = 0.$$

Since $\bar{m}(n)$ goes to infinity as n does, we have shown that for almost every $x \in [0,1)$, there exists an $N \in \mathbb{N}$, so that for all $n \geq N$, $P_n(x)$ is (n,η)-good and $Q_{\bar{m}(n)}(x)$ is $(\bar{m}(n),\eta)$-good and $x \notin D_n(\eta)$. In other words, for almost every $x \in [0,1)$, there exists an $N \in \mathbb{N}$, so that for all $n \geq N$, $P_n(x)$ is (n,η)-good and $Q_{\bar{m}(n)}(x)$ is $(\bar{m}(n),\eta)$-good and $P_n(x) \subset Q_{\bar{m}(n)}(x)$. Thus, for almost every $x \in [0,1)$, there exists an $N \in \mathbb{N}$, so that for all $n \geq N$,

$$m_{\mathcal{P},\mathcal{Q}}(n,x) \geq \bar{m}(n) = \left\lfloor (1-\varepsilon)\frac{c}{d}n \right\rfloor.$$

This proves that

$$\liminf_{n\to\infty} \frac{m_{\mathcal{P},\mathcal{Q}}(n,x)}{n} \geq (1-\varepsilon)\frac{c}{d} \quad \lambda - \text{a.e.}$$

Since $\varepsilon > 0$ was arbitrary, we have established the theorem. □

The above result allows us to compare any two well-known expansions of numbers. Since the *commonly used* expansions are usually performed for points in the unit interval, our underlying space will be $([0,1), \mathcal{B}, \lambda)$, where $\mathcal{B}$ is the Lebesgue σ-algebra, and λ the Lebesgue measure. Recall the definition of an NTFM from Definition 9.4.1. For an NTFM T with corresponding invariant probability measure μ and generating interval partition α, write $\alpha_n = \bigvee_{i=0}^{n-1} T^{-i}\alpha$ for the interval partition into cylinder sets of order n and $\alpha_n(x)$ for the partition element from α_n that contains the point x. The fact that each α_n is an interval partitions follows from property (a). We have the following theorem of which Lochs' Theorem is a specific instance.

Theorem 9.5.3. *Let T and S be two NTFM's on $[0,1)$ with corresponding invariant probability measures μ and ν and generating interval partitions α and β, respectively. Assume that both T and S are ergodic with respect to Lebesgue measure and that $h_\mu(T), h_\nu(S) > 0$. For $n \geq 1$ and $x \in [0,1)$ let*

$$m(n,x) = \sup\{m \geq 1 \,:\, \alpha_n(x) \subseteq \beta_m(x)\}.$$

Then,

$$\lim_{n\to\infty} \frac{m(n,x)}{n} = \frac{h_\mu(T)}{h_\nu(S)} \quad \lambda\text{-a.e.}$$

Proof. By the assumption that $H_\mu(\alpha), H_\nu(\beta) < \infty$ we can apply the Shannon-McMillan-Breiman Theorem to obtain that

$$\lim_{n\to\infty} -\frac{\log \mu(\alpha_n(x))}{n} = h_\mu(T) \quad \mu\text{-a.e.}$$

and similarly

$$\lim_{n\to\infty} -\frac{\log \nu(\beta_n(x))}{n} = h_\nu(S) \quad \nu\text{-a.e.}$$

Since μ and ν are equivalent to λ with bounded density, these statements also hold for λ-almost every $x \in [0,1)$. This means that the sequences of partitions (α_n) and (β_n) satisfy the conditions of Theorem 9.5.2, which gives the result. □

Exercise 9.5.1. Show that the $\times N$ transformation $T_N x = Nx \pmod 1$, $N \geq 2$ an integer, is an NTFM. From Theorem 9.5.1 one can deduce that typically regular continued fraction digits give us slightly more information on the precise value of a number x than its decimal digits. How large should we choose N so that the information obtained from knowing digits in base N is typically larger than the information given by the same number of regular continued fraction digits?

CHAPTER 10

The Variational Principle

Dynamical systems can have many different invariant probability measures each with their own metric entropy. Since entropy reflects the average amount of information gained by applications of the transformation T, one could wonder whether it is possible to find measures that maximize this amount. The search for such maximal measures is simplified when T is a continuous map on a compact metric space (X, d) and therefore, as in Chapter 7, we choose this setup here.

For continuous maps $T : X \to X$ one can define a topological analogue of the metric entropy by replacing the measurable partitions in the definition of metric entropy by open covers. The first section below includes two equivalent definitions of the notion of topological entropy. In the second section we prove the Variational Principle, which establishes a powerful relationship between topological entropy and metric entropy. In the last section we discuss measures for which the metric entropy is as large as possible.

10.1 TOPOLOGICAL ENTROPY

The definition of topological entropy comes in two flavors. Topological entropy was first introduced by R. L. Adler, A. G. Konheim and M. H. McAndrew in 1965 in [2] as an analogue of the successful concept of metric entropy. Their definition is in terms of open covers and is very similar to the definition of metric entropy from Chapter 9. It requires the space X to be a compact metric space. The second (and chronologically later) definition was first investigated by R. Bowen in [7] and E. I. Dinaburg in [15]. It uses (n, ε)-separating and spanning sets and only requires X to be a metric space. To ease the exposition we let (X, d) be a compact metric space in both cases and as we shall see both concepts then become equivalent.

Let (X, d) be a compact metric space and $T : X \to X$ a continuous map. An *open cover* of X, is a collection α of open subsets of X such that $X \subseteq \bigcup_{A \in \alpha} A$. Recall the definitions of refinement and common refinement of collections of sets from Definition 9.2.1.

Exercise 10.1.1. For a finite collection $(\alpha_i)_{i=1}^n$ of open covers of X and a continuous transformation $T : X \to X$ show that

$$\bigvee_{i=1}^{n} \alpha_i = \Big\{ \bigcap_{i=1}^{n} A_{j_i} \,:\, A_{j_i} \in \alpha_i \Big\}$$

and

$$T^{-1}\alpha = \{T^{-1}A \,:\, A \in \alpha\}$$

are again open covers of X. Moreover, show that $T^{-1}(\alpha \vee \beta) = T^{-1}(\alpha) \vee T^{-1}(\beta)$ and that $\alpha \leq \beta$ implies $T^{-1}\alpha \leq T^{-1}\beta$ (c.f. Exercise 9.2.1).

Let α be an open cover of X. The *diameter* of α is given by

$$diam_d(\alpha) := \sup_{A \in \alpha} diam_d(A) = \sup_{A \in \alpha} \sup_{x,y \in A} d(x, y).$$

By the compactness of X any open cover has a finite subcover. Let $N(\alpha)$ be the number of sets in a finite subcover of α of minimal cardinality and define the *entropy* of α to be $H_{top}(\alpha) = \log(N(\alpha))$. The following proposition summarizes some easy properties of $H_{top}(\alpha)$. The proof is left as an exercise.

Proposition 10.1.1. *Let α be an open cover of X. Then the following hold.*

(i) *$H_{top}(\alpha) \geq 0$ and $H_{top}(\alpha) = 0$ if and only if $N(\alpha) = 1$ if and only if $X \in \alpha$.*

(ii) *If $\alpha \leq \beta$, then $H_{top}(\alpha) \leq H_{top}(\beta)$.*

(iii) *$H_{top}(\alpha \vee \beta) \leq H_{top}(\alpha) + H_{top}(\beta)$.*

(iv) *For $T : X \to X$ continuous we have $H_{top}(T^{-1}\alpha) \leq H_{top}(\alpha)$. If T is surjective, then $H_{top}(T^{-1}\alpha) = H_{top}(\alpha)$.*

Exercise 10.1.2. Prove Proposition 10.1.1.

The topological entropy of T with respect to the open cover α is then defined as

$$h_{top}(\alpha, T) = \lim_{n\to\infty} \frac{1}{n} H_{top}\Big(\bigvee_{i=0}^{n-1} T^{-i}\alpha\Big).$$

The existence of the limit on the right-hand side follows as for the metric entropy in Proposition 9.2.3 using the subadditivity of the sequence $(H_{top}(\bigvee_{i=0}^{n-1} T^{-i}\alpha))$.

Exercise 10.1.3. Prove that $\lim_{n\to\infty} \frac{1}{n} H_{top}(\bigvee_{i=1}^{n-1} T^{-i}\alpha)$ exists.

Proposition 10.1.2. *The topological entropy $h_{top}(\alpha, T)$ of a continuous transformation T with respect to an open cover α satisfies the following properties.*

(i) $h_{top}(\alpha, T) \geq 0$.

(ii) *If* $\alpha \leq \beta$, *then* $h_{top}(\alpha, T) \leq h_{top}(\beta, T)$.

(iii) $h_{top}(\alpha, T) \leq H_{top}(\alpha)$.

Proof. These statements are easy consequences of Proposition 10.1.1. For example, for the third statement it follows from Proposition 10.1.1(iii) and (iv) that

$$H_{top}\Big(\bigvee_{i=0}^{n-1} T^{-i}\alpha\Big) \leq \sum_{i=0}^{n-1} H_{top}(T^{-i}\alpha) \leq nH_{top}(\alpha). \qquad \square$$

This brings us to the first definition of topological entropy.

Definition 10.1.1. Let $T : X \to X$ be a continuous transformation on a compact metric space (X, d). The *topological entropy* of T is

$$h_1(T) = \sup_{\alpha} h_{top}(\alpha, T),$$

where the supremum is taken over all open covers α of X.

The next result states that the topological entropy can be obtained from a suitable sequence of open covers.

Lemma 10.1.1. *Let (X, d) be a compact metric space. Suppose $(\alpha_n)_{n\geq 1}$ is a sequence of open covers of X such that* $\lim_{n\to\infty} diam_d(\alpha_n) = 0$. *Then* $\lim_{n\to\infty} h_{top}(\alpha_n, T) = h_1(T)$.

Proof. We will only prove the lemma for $h_1(T) < \infty$. Let $\varepsilon > 0$ be arbitrary and let β be an open cover of X such that $h_{top}(\beta, T) > h_1(T) - \varepsilon$. By Theorem 12.1.1(vi) there exists a Lebesgue number $\delta > 0$ for β and by assumption there exists an $N > 0$ such that $diam_d(\alpha_n) < \delta$ for $n \geq N$. So if $n \geq N$, then for any $A \in \alpha_n$ there exists a $B \in \beta$ such that $A \subseteq B$. In other words, $\beta \leq \alpha_n$. It follows from Proposition 10.1.2(ii) that

$$h_1(T) - \varepsilon < h_{top}(\beta, T) \leq h_{top}(\alpha_n, T) \leq h_1(T)$$

for all $n \geq N$ and the result follows. □

Exercise 10.1.4. Finish the proof of Lemma 10.1.1 by showing that if $h_1(T) = \infty$, then $\lim_{n\to\infty} h_{top}(\alpha_n, T) = \infty$.

Exercise 10.1.5. Let $T : X \to X$ be a continuous transformation on a compact metric space (X, d). For any subset $A \subseteq X$ and any open cover β let $N(A, \beta)$ denote the minimal cardinality that any subset of β covering A can have.

(a) Fix two open covers α and β of X. Prove that

$$N\Big(\bigvee_{i=0}^{n-1} T^{-i}\beta\Big) \leq N\Big(\bigvee_{i=0}^{n-1} T^{-i}\alpha\Big) + \max_{A\in\bigvee_{i=0}^{n-1} T^{-i}\alpha} N\Big(A, \bigvee_{i=0}^{n-1} T^{-i}\beta\Big).$$

(b) Deduce that for any open cover α of X,

$$h_{top}(T) \leq h_{top}(\alpha, T) + \sup_{\beta} \lim_{n\to\infty} \frac{1}{n} \log\Big(\max_{A\in\bigvee_{i=0}^{n-1} T^{-i}\alpha} N\Big(A, \bigvee_{i=0}^{n-1} T^{-i}\beta\Big)\Big),$$

where the supremum is taken over all open covers of X.

Let us now turn to the second definition of topological entropy. This approach was first explored by R. Bowen and E. I. Dinaburg in 1971 and is based on measuring the exponential growth rate of the number of essentially different initial parts of orbits. For each $n \geq 1$ define a new metric d_n on X by setting

$$d_n(x, y) = \max_{0\leq i\leq n-1} d(T^i x, T^i y).$$

Hence, d_n measures the maximal distance between the first n elements in the orbits of x and y.

For $n \geq 1$ and $\varepsilon > 0$ we say that a collection α of open subsets of X is (n, ε)-*covering* if α is an open cover of X and $diam_{d_n}(A) < \varepsilon$ for each $A \in \alpha$. Let $Co(n, \varepsilon, T)$ be the minimal cardinality that any covering of X by open sets of d_n-diameter less than ε can have. The compactness of X implies that $Co(n, \varepsilon, T) < \infty$. Moreover, if $0 < \varepsilon_1 < \varepsilon_2$, then $Co(n, \varepsilon_1, T) \geq Co(n, \varepsilon_2, T)$. So $Co(n, \varepsilon, T)$ decreases in ε.

Lemma 10.1.2. *For any $\varepsilon > 0$ the limit $\lim_{n\to\infty} \frac{1}{n} \log Co(n, \varepsilon, T)$ exists and is finite.*

Proof. Fix $n, m \geq 1$ and $\varepsilon > 0$. Let α and β be two open covers of X consisting of sets of d_m-diameter and d_n-diameter smaller than ε and with cardinalities $Co(m, \varepsilon, T)$ and $Co(n, \varepsilon, T)$, respectively. Pick any $A \in \alpha$ and $B \in \beta$. Then, for $x, y \in A \cap T^{-m}B$,

$$\begin{aligned} d_{m+n}(x, y) &= \max_{0\leq i\leq m+n-1} d(T^i x, T^i y) \\ &= \max\Big\{ \max_{0\leq i\leq m-1} d(T^i x, T^i y), \max_{m\leq j\leq m+n-1} d(T^j x, T^j y)\Big\} \\ &< \varepsilon. \end{aligned}$$

So $\alpha \vee T^{-m}\beta$ is an open cover of X of d_{m+n}-diameter less than ε. Moreover, the cardinality of $\alpha \vee T^{-m}\beta$ is at most $Co(m, \varepsilon, T) \cdot Co(n, \varepsilon, T)$. Hence,

$$Co(m + n, \varepsilon, T) \leq Co(m, \varepsilon, T) \cdot Co(n, \varepsilon, T).$$

Since log is an increasing function, the sequence (a_n) defined by $a_n = \log Co(n, \varepsilon, T)$ is subadditive and the result follows from Proposition 9.2.2. □

From the monotonicity of $Co(n, \varepsilon, T)$ in ε it follows that

$$h_2(T) := \lim_{\varepsilon\downarrow 0} \lim_{n\to\infty} \frac{1}{n} \log Co(n, \varepsilon, T)$$

exists, although it may be infinite. We will take this as our second definition of topological entropy. Before we prove that $h_1(T) = h_2(T)$, we give two alternative formulations of $h_2(T)$.

For $n \geq 1$ and $\varepsilon > 0$ a subset $A \subseteq X$ is called (n, ε)-*spanning* for X if for all $x \in X$ there exists a $y \in A$ such that $d_n(x, y) < \varepsilon$. Let $Sp(n, \varepsilon, T)$ denote the minimal cardinality that any (n, ε)-spanning set can have. The fact that $Sp(n, \varepsilon, T) < \infty$ follows from the compactness of X, so this minimal cardinality is well defined. A subset $A \subseteq X$ is called (n, ε)-*separated* if any $x, y \in A$ with $x \neq y$ satisfy $d_n(x, y) \geq \varepsilon$.

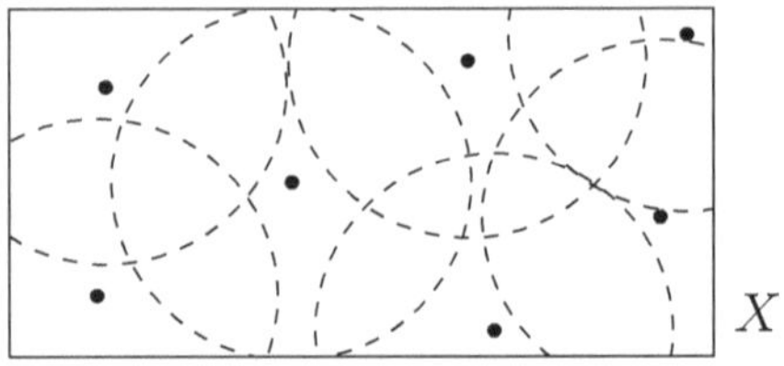

Figure 10.1 The dots indicate a subset $A \subseteq X$ and the dashed circles d_1-balls of radius ε. A is $(1, \varepsilon)$-spanning and $(1, \varepsilon)$-separated.

We define $Se(n, \varepsilon, T)$ to be the maximal cardinality that any (n, ε)-separated set can have. The fact that $Se(n, \varepsilon, T)$ is finite also follows from compactness, see Exercise 10.1.6. Hence, an (n, ε)-separated set of maximal cardinality exists. Both quantities $Sp(n, \varepsilon, T), Se(n, \varepsilon, T)$ are decreasing in ε. Figure 10.1 illustrates both concepts.

Exercise 10.1.6. Prove that $Se(n, \varepsilon, T) < \infty$.

We can also formulate the above in terms of open balls. If we use $B(x, r) = \{y \in X : d(x, y) < r\}$ to denote an open ball in the metric d, then the open ball with centre x and radius r in the metric d_n is given by

$$B_n(x, r) := \bigcap_{i=0}^{n-1} T^{-i} B(T^i x, r). \tag{10.1}$$

Hence, A is (n, ε)-spanning for X if

$$X = \bigcup_{a \in A} B_n(a, \varepsilon)$$

and A is (n, ε)-separated if

$$A \cap B_n(a, \varepsilon) = \{a\} \quad \text{for all } a \in A.$$

The following relation holds between $Co(n, \varepsilon, T)$, $Sp(n, \varepsilon, T)$ and $Se(n, \varepsilon, T)$.

Lemma 10.1.3. *For any $n \in \mathbb{Z}_{\geq 0}$ and $\varepsilon > 0$ it holds that*

$$Co(n, 3\varepsilon, T) \leq Sp(n, \varepsilon, T) \leq Se(n, \varepsilon, T) \leq Co(n, \varepsilon, T).$$

Proof. We will only prove the last two inequalities and leave the first one as an exercise to the reader. Let A be an (n,ε)-separated set of cardinality $Se(n,\varepsilon,T)$. Suppose A is not (n,ε)-spanning for X. Then there is some $x \in X$ such that $d_n(x,a) \geq \varepsilon$ for all $a \in A$. But then $A \cup \{x\}$ is an (n,ε)-separated set of cardinality larger than $Se(n,\varepsilon,T)$. This contradiction shows that A is an (n,ε)-spanning set for X. The second inequality now follows since the cardinality of A is at least as large as $Sp(n,\varepsilon,T)$.

To prove the third inequality, let A be an (n,ε)-separated set of cardinality $Se(n,\varepsilon,T)$. Note that if α is an open cover of d_n-diameter less than ε, then no element of α can contain more than one element of A. This holds in particular for an open cover of minimal cardinality, so the third inequality is proved. □

Exercise 10.1.7. Finish the proof of the above lemma by proving the first inequality.

From Lemma 10.1.2, the monotonicity of $Sp(n,\varepsilon,T)$ and $Se(n,\varepsilon,T)$ in ε and Lemma 10.1.3 we obtain that

$$\begin{aligned}\lim_{\varepsilon\downarrow 0}\lim_{n\to\infty}\frac{1}{n}\log Co(n,\varepsilon,T) &= \lim_{\varepsilon\downarrow 0}\liminf_{n\to\infty}\frac{1}{n}\log Sp(n,\varepsilon,T)\\ &= \lim_{\varepsilon\downarrow 0}\limsup_{n\to\infty}\frac{1}{n}\log Sp(n,\varepsilon,T)\end{aligned}$$

and

$$\begin{aligned}\lim_{\varepsilon\downarrow 0}\lim_{n\to\infty}\frac{1}{n}\log Co(n,\varepsilon,T) &= \lim_{\varepsilon\downarrow 0}\liminf_{n\to\infty}\frac{1}{n}\log Se(n,\varepsilon,T)\\ &= \lim_{\varepsilon\downarrow 0}\limsup_{n\to\infty}\frac{1}{n}\log Se(n,\varepsilon,T).\end{aligned}$$

This motivates the following definition.

Definition 10.1.2. The topological entropy of a continuous transformation $T: X \to X$ on a compact metric space (X,d) is given by

$$\begin{aligned}h_2(T) &= \lim_{\varepsilon\downarrow 0}\lim_{n\to\infty}\frac{1}{n}\log Co(n,\varepsilon,T)\\ &= \lim_{\varepsilon\downarrow 0}\liminf_{n\to\infty}\frac{1}{n}\log Sp(n,\varepsilon,T) = \lim_{\varepsilon\downarrow 0}\limsup_{n\to\infty}\frac{1}{n}\log Sp(n,\varepsilon,T)\\ &= \lim_{\varepsilon\downarrow 0}\liminf_{n\to\infty}\frac{1}{n}\log Se(n,\varepsilon,T) = \lim_{\varepsilon\downarrow 0}\limsup_{n\to\infty}\frac{1}{n}\log Se(n,\varepsilon,T).\end{aligned}$$

Note that there seems to be a concealed ambiguity in this definition, since $h_2(T)$ depends on the chosen metric d. Further on we see that from the compactness of (X, d) it follows that any metric that is topologically equivalent to d (induces the same topology) gives the same value for $h_2(T)$. First we show that Definitions 10.1.1 and 10.1.2 of $h_1(T)$ and $h_2(T)$ agree, justifying the use of $h_{top}(T)$ for both.

Theorem 10.1.1. *If $T : X \to X$ is a continuous transformation on a compact metric space (X, d), then $h_1(T) = h_2(T) =: h_{top}(T)$.*

Proof. We first show that $h_2(T) \leq h_1(T)$. Let (α_k) be the sequence of open covers of X defined by $\alpha_k = (B(x, \frac{1}{3k}) : x \in X)$. Then $diam_d(\alpha_k) = \frac{2}{3k} \to 0$ as $k \to \infty$, so by Lemma 10.1.1,

$$h_1(T) = \lim_{k\to\infty} h_{top}(\alpha_k, T) = \lim_{n\to\infty} \frac{1}{n} \log N\Big(\bigvee_{i=0}^{n-1} T^{-i}\alpha_k\Big).$$

If we take x, y in the same element of $\bigvee_{i=0}^{n-1} T^{-i}\alpha_k$, then for each $0 \leq i \leq n-1$ the points $T^i x$ and $T^i y$ are in the same ball of radius $\frac{1}{3k}$. Thus $d_n(x, y) < \frac{2}{3k} < \frac{1}{k}$, showing that the d_n-diameter of the open cover $\bigvee_{i=0}^{n-1} T^{-i}\alpha_k$ is less than $\frac{1}{k}$. Hence, $Co(n, \frac{1}{k}, T)$ is upper bounded by the cardinality of any subcover of $\bigvee_{i=0}^{n-1} T^{-i}\alpha_k$, so

$$Co\Big(n, \frac{1}{k}, T\Big) \leq N\Big(\bigvee_{i=0}^{n-1} T^{-i}\alpha_k\Big).$$

Subsequently taking the log, dividing by n, taking the limit for $n \to \infty$ and the limit for $k \to \infty$ gives the result.

To prove that $h_1(T) \leq h_2(T)$, let $(\beta_k)_{k=1}^{\infty}$ be the sequence of open covers of X with $\beta_k = (B(x, \frac{1}{k}) : x \in X)$. Then $diam_d(\beta_k) \to 0$ as $k \to \infty$. Fix an $n \geq 1$ and let $A \subseteq X$ be any $(n, \frac{1}{k})$-spanning set for X with cardinality $Sp(n, \frac{1}{k}, T)$. For any $a \in A$ and any $0 \leq i \leq n-1$ the open ball $B(T^i a, \frac{1}{k})$ is an element of β_k. Using the notation from (10.1), we have for each $a \in A$ that

$$\begin{aligned} B_n\Big(a, \frac{1}{k}\Big) &= B\Big(a, \frac{1}{k}\Big) \cap T^{-1}B\Big(Ta, \frac{1}{k}\Big) \cap \cdots \cap T^{-(n-1)}B\Big(T^{n-1}a, \frac{1}{k}\Big) \\ &\in \bigvee_{i=0}^{n-1} T^{-1}\beta_k. \end{aligned}$$

Since A is $(n, \frac{1}{k})$-spanning, we get

$$X \subseteq \bigcup_{a\in A} B_n\Big(a, \frac{1}{k}\Big),$$

so $\{B_n(a, \frac{1}{k}) : a \in A\}$ is a finite subcover of $\bigvee_{i=0}^{n-1} T^{-1}\beta_k$ of cardinality $Sp(n, \frac{1}{k}, T)$. Hence, $N(\bigvee_{i=0}^{n-1} T^{-1}\beta_k) \leq Sp(n, \frac{1}{k}, T)$, which gives

$$\lim_{n\to\infty} \frac{1}{n} \log N\Big(\bigvee_{i=0}^{n-1} T^{-1}\beta_k\Big) \leq \limsup_{n\to\infty} \frac{1}{n} \log Sp\Big(n, \frac{1}{k}, T\Big).$$

Using Lemma 10.1.1 we obtain,

$$\begin{aligned} h_1(T) &= \lim_{k\to\infty} \lim_{n\to\infty} \frac{1}{n} \log N\Big(\bigvee_{i=0}^{n-1} T^{-1}\beta_k\Big) \\ &\leq \lim_{k\to\infty} \limsup_{n\to\infty} \frac{1}{n} \log Sp\Big(n, \frac{1}{k}, T\Big) = h_2(T). \end{aligned}$$

Hence, $h_1(T) = h_2(T)$. □

The following theorem lists some elementary properties of the topological entropy.

Theorem 10.1.2. *Let (X, d) be a compact metric space and $T : X \to X$ continuous. Then the following hold.*

(i) *If a metric ρ generates the same topology on X as d, then $h_{top}(T)$ is the same under both metrics.*

(ii) *$h_{top}(T)$ is a conjugacy invariant.*

(iii) *For each $n \geq 1$ it holds that $h_{top}(T^n) = nh_{top}(T)$.*

(iv) *If T is a homeomorphism, then $h_{top}(T^{-1}) = h_{top}(T)$. Consequently, for each $n \in \mathbb{Z}$ we have $h_{top}(T^n) = |n|h_{top}(T)$.*

Proof. (i) Since d and ρ induce the same topology on X, both identity maps $i : (X, d) \to (X, \rho)$ and $j : (X, \rho) \to (X, d)$ are continuous, so by the compactness of X also uniformly continuous. Let $\varepsilon_1 > 0$. Then by the uniform continuity of j there is an $\varepsilon_2 > 0$ such that for all $x, y \in X$,

$$\rho(x, y) < \varepsilon_2 \Rightarrow d(x, y) < \varepsilon_1,$$

and then by the uniform continuity of i there is an $\varepsilon_3 > 0$ such that for all $x, y \in X$,

$$d(x, y) < \varepsilon_3 \Rightarrow \rho(x, y) < \varepsilon_2.$$

Let A be an (n, ε_2, ρ)-spanning set for T (where we have added the third argument to emphasize the metric). By definition this means that for each $x \in X$ there is an $a \in A$, such that

$$\rho_n(x, a) = \max_{0\leq i\leq n-1} \rho(T^i x, T^i a) < \varepsilon_2.$$

But then

$$d_n(x, a) = \max_{0 \le i \le n-1} d(T^i x, T^i a) < \varepsilon_1,$$

so A is an (n, ε_1, d)-spanning set. Hence, $Sp(n, \varepsilon_1, T, d) \le Sp(n, \varepsilon_2, T, \rho)$. Similarly we obtain $Sp(n, \varepsilon_2, T, \rho) \le Sp(n, \varepsilon_3, T, d)$, yielding

$$\begin{aligned} \liminf_{n\to\infty} \frac{1}{n} \log Sp(n, \varepsilon_1, T, d) &\le \liminf_{n\to\infty} \frac{1}{n} \log Sp(n, \varepsilon_2, T, \rho) \\ &\le \liminf_{n\to\infty} \frac{1}{n} \log Sp(n, \varepsilon_3, T, d). \end{aligned}$$

If we now let $\varepsilon_3 \downarrow 0$, $\varepsilon_2 \downarrow 0$ and then $\varepsilon_1 \downarrow 0$ we get

$$h_{top,d}(T) \le h_{top,\rho}(T) \le h_{top,d}(T).$$

(ii) Let (Y, ρ) be a compact metric space and $S : Y \to Y$ continuous and topologically conjugate to T with conjugacy $\psi : Y \to X$. Define another metric $\tilde{d}$ on Y by setting $\tilde{d}(x, y) = d(\psi(x), \psi(y))$. We claim that ρ and $\tilde{d}$ induce the same topology on Y. To see this, let $y \in Y$ and $\varepsilon > 0$ be arbitrary. Since ψ is a homeomorphism, the image $\psi(B_\rho(y, \varepsilon))$ of the open ball $B_\rho(y, \varepsilon)$ is an open set in X and thus contains an open ball $B_d(\psi(y), \tilde{\varepsilon})$ for some $\tilde{\varepsilon} > 0$. Then $\psi^{-1}(B_d(\psi(y), \tilde{\varepsilon})) \subseteq B_\rho(y, \varepsilon)$. Note that

$$z \in \psi^{-1}(B_d(\psi(y), \tilde{\varepsilon})) \Leftrightarrow d(\psi(z), \psi(y)) < \tilde{\varepsilon} \Leftrightarrow z \in B_{\tilde{d}}(y, \tilde{\varepsilon}).$$

Hence, $B_{\tilde{d}}(y, \tilde{\varepsilon}) \subseteq B_\rho(y, \varepsilon)$. Similarly we can show that any open ball $B_{\tilde{d}}(y, \varepsilon)$ contains an open ball $B_\rho(y, \tilde{\varepsilon})$, which gives the claim.

Let $x_1, x_2 \in X$. Fix an $n \ge 1$ and set $y_1 = \psi^{-1}(x_1)$, $y_2 = \psi^{-1}(x_2)$. Then for each $0 \le i \le n-1$,

$$\begin{aligned} d(T^i x_1, T^i x_2) &= d(T^i \psi(y_1), T^i \psi(y_2)) \\ &= d(\psi(S^i y_1), \psi(S^i y_2)) = \tilde{d}(S^i y_1, S^i y_2). \end{aligned}$$

This means that a collection α of open subsets of X is an (n, ε, d)-cover of X if and only if the collection $\psi(\alpha)$ is an $(n, \varepsilon, \tilde{d})$-cover of Y. Hence, $Co(n, \varepsilon, T, d) = Co(n, \varepsilon, S, \tilde{d})$ and thus $h_{top}(T) = h_{top}(S)$ by part (i).

(iii) Fix an $n \ge 1$. Observe that for each $m \ge 1$ and $x, y \in X$,

$$\begin{aligned} d_m(T^n x, T^n y) &= \max_{0 \le i \le m-1} d(T^{ni} x, T^{ni} y) \\ &\le \max_{0 \le j \le nm-1} d(T^j x, T^j y) = d_{nm}(x, y). \end{aligned}$$

Therefore $Sp(m, \varepsilon, T^n) \leq Sp(nm, \varepsilon, T)$, which implies that

$$\begin{aligned} h_{top}(T^n) &= \lim_{\varepsilon \downarrow 0} \liminf_{m \to \infty} \frac{1}{m} \log Sp(m, \varepsilon, T^n) \\ &\leq n \lim_{\varepsilon \downarrow 0} \limsup_{m \to \infty} \frac{1}{nm} \log Sp(nm, \varepsilon, T) \\ &\leq n \lim_{\varepsilon \downarrow 0} \limsup_{m \to \infty} \frac{1}{m} \log Sp(m, \varepsilon, T) = n h_{top}(T). \end{aligned} \tag{10.2}$$

For the other inequality, let $\varepsilon > 0$. By the uniform continuity of T on X, and thus of T^i for each $0 \leq i \leq n-1$, we can find a $\delta > 0$ such that for all $x, y \in X$ with $d(x, y) < \delta$ we have $d_n(x, y) < \varepsilon$. Let $m \geq 1$ and let A be an (m, δ)-spanning set for T^n. Then, by definition for all $x \in X$ there is an $a \in A$ such that

$$\max_{0 \leq i \leq m-1} d(T^{ni}x, T^{ni}a) < \delta$$

and by the above,

$$\max_{0 \leq j \leq nm-1} d(T^j x, T^j a) = \max_{0 \leq i \leq m-1} \max_{0 \leq k \leq n-1} d(T^{ni+k}x, T^{ni+k}a) < \varepsilon.$$

Thus, $Sp(mn, \varepsilon, T) \leq Sp(m, \delta, T^n)$ and it follows that $h_{top}(T^n) \geq n h_{top}(T)$.

(iv) Let $n \geq 1$ and $\varepsilon > 0$. Let A be an (n, ε)-separated set for T. Then, for any $x, y \in A$ with $x \neq y$ it holds that $d_n(x, y) > \varepsilon$. But then,

$$\max_{0 \leq i \leq n-1} d(T^{-i}(T^{n-1}x), T^{-i}(T^{n-1}y)) = \max_{0 \leq i \leq n-1} d(T^i x, T^i y) > \varepsilon.$$

So $T^{n-1}A$ is an (n, ε)-separated set for T^{-1} of the same cardinality. Conversely, every (n, ε)-separated set B for T^{-1} gives the (n, ε)-separated set $T^{-(n-1)}B$ for T. Thus, $Se(n, \varepsilon, T) = Se(n, \varepsilon, T^{-1})$ and the first statement follows. The second statement is a direct consequence of (iii). □

Exercise 10.1.8. Let (X, d), (Y, ρ) be compact metric spaces and $T : X \to X$ and $S : Y \to Y$ continuous. Show that $h_{top}(T \times S) = h_{top}(T) + h_{top}(S)$. (Hint: first show that the metric $\tilde{d} = max\{d, \rho\}$ induces the product topology on $X \times Y$).

Example 10.1.1. Let (X, d) be a compact metric space and $T : X \to X$ an isometry, so $d(Tx, Ty) = d(x, y)$ for each $x, y \in X$. Then T is obviously continuous and for each $n \geq 1$ we get $d_n(x, y) = d(x, y)$.

Hence, a set A is (n,ε)-covering if and only if it is $(1,\varepsilon)$-covering, giving for each $\varepsilon > 0$ that

$$\lim_{n\to\infty} \frac{1}{n} \log Co(n,\varepsilon,T) = \lim_{n\to\infty} \frac{1}{n} \log Co(1,\varepsilon,T) = 0.$$

This yields $h_{top}(T) = 0$.

In the proof of Proposition 7.3.3 we saw that any rotation $R_\theta : \mathbb{S}^1 \to \mathbb{S}^1$, $z \mapsto e^{2\pi i\theta} z$, $\theta \in [0,1)$, is an isometry with respect to the arc length distance function which induces the topology on $\mathbb{S}^1$. So, $h_{top}(R_\theta) = 0$ for any $\theta \in [0,1)$.

Example 10.1.2. Let $X = \{0,1,\dots,k-1\}^{\mathbb{Z}}$ (or $X = \{0,\dots,k-1\}^{\mathbb{N}}$) and let $T : X \to X$ be the left-shift. Recall from Example 7.1.1 that T is continuous with respect to the metric d from (7.1). There are more ways to put a metric on X.

Exercise 10.1.9. Define the metric ρ on X by

$$\rho(x,y) = \sum_{n=-\infty}^{\infty} \frac{|x_n - y_n|}{2^{|n|}}.$$

Prove that d and ρ are topologically equivalent.

Proposition 10.1.3. *Let $X = \{0,1,\dots,k-1\}^{\mathbb{Z}}$ or $X = \{0,\dots,k-1\}^{\mathbb{N}}$ with the metric d from (7.1) and let $T : X \to X$ be the left shift. The topological entropy of T is equal to $h_{top}(T) = \log k$.*

Proof. We will only prove the statement for the one-sided shift on X, since both proofs are similar. Let d be the metric from (7.1). Fix $0 < \varepsilon < 1$ and any $x = (x_i)_{i\ge1}$, $y = (y_i)_{i\ge1} \in X$. Notice that if at least one of the first n symbols of x and y differ, then

$$d_n(x,y) = \max_{0\le i\le n-1} d(T^i x, T^i y) = 1 > \varepsilon.$$

So, the set $A_n := \{a \in X : a_j = 0,\ j \ge n\}$ is (n,ε)-separated with $Se(n,\varepsilon,T) \ge k^n$. Hence,

$$h_{top}(T) = \lim_{\varepsilon\downarrow0} \limsup_{n\to\infty} \frac{1}{n} \log Se(n,\varepsilon,T) \ge \log k.$$

To prove the reverse inequality, take $\ell \ge 1$ such that $2^{-\ell} < \varepsilon$. Then $A_{n+\ell}$ is an (n,ε)-spanning set, since for every $x \in X$ there is an $a \in A_{n+\ell}$ for which the first $n+\ell$ digits coincide. In other words, $d_n(x,a) < 2^{-\ell} < \varepsilon$. Therefore,

$$h_{top}(T) = \lim_{\varepsilon \downarrow 0} \liminf_{n\to\infty} \frac{1}{n} \log Sp(n,\varepsilon,T) \leq \lim_{\varepsilon \downarrow 0} \lim_{n\to\infty} \frac{n+\ell}{n} \log k = \log k$$

and our proof is complete. □

As with the metric entropy it is often not so easy to compute the topological entropy of a specific system from the definition. In certain cases there is an easier way. The following theorem is a result by M. Misiurewicz and W. Szlenk from [43] from 1980 for piecewise monotone continuous interval maps. Let $T : [0,1] \to [0,1]$ be continuous. Then T is called *piecewise monotone* if there exists a finite interval partition (see also Section 6.3) α of $[0,1]$ so that the restriction of T to any element of α is monotone. The elements of such an interval partition of smallest cardinality are called the *fundamental intervals* of T, see Remark 2.3.1. We call the partition itself a *fundamental interval partition.* Note that a fundamental interval partition is not uniquely defined at the endpoints of the intervals. Note also that if T is a continuous piecewise monotone map, then so is any iterate T^n.

Theorem 10.1.3 (Misiurewicz and Szlenk). *Let $T : [0,1] \to [0,1]$ be a continuous piecewise monotone map. Then*

$$h_{top}(T) = \lim_{n\to\infty} \frac{1}{n} \log F_n,$$

where F_n is the number of fundamental intervals of T^n.

Before we prove the theorem, we prove the following result.

Proposition 10.1.4. *Let $T : [0,1] \to [0,1]$ be a piecewise monotone continuous interval map and α a fundamental interval partition. Then*

$$h_{top}(T) = h_{top}(\alpha, T).$$

Proof. Let ι be a covering of $[0,1]$ by finitely many intervals. Fix some $n \geq 1$ and $A \in \bigvee_{i=0}^{n-1} T^{-i}\alpha$. Then for each $0 \leq k \leq n$ the map T^k is monotone on A, so $A \cap T^{-k}I$ is an interval for any $I \in \iota$. Hence, if we let $\mathcal{E}_k$ denote the collection of all endpoints of the intervals in $A \cap T^{-k}\iota$, then $\#\mathcal{E}_k \leq 2\#\iota$. Any set

$$I_0 \cap T^{-1}I_1 \cap \cdots \cap T^{-(n-1)}I_{n-1} \in \bigvee_{i=0}^{n-1} T^{-i}\iota$$

is an interval with endpoints in the set $\bigcup_{i=1}^{n} \mathcal{E}_i$, which has cardinality at most $2n\#\iota$. So the collection $A \cap \bigvee_{i=0}^{n-1} T^{-i}\iota$ specifies at most $2n\#\iota$ possible endpoints in A. Since each interval has two endpoints and for each interval there are four possibilities to have the endpoints either open or closed, with this number of endpoints we can form at most $4(2n\#\iota)^2$ intervals. Hence,

$$N\Big(A \cap \bigvee_{i=0}^{n-1} T^{-i}\iota\Big) \leq 4(2n\#\iota)^2.$$

From this it follows that

$$\lim_{n\to\infty} \frac{1}{n} \log\Big(\max_{A\in\bigvee_{i=0}^{n-1} T^{-i}\alpha} N\Big(A \cap \bigvee_{i=0}^{n-1} T^{-i}\iota\Big)\Big) \leq \lim_{n\to\infty} \frac{1}{n} \log(4(2n\#\iota)^2) = 0.$$

Hence,

$$\sup_{\iota} \lim_{n\to\infty} \frac{1}{n} \log\Big(\max_{A\in\bigvee_{i=0}^{n-1} T^{-i}\alpha} N\Big(A \cap \bigvee_{i=0}^{n-1} T^{-i}\iota\Big)\Big) = 0,$$

where the supremum is taken over all finite covers of $[0,1]$ by intervals. Since any open subset $B \subseteq [0,1]$ can be written as a countable disjoint union of open intervals, we can find for each open cover β of $[0,1]$ a finite cover ι of $[0,1]$ by intervals such that

$$\max_{A\in\bigvee_{i=0}^{n-1} T^{-i}\alpha} N\Big(A \cap \bigvee_{i=0}^{n-1} T^{-i}\beta\Big) \leq \max_{A\in\bigvee_{i=0}^{n-1} T^{-i}\alpha} N\Big(A \cap \bigvee_{i=0}^{n-1} T^{-i}\iota\Big).$$

The result then follows from Exercise 10.1.5. □

Proof of Theorem 10.1.3. For each $n \geq 1$, let α_n be a fundamental interval partition for T^n. Let $m, k \geq 1$ be given. Then the collection $T^{-k}\alpha_m \vee \alpha_k$ consists of finitely many intervals on which the map T^{m+k} is monotone. Hence,

$$F_{m+k} \leq \#(T^{-k}\alpha_m \vee \alpha_k) \leq F_m \cdot F_k.$$

This implies that $(\log F_n)$ is a subadditive sequence, and hence by Proposition 9.2.2 the limit $\lim_{n\to\infty} \frac{1}{n} \log F_n$ exists.

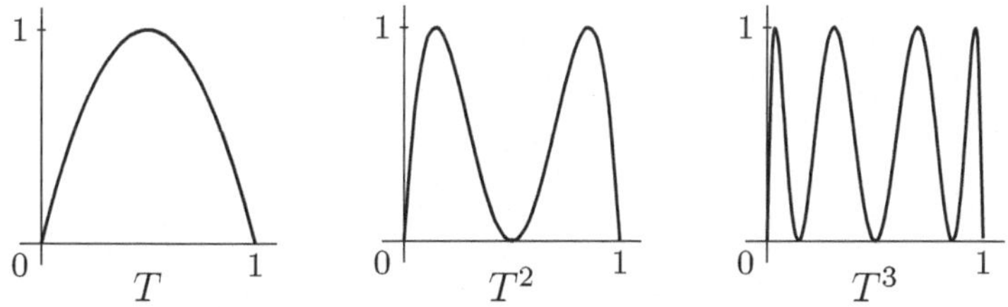

Figure 10.2 The first three iterations of the logistic map from Example 10.1.3.

Now fix some $n \geq 1$. By Theorem 10.1.2(iii), Proposition 10.1.4 and Lemma 10.1.2(iii) we have

$$h_{top}(T) = \frac{1}{n} h_{top}(T^n) = \frac{1}{n} h_{top}(\alpha_n, T^n) \leq \frac{1}{n} H_{top}(\alpha_n) = \frac{1}{n} \log F_n.$$

Hence,

$$h_{top}(T) \leq \lim_{n\to\infty} \frac{1}{n} \log F_n.$$

For the other direction, again fix some $n \geq 1$. For any $k \geq 1$ the collection $\bigvee_{i=0}^{k-1} T^{-ni}\alpha_n$ is a finite collection of intervals, such that the restriction of T^{nk} to each of these intervals is monotone. Hence, $F_{nk} \leq N(\bigvee_{i=0}^{k-1} T^{-ni}\alpha_n)$, which by Proposition 10.1.4 and Theorem 10.1.2(iii) leads to

$$\begin{aligned}
\lim_{k\to\infty} \frac{1}{k} \log F_k &= \lim_{k\to\infty} \frac{1}{nk} \log F_{nk} \\
&\leq \lim_{k\to\infty} \frac{1}{nk} \log N\Big(\bigvee_{i=0}^{k-1} T^{-ni}\alpha_n\Big) \\
&= \frac{1}{n} h_{top}(\alpha_n, T^n) = \frac{1}{n} h_{top}(T^n) = h_{top}(T).
\end{aligned}$$

This finishes the proof. □

Example 10.1.3. Consider the logistic map $T : [0,1] \to [0,1]$, $x \mapsto 4x(1-x)$ we saw in Example 6.3.1. See Figure 10.2 for the first three iterations. The fundamental intervals of T are $[0, \frac{1}{2}]$ and $[\frac{1}{2}, 1]$ and since both intervals are mapped to $[0,1]$ by T, one immediately sees that T^n has 2^n fundamental intervals. Hence, $h_{top}(T) = \log 2$.

Exercise 10.1.10. Let $a > 1$. Compute the topological entropy of the transformation

$$T : [0, 1] \to [0, 1],\ x \mapsto \begin{cases} ax, & \text{if } 0 \leq x \leq \frac{1}{a}, \\ \frac{a+1}{a} - x, & \text{if } \frac{1}{a} < x \leq 1. \end{cases}$$

10.2 PROOF OF THE VARIATIONAL PRINCIPLE

The *Variational Principle* is a striking result linking the notions of topological and metric entropy. It was originally obtained by E. I. Dinaburg, T. Goodman and L. Goodwyn, see [15, 18, 19] and states the following: if $T : X \to X$ is a continuous transformation on a compact metric space (X, d), then

$$h_{top}(T) = \sup\{h_\mu(T) \,:\, \mu \in M(X, T)\},$$

where $M(X, T)$ is the set of all T-invariant Borel probability measures as in Chapter 7. To prove this statement, we will proceed along the shortest and most popular route to victory, provided by M. Misiurewicz in [42] from 1976. The first part of the proof uses only some properties of metric entropy that we have listed in the following lemma. For any collection α of subsets of X and any $\mu \in M(X)$, let $N_\mu(\alpha)$ be the number of sets in α of positive μ-measure. Throughout the section and as before we will use Φ to denote the function

$$\Phi : [0, \infty) \to \mathbb{R},\ x \mapsto \begin{cases} -x \log x, & \text{if } x \neq 0, \\ 0, & \text{if } x = 0. \end{cases} \tag{10.3}$$

Recall that Φ is concave.

Lemma 10.2.1. *Let α, β be finite, measurable partitions of X and T a measure preserving transformation on the probability space $(X, \mathcal{F}, \mu)$. Then the following hold.*

(i) *$H_\mu(\alpha) \leq \log(N_\mu(\alpha))$. Equality holds if and only if $\mu(A) = \frac{1}{N_\mu(\alpha)}$ for all $A \in \alpha$ with $\mu(A) > 0$.*

(ii) *If $\alpha \leq \beta$, then $h_\mu(\alpha, T) \leq h_\mu(\beta, T)$.*

Proof. (i) Recall that $H_\mu(\alpha) = \sum_{A\in\alpha} \Phi(\mu(A))$ for the concave function Φ from (10.3). Applying Jensen's Inequality then gives

$$H_\mu(\alpha) \leq N_\mu(\alpha)\Phi\left(\frac{1}{N_\mu(\alpha)} \sum_{A\in\alpha} \mu(A)\right) = \log(N_\mu(\alpha)).$$

(ii) From $\alpha \le \beta$ we get that for every $B \in \beta$, there is some $A \in \alpha$ such that $T^{-i}B \subseteq T^{-i}A$ for any $i \in \mathbb{Z}_{\ge 0}$. Therefore, for $n \ge 1$

$$\bigvee_{i=0}^{n-1} T^{-i}\alpha \le \bigvee_{i=0}^{n-1} T^{-i}\beta.$$

The result now follows from Proposition 9.2.1(v), division by n and taking the limit as $n \to \infty$. □

We are now ready to prove the first part of the Variational Principle.

Theorem 10.2.1. *Let $T : X \to X$ be a continuous transformation on a compact metric space (X, d). Then $h_{top}(T) \ge \sup\{h_\mu(T) : \mu \in M(X,T)\}$.*

Proof. We start by taking an arbitrary measure $\mu \in M(X,T)$ and a finite partition $\alpha = \{A_1, \ldots, A_N\}$ of X into measurable sets. Our goal is to show that $h_\mu(\alpha, T) < h_{top}(T)$ by finding an open cover of X that is close to α. To do so, first pick an $0 < \varepsilon < \frac{1}{N \log N}$. By Theorem 12.6.1, μ is regular, so for each $i = 1, \ldots, N$ we can find a closed set $C_i \subseteq A_i$ such that $\mu(A_i \setminus C_i) < \varepsilon$. Set $C_0 = X \setminus \bigcup_{i=1}^N C_i$ and let γ be the partition $\gamma = \{C_0, \ldots, C_N\}$. If $\mu(C_0) > 0$, then since $C_i \subseteq A_i$ for each $i = 1, \ldots, N$ we get from Jensen's Inequality that

$$\begin{aligned}
H_\mu(\alpha|\gamma) &= \sum_{i=0}^{N}\sum_{j=1}^{N} \mu(C_i)\Phi\Big(\frac{\mu(C_i \cap A_j)}{\mu(C_i)}\Big) \\
&= \mu(C_0)\sum_{j=1}^{N} \Phi\Big(\frac{\mu(C_0 \cap A_j)}{\mu(C_0)}\Big) \\
&\le N\mu(C_0)\Phi\left(\frac{1}{N}\sum_{j=1}^{N} \frac{\mu(C_0 \cap A_j)}{\mu(C_0)}\right) \\
&= \mu(C_0)\log N \\
&< N\varepsilon \log N < 1.
\end{aligned}$$

From Exercise 9.3.1(b) we see that

$$h_\mu(\alpha, T) \le h_\mu(\gamma, T) + H_\mu(\alpha|\gamma) < h_\mu(\gamma, T) + 1. \tag{10.4}$$

If $\mu(C_0) = 0$, then we have $h_\mu(\alpha, T) = h_\mu(\gamma, T)$, which also yields (10.4).

For each $1 \le i \le N$ the set $C_0 \cup C_i = X \setminus \bigcup_{j \ne i} C_j$ is an open set, so the collection $\beta = \{C_0 \cup C_1, \ldots, C_0 \cup C_N\}$ is an open cover of X. For any $B \in \bigvee_{i=0}^{n-1} T^{-i}\beta$ there are some some $i_j \in \{1, \ldots, N\}$, such that B can be written as

$$B = (C_0 \cup C_{i_0}) \cap T^{-1}(C_0 \cup C_{i_1}) \cap \ldots \cap T^{-(n-1)}(C_0 \cup C_{i_{n-1}})$$
$$= \bigcup_{m=0}^{n-1} \bigcup_{\substack{S \subseteq \{0,1,\ldots,n-1\}: \\ \#S = m}} \Big(\bigcap_{j \in S} T^{-j} C_0 \cap \bigcap_{j \notin S} T^{-j} C_{i_j} \Big). \tag{10.5}$$

Here the sets $\mathcal{S}$ specify the positions in which we choose C_0 instead of C_{i_j}. We see that each element of $\bigvee_{i=0}^{n-1} T^{-i}\beta$ can be written as a disjoint union of 2^n elements from $\bigvee_{i=0}^{n-1} T^{-i}\gamma$, some of which might be empty. Recall that $N(\bigvee_{i=0}^{n-1} T^{-i}\beta)$ denotes the minimal cardinality that any subcover of $\bigvee_{i=0}^{n-1} T^{-i}\beta$ can have. Let $\tilde{\beta}$ be a subcover of $\bigvee_{i=0}^{n-1} T^{-i}\beta$ of cardinality $N(\bigvee_{i=0}^{n-1} T^{-i}\beta)$. Then every element of $\tilde{\beta}$ contains at most 2^n elements from $\bigvee_{i=0}^{n-1} T^{-i}\gamma$. Moreover, $\tilde{\beta}$ is a cover of X, so each element from $\bigvee_{i=0}^{n-1} T^{-i}\gamma$ is contained in at least one element from $\tilde{\beta}$. This leads to the estimate

$$N_\mu\Big(\bigvee_{i=0}^{n-1} T^{-i}\gamma\Big) \le 2^n N\Big(\bigvee_{i=0}^{n-1} T^{-i}\beta\Big).$$

By Lemma 10.2.1(ii) we then obtain that

$$\begin{aligned} h_\mu(\gamma, T) = \lim_{n\to\infty} \frac{1}{n} H_\mu(\gamma, T) &\le \lim_{n\to\infty} \frac{1}{n} \log N_\mu\Big(\bigvee_{i=0}^{n-1} T^{-i}\gamma\Big) \\ &\le \lim_{n\to\infty} \frac{1}{n} \log\Big(2^n N\Big(\bigvee_{i=0}^{n-1} T^{-i}\beta\Big)\Big) \\ &= h_{top}(\beta, T) + \log 2, \end{aligned}$$

so that $h_\mu(\alpha, T) < h_{top}(\beta, T) + \log 2 + 1 \le h_{top}(T) + \log 2 + 1$ and taking the supremum over all finite measurable partitions of X gives

$$h_\mu(T) < h_{top}(T) + \log 2 + 1. \tag{10.6}$$

We now use the results on the entropies of T^n from Exercise 9.3.3 and Theorem 10.1.2. Any T-invariant measure is automatically T^n-invariant. So (10.6) also holds with T replaced by T^n for any $n \ge 1$. Then Exercise 9.3.3(d) and Theorem 10.1.2(iii) lead to

$$nh_\mu(T) = h_\mu(T^n) < h_{top}(T^n) + \log 2 + 1 = nh_{top}(T) + \log 2 + 1.$$

Dividing by n, letting $n \to \infty$ and taking the supremum over all $\mu \in M(X, T)$ gives the result. □

To finish the proof of the Variational Principle the opposite inequality still remains. We need a few lemmas.

Lemma 10.2.2. *Let (X, d) be a compact metric space and α a finite measurable partition of X. Then, for any $\mu, \nu \in M(X, T)$ and $p \in [0, 1]$ we have $H_{p\mu+(1-p)\nu}(\alpha) \geq pH_\mu(\alpha) + (1-p)H_\nu(\alpha)$.*

Proof. The concavity of the function Φ from (10.3) gives for any measurable set A that

$$0 \leq \Phi(p\mu(A) + (1-p)\nu(A)) - p\Phi(\mu(A)) - (1-p)\Phi(\nu(A)).$$

The result now follows easily. □

Exercise 10.2.1. (a) Suppose (X, d) is a compact metric space and $T : X \to X$ a continuous transformation. Use (the proof of) Lemma 10.2.2 to show that for any $\mu, \nu \in M(X, T)$ and $p \in [0, 1]$ we have $h_{p\mu+(1-p)\nu}(T) \geq ph_\mu(T) + (1-p)h_\nu(T)$.
(b) Improve the result in part (a) by showing that we can replace the inequality by an equality sign, i.e., for any $\mu, \nu \in M(X, T)$ and $p \in [0, 1]$ we have $h_{p\mu+(1-p)\nu}(T) = ph_\mu(T) + (1-p)h_\nu(T)$.

Recall that the boundary of a set A is defined by $\partial A = \overline{A} \setminus A^o$, where A^o denotes the interior of A.

Lemma 10.2.3. *Let (X, d) be a compact metric space. The following hold.*

(i) *Let $\mu \in M(X)$. For any $\delta > 0$, there is a finite, measurable partition $\alpha = \{A_1, \ldots, A_n\}$ of X such that $\text{diam}_d(A_j) < \delta$ and $\mu(\partial A_j) = 0$ for all j.*

(ii) *Let $T : X \to X$ be continuous and $\mu \in M(X, T)$. Suppose that for each $j = 0, \ldots, n-1$ we have $A_j \in \mathcal{B}$ satisfying $\mu(\partial A_j) = 0$. Then $\mu(\partial(\bigcap_{j=0}^{n-1} T^{-j} A_j)) = 0$.*

Proof. (i) First note that for every $x \in X$ and $\delta > 0$ we can find an $0 < \eta < \delta$ such that the open ball $B(x, \eta)$ satisfies $\mu(\partial B(x, \eta)) = 0$. To see this, suppose that the opposite is true. Then there exists an $x \in X$ and a $\delta > 0$ such that for all $0 < \eta < \delta$ it holds that $\mu(\partial B(x, \eta)) > 0$. This gives an uncountable collection of disjoint subsets of X with positive measure, contradicting the fact that μ is a probability measure. To prove the statement, fix $\delta > 0$ and for each $x \in X$ let $0 < \eta_x < \delta/2$ be such that $\mu(\partial B(x, \eta_x)) = 0$. The collection $\{B(x, \eta_x) : x \in X\}$ forms an open cover of X, so by compactness there exists a finite subcover which we denote by $\beta = \{B_1, \ldots, B_n\}$. Define α by letting $A_1 = \overline{B_1}$ and let $A_j = \overline{B_j} \setminus (\bigcup_{k=1}^{j-1} \overline{B_k})$ for $0 < j \leq n$. Then α is a partition of X into Borel measurable set with $diam_d(A_j) \leq diam_d(B_j) < \delta$ and $\mu(\partial A_j) \leq \mu(\bigcup_{i=1}^{n} \partial B_i) = 0$.

(ii) Let $x \in \partial(\bigcap_{j=0}^{n-1} T^{-j}A_j)$. Then $x \in \overline{\bigcap_{j=0}^{n-1} T^{-j}A_j}$, but $x \notin (\bigcap_{j=0}^{n-1} T^{-j}A_j)^o$. That is, every open neighborhood of x intersects every $T^{-j}A_j$, but there is a $0 \leq k \leq n-1$ for which $x \notin (T^{-k}A_k)^o$. Hence, $x \in \partial T^{-k}A_k$ and by continuity of T it then also follows that $x \in T^{-k}\partial A_k$ (see Lemma 12.1.1). Hence, $\partial(\bigcap_{j=0}^{n-1} T^{-j}A_j) \subseteq \bigcup_{j=0}^{n-1} T^{-j}\partial A_j$. The statement follows since μ is T-invariant. □

Lemma 10.2.4. *Let q, n be integers such that $1 < q < n$. Define for $0 \leq j \leq q-1$ the numbers $a(j) = \lfloor \frac{n-j}{q} \rfloor$, where $\lfloor \cdot \rfloor$ denotes the integer part. Then the following hold.*

(i) $a(0) \geq a(1) \geq \cdots \geq a(q-1)$.

(ii) *Fix $0 \leq j \leq q-1$ and let $S_j = \{0, 1, \ldots, j-1, j+a(j)q, j+a(j)q+1, \ldots, n-1\}$. Then*

$$\{0, 1, \ldots, n-1\} = \{j+rq+i : 0 \leq r \leq a(j)-1,\ 0 \leq i \leq q-1\} \cup S_j$$

and $\#S_j \leq 2q$.

(iii) *For each $0 \leq j \leq q-1$, $(a(j)-1)q + j \leq \lfloor \frac{n-j}{q} - 1 \rfloor q + j \leq n-q$. The numbers in the set $\{j + rq : 0 \leq j \leq q-1,\ 0 \leq r \leq a(j)-1\}$ are all distinct and do not exceed $n-q$.*

The number $a(j)$ represents the amount of times you have to add q to j to get a number between $n-q$ and n. The proof of these three statements is left to the reader.

To prove the main result of this section we construct a Borel probability measure μ with metric entropy $h_\mu(T) \geq \liminf_{n\to\infty} \frac{1}{n} \log Se(n, \varepsilon, T)$. To do this, we first find a sequence of measures ν_n with

$$H_{\nu_n}\Big(\bigvee_{i=0}^{n-1} T^{-i}\alpha\Big) = \log Se(n, \varepsilon, T)$$

for a suitably chosen partition α. Theorem 7.1.1 is then used to obtain a suitable $\mu \in M(X,T)$ and the appropriate estimate for H_μ then comes from applying Lemma 10.2.4 to the tails $S(j)$ of the partition $\bigvee_{i=0}^{n-1} T^{-i}\alpha$. Lemma 10.2.3 fills in the remaining technicalities.

Theorem 10.2.2. *Let $T : X \to X$ be a continuous transformation on a compact metric space (X, d). Then for each $\varepsilon > 0$ there is a measure $\mu \in M(X,T)$ that satisfies*

$$h_\mu(T) \geq \liminf_{n\to\infty} \frac{1}{n} \log Se(n,\varepsilon,T).$$

Proof. Fix $\varepsilon > 0$. For each n, let E_n be an (n,ε)-separated set of cardinality $Se(n,\varepsilon,T)$. Define $\nu_n \in M(X)$ by $\nu_n = \frac{1}{Se(n,\varepsilon,T)} \sum_{x\in E_n} \delta_x$, where δ_x is the Dirac measure concentrated at x. Define $\mu_n \in M(X)$ by $\mu_n = \frac{1}{n}\sum_{i=0}^{n-1} \nu_n \circ T^{-i}$. By Theorem 12.6.4, $M(X)$ is compact, hence there exists a subsequence (n_j) such that (μ_{n_j}) converges weakly in $M(X)$ to some $\mu \in M(X)$. By Theorem 7.1.1, $\mu \in M(X,T)$. We will show that this measure μ satisfies the statement from the theorem.

By Lemma 10.2.3(i), we can find a μ-measurable partition $\alpha = \{A_1, \dots, A_k\}$ of X such that $diam_d(A_j) < \varepsilon$ and $\mu(\partial A_j) = 0$ for all $j = 1, \dots, k$. Since E_n is (n,ε)-separated any set $A \in \bigvee_{i=0}^{n-1} T^{-i}\alpha$ can contain at most one element from E_n. Hence, either $\nu_n(A) = 0$ or $\nu_n(A) = \frac{1}{Se(n,\varepsilon,T)}$. Since $E_n \subseteq \bigcup_{j=1}^k A_j$, we see that $H_{\nu_n}(\bigvee_{i=0}^{n-1} T^{-i}\alpha) = \log(Se(n,\varepsilon,T))$. Fix integers q, n with $1 < q < n$ and define for each $0 \leq j \leq q-1$ the numbers $a(j)$ as in Lemma 10.2.4. Fix $0 \leq j \leq q-1$. Since

$$\bigvee_{i=0}^{n-1} T^{-i}\alpha = \Big(\bigvee_{r=0}^{a(j)-1} T^{-(rq+j)}\Big(\bigvee_{i=0}^{q-1} T^{-i}\alpha\Big)\Big) \vee \Big(\bigvee_{i\in S_j} T^{-i}\alpha\Big),$$

we find using Proposition 9.2.1(i) and (iv) and Lemma 10.2.1(ii) that

$$\begin{aligned}
\log Se(n,\varepsilon,T) &= H_{\nu_n}\Big(\bigvee_{i=0}^{n-1} T^{-i}\alpha\Big)\\
&\leq \sum_{r=0}^{a(j)-1} H_{\nu_n}\Big(T^{-(rq+j)}\Big(\bigvee_{i=0}^{q-1} T^{-i}\alpha\Big)\Big) + \sum_{i\in S_j} H_{\nu_n}(T^{-i}\alpha)\\
&\leq \sum_{r=0}^{a(j)-1} H_{\nu_n\circ T^{-(rq+j)}}\Big(\bigvee_{i=0}^{q-1} T^{-i}\alpha\Big) + 2q\log k.
\end{aligned}$$

Now if we sum the above inequality over j and divide both sides by n, we obtain by Lemmas 10.2.4(iii) and 10.2.2 that

$$\begin{aligned}
\frac{q}{n}\log Se(n,\varepsilon,T) &\leq \frac{1}{n}\sum_{\ell=0}^{n-1} H_{\nu_n\circ T^{-\ell}}\Big(\bigvee_{i=0}^{q-1} T^{-i}\alpha\Big) + \frac{2q^2}{n}\log k\\
&\leq H_{\mu_n}\Big(\bigvee_{i=0}^{q-1} T^{-i}\alpha\Big) + \frac{2q^2}{n}\log k.
\end{aligned}$$

By Lemma 10.2.3(ii) each atom A of $\bigvee_{i=0}^{q-1} T^{-i}\alpha$ has boundary of μ-measure zero. Together with the weak convergence of μ_{n_j} to μ in $M(X)$ this implies that $\lim_{j\to\infty} \mu_{n_j}(A) = \mu(A)$ (see Theorem 12.6.5). Hence,

$$\lim_{j\to\infty} H_{\mu_{n_j}}\Big(\bigvee_{i=0}^{q-1} T^{-i}\alpha\Big) + \frac{2q^2}{n_j}\log k = H_\mu\Big(\bigvee_{i=0}^{q-1} T^{-i}\alpha\Big),$$

giving that

$$\liminf_{n\to\infty} \frac{1}{n}\log Se(n,\varepsilon,T) \le \frac{1}{q} H_\mu\Big(\bigvee_{i=0}^{q-1} T^{-i}\alpha\Big).$$

Taking the limit as $q \to \infty$ now yields the desired inequality. □

Corollary 10.2.1 (Variational Principle). *The topological entropy of a continuous transformation $T: X \to X$ on a compact metric space (X,d) is given by*

$$h_{top}(T) = \sup\{h_\mu(T) \,:\, \mu \in M(X,T)\}.$$

Proof. By Theorem 10.2.1 we have $h_{top}(T) \ge \sup\{h_\mu(T) \,:\, \mu \in M(X,T)\}$. From Theorem 10.2.2 we get for each $\varepsilon > 0$ the existence of a measure $\mu = \mu_\varepsilon \in M(X,T)$ with

$$\sup\{h_\mu(T) \,:\, \mu \in M(X,T)\} \ge h_{\mu_\varepsilon}(T) \ge \liminf_{n\to\infty} \frac{1}{n}\log Se(n,\varepsilon,T).$$

The other inequality then follows from Definition 10.1.2. □

To get a taste of the power of this statement, let us revisit our proof of the invariance of topological entropy under conjugacy given in Theorem 10.1.2(ii). Let (X,d), (Y,ρ) be two compact metric spaces with continuous transformations $T: X \to X$ and $S: Y \to Y$ and let $\psi: X \to Y$ be a conjugacy. Note that $\mu \in M(X,T)$ if and only if $\mu\circ\psi^{-1} \in M(Y,S)$. With these measures the maps T and S are measure preservingly isomorphic according to Definition 5.1.1 and thus by Theorem 9.2.2 we obtain that $h_\mu(T) = h_{\mu\circ\psi^{-1}}(S)$. It then follows from the Variational Principle that $h_{top}(T) = h_{top}(S)$.

10.3 MEASURES OF MAXIMAL ENTROPY

The Variational Principle suggests an educated way for choosing a Borel probability measure on X, namely one that maximizes the entropy of T.

Definition 10.3.1. Let (X, d) be a compact metric space and $T : X \to X$ continuous. A measure $\mu \in M(X, T)$ is called a *measure of maximal entropy* if $h_\mu(T) = h_{top}(T)$. Let $M_{max}(X, T) = \{\mu \in M(X, T) : h_\mu(T) = h_{top}(T)\}$. If $M_{max}(X, T) = \{\mu\}$, then μ is called the *unique measure of maximal entropy.*

Example 10.3.1. Recall that for any $\theta \in [0, 1)$ the circle rotation R_θ has topological entropy $h_{top}(R_\theta) = 0$. We know from Chapter 7 that Haar measure λ is the unique invariant probability measure for R_θ. Since $h_\lambda(R_\theta) \geq 0$, it immediately follows from the Variational Principle that λ is the unique measure of maximal entropy for R_θ with $h_\lambda(R_\theta) = 0$.

Exercise 10.3.1. Consider the set

$$K = \{f : [0,1] \to [0,1] \, : \, |f(x) - f(y)| \leq |x - y| \text{ for all } x, y \in [0,1]\}$$

of Lipschitz continuous functions on $[0, 1]$ with Lipschitz constant 1. Endowed with the uniform distance metric

$$d(f, g) = \sup_{x \in [0,1]} |f(x) - g(x)|$$

the set K becomes a compact metric space. Let $g \in K$ be a bijection. Define the mapping $T : K \to K$ by $T \circ f = f \circ g$.

(a) Prove that $T \circ f \in K$ for all $f \in K$.

(b) Prove that $h_\mu(T) = 0$ for all $\mu \in M(K, T)$.

Measures of maximal entropy are closely connected to (uniquely) ergodic measures, as will become apparent from the following theorem.

Theorem 10.3.1. *Let (X, d) be a compact metric space and $T : X \to X$ continuous. Then the following hold.*

(i) *$M_{max}(X, T)$ is a convex set.*

(ii) *If $h_{top}(T) < \infty$, then the extreme points of $M_{max}(X, T)$ are precisely the ergodic members of $M_{max}(X, T)$.*

(iii) *If $h_{top}(T) = \infty$, then $M_{max}(X, T) \neq \emptyset$. If, moreover, T has a unique measure of maximal entropy, then T is uniquely ergodic.*

Proof. (i) Let $p \in [0, 1]$ and $\mu, \nu \in M_{max}(X, T)$. Then, by Exercise 10.2.1,

$$\begin{aligned} h_{p\mu+(1-p)\nu}(T) &= p h_\mu(T) + (1 - p) h_\nu(T) \\ &= p h_{top}(T) + (1 - p) h_{top}(T) = h_{top}(T). \end{aligned}$$

Hence, $p\mu + (1-p)\nu \in M_{max}(X,T)$.

(ii) Suppose $\mu \in M_{max}(X,T)$ is ergodic. Then, by Theorem 7.1.2, μ cannot be written as a non-trivial convex combination of elements of $M(X,T)$. Since $M_{max}(X,T) \subseteq M(X,T)$, μ is an extreme point of $M_{max}(X,T)$. Conversely, suppose μ is an extreme point of $M_{max}(X,T)$ and suppose there is a $p \in (0,1)$ and $\nu_1, \nu_2 \in M(X,T)$ such that $\mu = p\nu_1 + (1-p)\nu_2$. By Exercise 10.2.1, $h_{top}(T) = h_\mu(T) = ph_{\nu_1}(T) + (1-p)h_{\nu_2}(T)$. Since $h_{\nu_1}(T), h_{\nu_2}(T) \leq h_{top}(T) < \infty$, we must have $h_{top}(T) = h_{\nu_1}(T) = h_{\nu_2}(T)$. Thus, $\nu_1, \nu_2 \in M_{max}(X,T)$. Since μ is an extreme point of $M_{max}(X,T)$, we must have $\mu = \nu_1 = \nu_2$. Therefore, μ is also an extreme point of $M(X,T)$ and we conclude that μ is ergodic.

(iii) By the Variational Principle we can find for any $n \geq 0$ a measure $\mu_n \in M(X,T)$ with $h_{\mu_n}(T) > 2^n$. Define $\mu \in M(X,T)$ by $\mu = \sum_{n\geq 1} \frac{\mu_n}{2^n}$. Then $\mu = \sum_{n=1}^{N} \frac{\mu_n}{2^n} + \sum_{n\geq N+1} \frac{\mu_n}{2^n}$ for any $N \geq 1$. But then Exercise 10.2.1 implies that

$$h_\mu(T) \geq \sum_{n=1}^{N} \frac{h_{\mu_n}}{2^n} > N.$$

Since this holds for arbitrary $N \geq 1$, we obtain $h_\mu(T) = h_{top}(T) = \infty$ and μ is a measure of maximal entropy. Now suppose that μ is the unique measure of maximal entropy for T. Then, for any $\nu \in M(X,T)$, $h_{\mu/2+\nu/2}(T) = \frac{1}{2}h_\mu(T) + \frac{1}{2}h_\nu(T) = \infty$. Hence, $\mu = \nu$ and $M(X,T) = \{\mu\}$. □

Corollary 10.3.1. *Let (X,d) be a compact metric space and $T : X \to X$ continuous. If T has a unique measure of maximal entropy, then it is ergodic. Conversely, if T is uniquely ergodic, then T has a measure of maximal entropy.*

Proof. The first statement follows from Theorem 10.3.1(ii) for $h_{top}(T) < \infty$ and (iii) for $h_{top}(T) = \infty$. If T is uniquely ergodic, then $M(X,T) = \{\mu\}$ for some μ. By the Variational Principle, $h_\mu(T) = h_{top}(T)$, hence μ is a measure of maximal entropy. □

Exercise 10.3.2. Let $X = \{0, 1, \ldots, k-1\}^{\mathbb{Z}}$ and let $T : X \to X$ be the left shift on X. Use Proposition 10.1.3 to show that the uniform product measure is the unique measure of maximal entropy for T.

We end this section with a generalization of the above exercise. A homeomorphism $T : X \to X$ is called *expansive* with constant $\delta > 0$

if for all $x, y \in X$ with $x \neq y$ there is a $k = k_{x,y} \in \mathbb{Z}$ such that $d(T^k x, T^k y) > \delta$.

Example 10.3.2. Let T be the left shift on $X = \{0, 1, \ldots, k-1\}^{\mathbb{Z}}$ and as in Example 7.1.1 consider the metric

$$d(x, y) = 2^{-\min\{|i| \,:\, x_i \neq y_i\}}.$$

We already saw in Example 7.1.1 that T is a homeomorphism on (X, d). If $x, y \in X$ with $x \neq y$, then there is an $i \in \mathbb{Z}$ with $x_i \neq y_i$. Then $d(T^i x, T^i y) = 1$ and hence T is expansive with any constant $\delta < 1$.

Proposition 10.3.1. *Every expansive homeomorphism of a compact metric space has a measure of maximal entropy.*

Proof. Let $T : X \to X$ be an expansive homeomorphism, and let $\delta > 0$ be an expansive constant for T. Fix $0 < \varepsilon < \delta$ and let $\mu \in M(X, T)$ be the measure given by Theorem 10.2.2, so that

$$h_\mu(T) \geq \liminf_{n \to \infty} \frac{1}{n} \log Se(n, \varepsilon, T).$$

We will show that $h_{top}(T) = \liminf_{n \to \infty} \frac{1}{n} \log Se(n, \varepsilon, T)$, from which it then immediately follows that $\mu \in M_{max}(X, T)$.

Pick any $0 < \eta < \varepsilon$, $n \geq 1$ and let A be an (n, η)-separated set of cardinality $Se(n, \eta, T)$. By expansiveness we can find for any $x, y \in A$ some $k = k_{x,y} \in \mathbb{Z}$ such that

$$d(T^k x, T^k y) > \varepsilon.$$

Since A is finite, $\ell := \max\{|k_{x,y}| \,:\, x, y \in A\}$ exists. For any pair of points $x, y \in T^{-\ell} A$ with $x \neq y$ we have $T^\ell x, T^\ell y \in A$, so there is a $k \in \mathbb{Z}$ with $|k| \leq \ell$ such that $d(T^{k+\ell} x, T^{k+\ell} y) > \varepsilon$. Hence,

$$d_{2\ell+n}(x, y) \geq \max_{0 \leq i \leq 2\ell} d(T^i x, T^i y) > \varepsilon.$$

This implies that $T^{-\ell} A$ is a $(2\ell + n, \varepsilon, T)$-separated set and $\# T^{-\ell} A = Se(n, \eta, T)$. Hence, $Se(2\ell + n, \varepsilon, T) \geq Se(n, \eta, T)$ and

$$\begin{aligned}
\liminf_{n \to \infty} \frac{1}{n} \log Se(n, \eta, T) &\leq \liminf_{n \to \infty} \frac{1}{n} \log Se(2\ell + n, \varepsilon, T) \\
&= \liminf_{n \to \infty} \frac{1}{2\ell + n} \log Se(2\ell + n, \varepsilon, T) \\
&= \liminf_{n \to \infty} \frac{1}{n} \log Se(n, \varepsilon, T).
\end{aligned}$$

Conversely, any (n,ε)-separated set is also (n,η)-separated, so that $Se(n,\varepsilon,T) \leq Se(n,\eta,T)$. We conclude that

$$\liminf_{n\to\infty} \frac{1}{n} \log Se(n,\eta,T) = \liminf_{n\to\infty} \frac{1}{n} \log Se(n,\varepsilon,T).$$

Since $0 < \eta < \varepsilon$ was arbitrary, this shows that

$$h_{top}(T) = \lim_{\eta\downarrow 0} \liminf_{n\to\infty} \frac{1}{n} \log Se(n,\eta,T) = \liminf_{n\to\infty} \frac{1}{n} \log Se(n,\varepsilon,T). \quad \square$$

From the proof we extract the following corollary.

Corollary 10.3.2. *Let $T : X \to X$ be an expansive homeomorphism with constant δ. Then for any $0 < \varepsilon < \delta$,*

$$h_{top}(T) = \liminf_{n\to\infty} \frac{1}{n} \log Se(n,\varepsilon,T).$$

CHAPTER 11

Infinite Ergodic Theory

In large parts of the book we have only considered measure preserving transformations $T : X \to X$ on a probability space $(X, \mathcal{F}, \mu)$ and in fact several fundamental results we have seen, the Ergodic Theorems in particular, do not work in case the underlying measure space is infinite. In this chapter we discuss the dynamics on infinite measure systems in more detail. We first consider some more examples and elaborate on the recurrence behavior of infinite measure systems. We then introduce jump and induced transformations. For these transformations, one considers the dynamics only at the time steps in which the system visits a well chosen finite measure subset of the space. Then many tools developed for finite measure systems in the previous chapters become available nevertheless. At the end of the chapter we discuss Ergodic Theorems in relation to infinite measure systems.

11.1 EXAMPLES

In Example 1.3.1 and Example 1.3.4 we already encountered two transformations that were measure preserving with respect to Lebesgue measure on $\mathbb{R}$. Here we introduce two examples of transformations on the unit interval $[0, 1]$ that are non-singular with respect to Lebesgue measure λ, but have an invariant measure that is infinite and absolutely continuous with respect to λ. Contrary to Examples 1.3.1 and 1.3.4 the infiniteness of the measure here is not caused by the Lebesgue measure of the domain, but by the presence of a *neutral fixed point*, i.e., a point c in the space for which $Tc = c$ and the derivative of T at c equals 1.

Example 11.1.1. The *Farey map* is the transformation $T : [0,1] \to [0,1]$ defined by

$$Tx = \begin{cases} \frac{x}{1-x}, & \text{if } 0 \le x < \frac{1}{2}, \\ \frac{1-x}{x}, & \text{if } \frac{1}{2} \le x \le 1. \end{cases}$$

See Figure 11.1(a) for the graph. Note that 0 is a neutral fixed point for T. Let $\mathcal{B}$ denote the Lebesgue σ-algebra on $[0,1]$ and define the measure μ on $([0,1], \mathcal{B})$ by

$$\mu(A) = \int_A \frac{1}{x}\, d\lambda(x) \quad \text{for all } A \in \mathcal{B}. \tag{11.1}$$

We check that μ is T-invariant. For any $b \in [0,1]$ it holds that $\mu((0,b)) = \infty$ and since T0=0, $T^{-1}(0,b)$ contains an interval of the form $(0,c)$. Hence, $\mu(T^{-1}(0,b)) = \infty = \mu((0,b))$. Let $0 < a < b \le 1$. Then

$$T^{-1}(a,b) = \left(\frac{a}{a+1}, \frac{b}{b+1}\right) \cup \left(\frac{1}{b+1}, \frac{1}{a+1}\right),$$

where the union is disjoint. Hence,

$$\begin{aligned} \mu(T^{-1}(a,b)) &= \log\left(\frac{b}{b+1}\right) - \log\left(\frac{a}{a+1}\right) \\ &\quad + \log\left(\frac{1}{a+1}\right) - \log\left(\frac{1}{b+1}\right) \\ &= \log b - \log a = \mu((a,b)). \end{aligned}$$

The same holds for any other type of interval, so the invariance of μ follows from Theorem 1.2.1.

Example 11.1.2. Define the transformation $T : [0,1] \to [0,1]$ by

$$Tx = \begin{cases} \frac{x}{1-x}, & \text{if } 0 \le x < \frac{1}{2}, \\ 2x - 1, & \text{if } \frac{1}{2} \le x \le 1. \end{cases}$$

This map equals the Farey map on the interval $[0, \frac{1}{2})$. The graph is shown in Figure 11.1(b). Let μ be the measure from (11.1). As in the previous example for any interval $(0,b)$ with $0 < b \le 1$ it holds that $\mu((0,b)) = \infty = \mu(T^{-1}(0,b))$. Let $0 < a < b \le 1$. Then

$$\begin{aligned} \mu(T^{-1}(a,b)) &= \log\left(\frac{b}{b+1}\right) - \log\left(\frac{a}{a+1}\right) \\ &\quad + \log\left(\frac{b+1}{2}\right) - \log\left(\frac{a+1}{2}\right) \\ &= \log b - \log a = \mu((a,b)). \end{aligned}$$

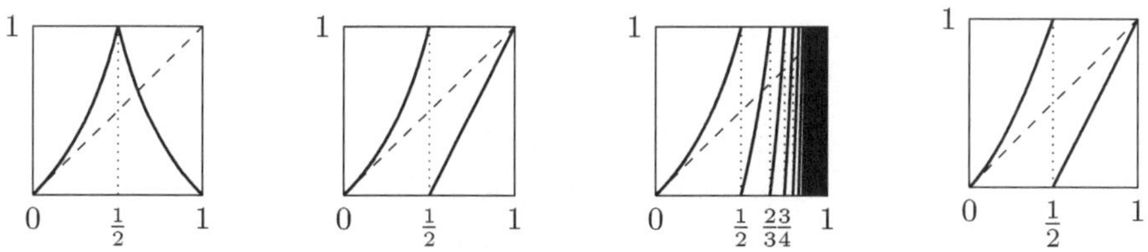

(a) Farey (b) T from Example 11.1.2 (c) Rényi (d) LSV

Figure 11.1 The graph of the Farey map from Example 11.1.1 in (a), of the transformation from Example 11.1.2 in (b), of the Rényi map from Exercise 11.1.1 in (c) and of an LSV maps from Exercise 11.1.2 in (d).

Hence also this transformation T is measure preserving with respect to μ.

In Examples 11.1.1 and 11.1.2, once an orbit lands near the neutral fixed point 0 it will stay close to 0 for a long time. The invariant measure μ from (11.1) reflects the amount of time typical orbits spend in different parts of the state space. It therefore has an accumulation of mass at 0 and in light of that what happens on the interval $\left[\frac{1}{2}, 1\right]$ has no significant effect.

Exercise 11.1.1. The *Rényi map* or *backwards continued fraction map* $T : [0, 1) \to [0, 1)$ is defined by

$$Tx = \frac{x}{1 - x} \pmod 1.$$

The graph is shown in Figure 11.1(c).

(a) Verify that for Lebesgue a.e. $x \in [0, 1)$ it holds that $Tx = S(1 - x)$, where S is the Gauss map from Example 1.3.7. Also find an invertible map $\varphi : [0, 1) \to [-1, 0)$, such that $T = \varphi^{-1} \circ T_0 \circ \varphi$, where $T_0 : [-1, 0) \to [-1, 0)$ is the α-continued fraction transformation from Example 6.3.3 with $\alpha = 0$.

(b) Find an invariant measure for T that is absolutely continuous with respect to the Lebesgue measure.

Exercise 11.1.2. Let $\alpha > 0$ and define the map $T_\alpha : [0, 1) \to [0, 1)$ by

$$T_\alpha x = \begin{cases} x(1 + 2^\alpha x^\alpha), & \text{if } 0 \leq x < \frac{1}{2}, \\ 2x - 1, & \text{if } \frac{1}{2} < x < 1. \end{cases}$$

Verify that the measure μ from (11.1) is not invariant for any LSV map T_α.

These maps T_α, which are now commonly called LSV maps due to the article [36] by C. Liverani, B. Saussol and S. Vaienti, are a modification of the so-called *Pomeau Manneville maps*, see [53, 52]. Figure 11.1(d) shows the graph of one of them. T_α has a σ-finite invariant measure that is absolutely continuous with respect to Lebesgue measure λ for any $\alpha > 0$. This is a probability measure for $\alpha \in (0,1)$ and an infinite measure for $\alpha > 1$.

11.2 CONSERVATIVE AND DISSIPATIVE PART

Recall from the statement of Halmos' Recurrence Theorem that a conservative dynamical system is characterized by the property that for any measurable set B almost all points in B return to B infinitely often. In that sense conservative systems are those infinite measure systems that behave most like finite measure systems. Any infinite measure system can be split into two parts according to its recurrence behavior: the conservative and the dissipative part. To make this precise, we introduce the following notions.

Definition 11.2.1. Let $(X, \mathcal{F}, \mu)$ be a measure space.

(i) A collection $\mathcal{H} \subseteq \mathcal{F}$ is called *hereditary* if $C \in \mathcal{H}$, $B \subseteq C$ and $B \in \mathcal{F}$ imply that $B \in \mathcal{H}$.

(ii) A set $U \in \mathcal{F}$ is called a *cover* of $\mathcal{H}$ if $\mu(A \setminus U) = 0$ for all $A \in \mathcal{H}$.

(iii) A hereditary collection $\mathcal{H}$ is said to *saturate* a set $A \in \mathcal{F}$ if for all $B \in \mathcal{F}$ with $B \subseteq A$ and $\mu(B) > 0$ there is a $C \in \mathcal{H}$ with $C \subseteq B$ and $\mu(C) > 0$.

(iv) A set $U \in \mathcal{F}$ is called the *measurable union* of a hereditary collection $\mathcal{H}$ if U is a cover of $\mathcal{H}$ and $\mathcal{H}$ saturates U.

Measurable unions are unique modulo sets of measure zero. To see this, let U, U' be two measurable unions for the same hereditary collection $\mathcal{H}$ and assume $\mu(U \Delta U') > 0$. Then $\mu(U' \setminus U) > 0$ or $\mu(U \setminus U') > 0$. Assume with no loss of generality that $\mu(U \setminus U') > 0$. Then, since $\mathcal{H}$ saturates U, there is a $C \in \mathcal{H}$ with $C \subseteq U \setminus U'$ and $\mu(C) > 0$ and since U' covers $\mathcal{H}$, $\mu(C \setminus U') = 0$. This gives a contradiction. Hence, $\mu(U \setminus U') = 0$ and by symmetry also $\mu(U' \setminus U) = 0$. The existence of a measurable union is given by Proposition 12.2.1 from the Appendix. We denote the measurable union of $\mathcal{H}$ by $U(\mathcal{H})$.

Exercise 11.2.1. Recall from Definition 2.1.1 that a set $W \in \mathcal{F}$ is called wandering for a transformation T if $\mu(T^{-n}W \cap T^{-m}W) = 0$ for all $0 \leq n < m$. Prove that the collection of wandering sets $\mathcal{W}_T$ is hereditary, so that the measurable union $U(\mathcal{W}_T)$ exists.

Definition 11.2.2. Let $T : X \to X$ be non-singular. The *dissipative part* of T is $\mathcal{D}(T) := U(\mathcal{W}_T)$. The *conservative part* is the complement: $\mathcal{C}(T) = X \setminus \mathcal{D}(T)$. The partition $\{\mathcal{C}(T), \mathcal{D}(T)\}$ of X is called the *Hopf decomposition* of X.

The set $U(\mathcal{W}_T)$ is the smallest measurable set containing all wandering sets in the sense that if $A \in \mathcal{F}$ has $\mu(A \cap U(\mathcal{W}_T)) > 0$, then A contains a wandering set W with $\mu(W) > 0$, see Definition 11.2.1(iii). In particular, if $\mu(U(\mathcal{W}_T)) > 0$, then there is a set $W \in \mathcal{W}_T$ with $\mu(W) > 0$ and T is not conservative. On the other hand, if there is a wandering set W for T with $\mu(W) > 0$, then $\mu(U(\mathcal{W}_T)) > 0$. So T is conservative if and only if $\mu(\mathcal{D}(T)) = 0$. On the other end of the spectrum we have totally dissipative systems.

Definition 11.2.3. A transformation $T : X \to X$ on a measure space $(X, \mathcal{F}, \mu)$ is called *totally dissipative* if $\mu(\mathcal{C}(T)) = 0$.

Example 11.2.1. Recall the dynamical system $(\mathbb{R}, \mathcal{B}, \lambda, T)$ from Example 1.3.1 where T is a translation on $\mathbb{R}$ by $t \in \mathbb{R}$. For $t \neq 0$ each of the intervals $[tn, t(n+1))$, $n \in \mathbb{Z}$, is a wandering set for T, so $\lambda([tn, t(n+1)) \setminus \mathcal{D}(T)) = 0$. Hence also

$$\lambda(\mathbb{R} \setminus \mathcal{D}(T)) = \sum_{n \in \mathbb{Z}} \lambda([tn, t(n+1)) \setminus \mathcal{D}(T)) = 0.$$

This implies that $\lambda(\mathcal{C}(T)) = 0$, and thus $(\mathbb{R}, \mathcal{B}, \lambda, T)$ is totally dissipative.

Since $T^{-1}\mathcal{W}_T \subseteq \mathcal{W}_T$, it follows that $\mu(T^{-1}\mathcal{D}(T) \setminus \mathcal{D}(T)) = 0$. This implies that points in the conservative part of the space typically stay within the conservative part, but it makes no claim about what happens on the dissipative part. In case T is invertible and ergodic, we can say more.

Proposition 11.2.1. *Let $T : X \to X$ be an invertible transformation on a measure space $(X, \mathcal{F}, \mu)$. Then there is a wandering set W, such that*

$$\mu\Big(\mathcal{D}(T) \Delta \bigcup_{n \in \mathbb{Z}} T^n W\Big) = 0.$$

Proof. Let $\mathcal{I} = \{A \in \mathcal{F} : T^{-1}A = A\}$ be the sub-σ-algebra of $\mathcal{F}$ of T-invariant sets. Define the collection

$$\mathcal{V} = \Big\{ \bigcup_{n \in \mathbb{Z}} T^n W : W \in \mathcal{W}_T \Big\}.$$

Then obviously $\mathcal{V} \subseteq \mathcal{I}$. We show that $\mathcal{V}$ is hereditary in $\mathcal{I}$. To this end, let $W \in \mathcal{W}_T$ and consider $A = \bigcup_{n \in \mathbb{Z}} T^n W \in \mathcal{V}$. Let $B \subseteq A$ with $B \in \mathcal{I}$. Then $B \cap W$ is a wandering set and

$$\bigcup_{n \in \mathbb{Z}} T^n(B \cap W) = \bigcup_{n \in \mathbb{Z}} B \cap T^n W = B.$$

So, $B \in \mathcal{V}$ and $\mathcal{V}$ is hereditary. By Proposition 12.2.1 there then exists a countable collection (A_k) of disjoint sets in $\mathcal{V}$ and with that a countable collection (W_k) of disjoint wandering sets, such that

$$U(\mathcal{V}) = \bigcup_{k \geq 1} A_k = \bigcup_{k \geq 1} \bigcup_{n \in \mathbb{Z}} T^n W_k.$$

Note that $\mathcal{D}(T) = U(\mathcal{W}_T) \subseteq U(\mathcal{V})$. Since the sets $\bigcup_{n \in \mathbb{Z}} T^n W_k$, $k \geq 1$, are all disjoint, for each k, ℓ, m, n we have

$$\mu(T^n W_k \cap T^m W_\ell) = 0.$$

Hence, the set $W = \bigcup_{k \geq 1} W_k$ is a wandering set and

$$U(\mathcal{V}) = \bigcup_{n \in \mathbb{Z}} T^n W \subseteq U(\mathcal{W}_T). \qquad \square$$

Exercise 11.2.2. Let $T : X \to X$ be an invertible and ergodic transformation on a measure space $(X, \mathcal{F}, \mu)$. Prove that if μ is non-atomic, then T is conservative.

The next proposition gives a characterization of the conservative part.

Proposition 11.2.2. *Let $(X, \mathcal{F}, \mu, T)$ be a dynamical system with $T : X \to X$ measure preserving.*

(i) *For any $f \in L^1(X, \mathcal{F}, \mu)$ with $f \geq 0$ it holds that*

$$\mu\Big(\Big\{x \in X : \sum_{n \geq 1} f(T^n x) = \infty\Big\} \setminus \mathcal{C}(T)\Big) = 0.$$

(ii) *For any* $f \in L^1(X, \mathcal{F}, \mu)$ *with* $f > 0$ μ*-a.e. it holds that*

$$\mu\Big(\mathcal{C}(T) \Delta \Big\{x \in X : \sum_{n \geq 1} f(T^n x) = \infty\Big\}\Big) = 0.$$

Proof. For (i) let $f \in L^1(X, \mathcal{F}, \mu)$ with $f \geq 0$ be given and let W be a wandering set. By the change of variable formula (6.6) together with the fact that T is measure preserving, we have for each $k \geq 1$ and each $g \in L^1(X, \mathcal{F}, \mu)$ that $\int_W g \, d\mu = \int_{T^{-k}W} g \circ T^k \, d\mu$. Thus, for each $n \geq 1$ we obtain

$$\begin{aligned}
\int_W \sum_{k=1}^{n} f \circ T^k \, d\mu &= \sum_{k=0}^{n-1} \int_W f \circ T^{n-k} \, d\mu = \sum_{k=0}^{n-1} \int_{T^{-k}W} f \circ T^n \, d\mu \\
&= \int_{\bigcup_{k=0}^{n-1} T^{-k}W} f \circ T^n \, d\mu \leq \int_X f \circ T^n d\mu = \int_X f \, d\mu.
\end{aligned}$$

Since $f \geq 0$ we can apply the Monotone Convergence Theorem to get

$$\int_W \sum_{k \geq 1} f \circ T^k \, d\mu \leq \int_X f \, d\mu < \infty,$$

so that $\mu(W \cap \{x \in X : \sum_{n \geq 1} f(T^n x) = \infty\}) = 0$. Since W was an arbitrary wandering set, we obtain

$$\mu\Big(U(\mathcal{W}_T) \cap \Big\{x \in X : \sum_{n \geq 1} f(T^n x) = \infty\Big\}\Big) = 0,$$

which gives the result.

To prove (ii), assume that $f > 0$ μ-a.e. Let $A \in \mathcal{F} \cap \mathcal{C}(T)$ satisfy $\mu(A) > 0$. Then $\mu(A \cap W) = 0$ for all wandering sets W. Set $E = A \cap \{x \in X : f(x) > 0\}$ and define for each $k \geq 1$ the set

$$A_k = A \cap \Big\{x \in X : f(x) > \frac{1}{k}\Big\}.$$

Then $\mu(E) = \mu(A) > 0$ and the sequence (A_k) increases to E. So by continuity of measures we can find a $k \geq 1$, such that $\mu(A_k) > 0$ and this implies that $f|_{A_k} \geq \frac{1}{k} 1_{A_k}$. Since $A_k \subseteq A$, we can apply Halmos' Recurrence Theorem to A and A_k to get for μ-almost every $x \in A_k$ that $\sum_{n \geq 1} 1_{A_k} \circ T^n(x) = \infty$. Hence,

$$\int_A \sum_{n \geq 1} f \circ T^n \, d\mu \geq \frac{1}{k} \int_{A_k} \sum_{n \geq 1} 1_{A_k} \circ T^n \, d\mu = \infty.$$

Since this holds for all positive measure sets $A \in \mathcal{F} \cap \mathcal{C}(T)$, it follows that

$$\mu\Big(\mathcal{C}(T) \setminus \Big\{x \in X : \sum_{n \geq 1} f(T^n x) = \infty\Big\}\Big) = 0.$$

Combining this with (i) then yields (ii). □

The following theorem, due to D. Maharam (see [40] from 1964), gives a way to check whether a system is conservative. Recall from Proposition 2.2.2 that in a conservative and ergodic system any set $A \in \mathcal{F}$ with $\mu(A) > 0$ is visited eventually by μ-a.e. $x \in X$. A set A of finite, positive measure with this property is called a sweep-out set.

Definition 11.2.4. A set $A \in \mathcal{F}$ is called a *sweep-out set* if $0 < \mu(A) < \infty$ and $\mu(X \setminus \bigcup_{n \geq 0} T^{-n} A) = 0$.

As in Remark 2.2.1 it follows that if A is a sweep-out set, then in fact μ-a.e. $x \in X$ will visit A infinitely many times under T.

Theorem 11.2.1 (Maharam's Recurrence Theorem). *Let $(X, \mathcal{F}, \mu)$ be a measure space and $T : X \to X$ a measure preserving transformation. If there exists a sweep-out set for T, then T is conservative.*

Proof. Let A be a sweep-out set for T. Then by Remark 2.2.1,

$$\mu\Big(X \setminus \Big\{x \in X : \sum_{n \geq 1} 1_A(T^n x) = \infty\Big\}\Big) = 0.$$

From Proposition 11.2.2(i) we obtain that

$$\mu\Big(\Big\{x \in X : \sum_{n \geq 1} 1_A(T^n x) = \infty\Big\} \setminus \mathcal{C}(T)\Big) = 0.$$

Hence, $\mu(X \setminus \mathcal{C}(T)) = 0$ and T is conservative. □

Example 11.2.2. Recall Boole's transformation from Example 1.3.4 and consider the set $A = \big[-\frac{1}{\sqrt{2}}, \frac{1}{\sqrt{2}}\big]$. Then for Lebesgue measure λ we have $\lambda(A) = \sqrt{2} > 0$. The endpoints of A are in the same period 2 orbit:

$$T\Big(-\frac{1}{\sqrt{2}}\Big) = \frac{1}{\sqrt{2}} \quad \text{and} \quad T\Big(\frac{1}{\sqrt{2}}\Big) = -\frac{1}{\sqrt{2}}.$$

Define the sequence of points $(x_n^+)_{n \geq 0} \subseteq (0, \infty)$ by $x_0^+ = \frac{1}{\sqrt{2}}$ and $Tx_n^+ = x_{n-1}^+$ for $n \geq 1$. Then for each $n \geq 1$ it holds that $T^n[x_{n-1}^+, x_n^+] =$

A. We can similarly define a sequence of points $(x_n^-)_{n\geq 0} \subsetneq (-\infty, 0)$ and deduce from $\lim_{n\to\infty} x_n^+ = \infty$ and $\lim_{n\to\infty} x_n^- = -\infty$ that $X = \bigcup_{n\geq 0} T^{-n}A$. So A is a sweep-out set. From Theorem 11.2.1 it follows that T is conservative.

Example 11.2.3. Let T be any of the transformations from Example 11.1.1, Example 11.1.2 or Exercise 11.1.1. Take $A = \left[\frac{1}{2}, 1\right)$. Then

$$\mu(A) = \int_{[\frac{1}{2},1)} \frac{1}{x}\, d\lambda(x) = \log 2 \in (0, \infty).$$

Notice that for each $n \geq 2$ we have $T\frac{1}{n+1} = \frac{1}{n}$, so $T^{n-1}\left[\frac{1}{n+1}, \frac{1}{n}\right) = A$. Hence,

$$\bigcup_{n\geq 0} T^{-n}A = \bigcup_{n\geq 1} \left[\frac{1}{n+1}, \frac{1}{n}\right) = (0, 1),$$

which by Theorem 11.2.1 implies that T is conservative.

11.3 INDUCED SYSTEMS

A powerful technique to study infinite measure dynamical systems is inducing. Here the system is restricted to a subset of the state space X and the dynamics is only observed when it passes through this set. To be sure that almost all points return to this part, we take a conservative system as a starting point.

Let $(X, \mathcal{F}, m)$ be a measure space and $T : X \to X$ a conservative transformation. Fix a set $A \in \mathcal{F}$ with $0 < m(A) < \infty$. For $x \in A$, set

$$r(x) := \inf\{n \geq 1 : T^n x \in A\}.$$

We call $r(x)$ the *first return time* of x to A. By Corollary 2.1.1 r is finite m-a.e. on A. In the sequel we remove from A the zero measure set of points that do not return to A infinitely often, i.e., the set of points $x \in A$ for which there is a $k \geq 1$ with $r(T^k x) = \infty$, and we denote the new set again by A. Consider the σ-algebra $\mathcal{F} \cap A$ on A, which is the restriction of $\mathcal{F}$ to A.

Exercise 11.3.1. Show that r is measurable with respect to $\mathcal{F} \cap A$.

Let m_A be the probability measure on $(A, \mathcal{F} \cap A)$ defined by

$$m_A(B) = \frac{m(B)}{m(A)} \quad \text{for } B \in \mathcal{F} \cap A,$$

so that $(A, \mathcal{F} \cap A, m_A)$ is a probability space. Define the *induced transformation* $T_A : A \to A$ by

$$T_A x = T^{r(x)} x.$$

What kind of a transformation is T_A?

Exercise 11.3.2. Show that T_A is measurable with respect to $\mathcal{F} \cap A$.

For $k \geq 1$, let

$$A_k = \{x \in A : r(x) = k\}, \tag{11.2}$$

so that $A = \bigcup_{k \geq 1} A_k$. We can use these sets to prove the following.

Proposition 11.3.1. *Let $(X, \mathcal{F}, m)$ be a measure space and $T : X \to X$ a conservative transformation. Then for any $A \in \mathcal{F}$ with $0 < m(A) < \infty$ the induced transformation $T_A : A \to A$ is non-singular with respect to the measure m_A.*

Proof. Let $C \in \mathcal{F} \cap A$. Then we can write $T_A^{-1} C$ as a countable disjoint union according to the first return times, i.e.,

$$T_A^{-1} C = \bigcup_{k \geq 1} (A_k \cap T^{-k} C). \tag{11.3}$$

Assume that $m_A(C) = 0$. Then $m(C) = 0$ and the non-singularity of T gives

$$0 \leq m(T_A^{-1} C) = \sum_{k \geq 1} m(T^{-k} C \cap A_k) \leq \sum_{k \geq 1} m(T^{-k} C) = 0.$$

So, $m_A(T_A^{-1} C) = 0$. For the other direction, assume $m_A(C) > 0$. Then $m(C) > 0$ and by the conservativity of T there is an $n \geq 1$, such that $m(A \cap T^{-n} C) \geq m(C \cap T^{-n} C) > 0$. Let n_0 be the least integer with this property. Since $C \subseteq A$, there exists a $k \leq n_0$ such that $m(A_k \cap T^{-n_0} C) > 0$. We claim $k = n_0$. Suppose $k < n_0$ and note that $A_k \cap T^{-n_0} C \subseteq T^{-k}\big(A \cap T^{-(n_0-k)} C\big)$. Hence, $m\Big(T^{-k}\big(A \cap T^{-(n_0-k)} C\big)\Big) > 0$ and the non-singularity of T gives $m(A \cap T^{-(n_0-k)} C) > 0$, contradicting the minimality of n_0. Thus, $m(A_{n_0} \cap T^{-n_0} C) > 0$ implying $m(T_A^{-1} C) > 0$ and then also $m_A(T_A^{-1} C) > 0$. The non-singularity of T_A follows. □

This proposition, together with Exercise 11.3.2, implies that $(A, \mathcal{F} \cap A, m_A, T_A)$ is a dynamical system, which justifies the following definition.

Definition 11.3.1. Let $(X, \mathcal{F}, m, T)$ be a dynamical system with $T : X \to X$ conservative and let $A \in \mathcal{F}$ satisfy $0 < m(A) < \infty$. Then the dynamical system $(A, \mathcal{F} \cap A, m_A, T_A)$ is called the *induced system* of $(X, \mathcal{F}, m, T)$ on A.

Example 11.3.1. Consider again the transformation $T : [0, 1) \to [0, 1)$ from Example 11.1.2 given by

$$Tx = \begin{cases} \frac{x}{1-x}, & \text{if } 0 \le x < \frac{1}{2}, \\ 2x - 1, & \text{if } \frac{1}{2} \le x < 1. \end{cases}$$

We construct the induced transformation T_A for $A = [\frac{1}{2}, 1)$. Obviously, any $x \in [\frac{3}{4}, 1)$ has first return time $r(x) = 1$. Furthermore, for each $n \ge 2$ we have $T\frac{1}{n+1} = \frac{1}{n}$, so $T^{n-1}[\frac{1}{n+1}, \frac{1}{n}) = A$. The point $\frac{n+1}{2n}$ lies in A and satisfies $T\frac{n+1}{2n} = \frac{1}{n}$. Hence, using the notation from (11.2),

$$A_n = \left[\frac{n+2}{2(n+1)}, \frac{n+1}{2n}\right), \quad n \ge 1,$$

or in other words $r(x) = n$ precisely when $x \in A_n$. See Figure 11.2(a). Write $T_1 : x \mapsto \frac{x}{1-x}$ for the first branch of T. Then for each $n \ge 1$,

$$T_1^n x = \frac{x}{1 - nx}. \tag{11.4}$$

Hence, on A_n it holds that

$$T_A x = T_1^{n-1}(2x - 1) = \frac{2x - 1}{n - (n-1)2x}.$$

The graph of T_A is shown in Figure 11.2(b).

Proposition 11.3.2. *Let $(X, \mathcal{F}, \mu, T)$ be a dynamical system with $T : X \to X$ conservative and let $A \in \mathcal{F}$ satisfy $0 < \mu(A) < \infty$. Then $T_A : A \to A$ is conservative on $(A, \mathcal{F} \cap A, \mu_A)$. Moreover, if T is measure preserving with respect to μ, then so is T_A with respect to μ_A.*

Proof. Recall the definition of the sets A_k from (11.2). To show that T_A is conservative, let $C \in \mathcal{F} \cap A$. Note that for each $x \in A$, since $C \subseteq A$,

$$\sum_{n \ge 1} 1_C(T_A^n x) = \infty \quad \Leftrightarrow \quad \sum_{n \ge 1} 1_C(T^n x) = \infty.$$

T is conservative, so by Corollary 2.1.1

$$\mu\Big(\Big\{x \in C : \sum_{n \ge 1} 1_C(T^n x) = \infty\Big\}\Big) = \mu(C)$$

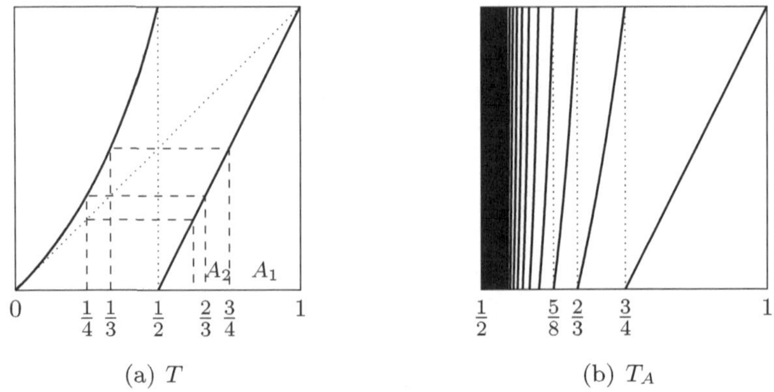

Figure 11.2 The graph of the map T from Example 11.3.1 on the left and its induced transformation on the interval $A = [\frac{1}{2}, 1)$ on the right.

and hence

$$\mu_A\Big(\Big\{x \in C : \sum_{n \geq 1} 1_C(T_A^n x) = \infty\Big\}\Big)$$

$$= \frac{\mu\Big(\Big\{x \in C : \sum_{n\geq 1} 1_C(T^n x) = \infty\Big\}\Big)}{\mu(A)} = \mu_A(C).$$

Since $C \in \mathcal{F} \cap A$ was arbitrary, Corollary 2.1.1 implies that T_A is conservative.

Now assume that T is measure preserving with respect to μ and let $C \in \mathcal{F} \cap A$. Since $\mu(C) = \mu(T^{-1}C)$, for the measure preservingness of T_A it is enough to show that $\mu(T_A^{-1}C) = \mu(T^{-1}C)$. For each $k \geq 1$ define

$$B_k = \{x \in A^c : Tx, \ldots, T^{k-1}x \notin A, T^k x \in A\}.$$

Notice that

$$T^{-1}A = A_1 \cup B_1 \quad \text{and} \quad T^{-1}B_n = A_{n+1} \cup B_{n+1}. \tag{11.5}$$

See Figure 11.3 for an illustration of where T maps the sets A_k and B_k.

From (11.3) we know that

$$\mu(T_A^{-1}C) = \sum_{n \geq 1} \mu(A_n \cap T^{-n}C).$$

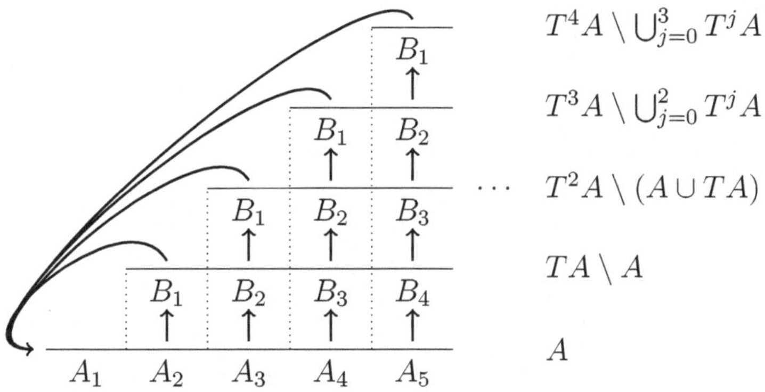

Figure 11.3 A tower.

On the other hand, using (11.5) repeatedly one gets for any $n \geq 1$,

$$\begin{aligned}
\mu(T^{-1}C) &= \mu(A_1 \cap T^{-1}C) + \mu(B_1 \cap T^{-1}C) \\
&= \mu(A_1 \cap T^{-1}C) + \mu(T^{-1}(B_1 \cap T^{-1}C)) \\
&= \mu(A_1 \cap T^{-1}C) + \mu(A_2 \cap T^{-2}C) + \mu(B_2 \cap T^{-2}C) \\
&\ \ \vdots \\
&= \sum_{k=1}^{n} \mu(A_k \cap T^{-k}C) + \mu(B_n \cap T^{-n}C).
\end{aligned}$$

We claim that $\lim_{n\to\infty} \mu(B_n) = 0$. This follows from the fact that if we replace C by A in the above calculations, we get for each n

$$\mu(A) = \mu(T^{-1}A) = \sum_{k=1}^{n} \mu(A_k) + \mu(B_n).$$

Taking limits, and using the fact that $\mu(A) = \sum_{k=1}^{\infty} \mu(A_k)$ as well as $\mu(A) < \infty$, we see that $\lim_{n\to\infty} \mu(B_n) = 0$. Since $\mu(B_n \cap T^{-n}C) \leq \mu(B_n)$, we have $\lim_{n\to\infty} \mu(B_n \cap T^{-n}C) = 0$. Thus,

$$\mu(C) = \mu(T^{-1}C) = \sum_{n\geq 1} \mu(A_n \cap T^{-n}C) = \mu(T_A^{-1}C).$$

This gives the result. □

Exercise 11.3.3. Let $(X, \mathcal{F}, \mu, T)$ be a dynamical system and assume that $T : X \to X$ is conservative, measure preserving and invertible. Show that

$$\mu_A(C) = \mu_A(T_A C)$$

for all $C \in \mathcal{F} \cap A$ without using Proposition 11.3.2.

Exercise 11.3.4. Let $(X, \mathcal{F}, m, T)$ be a dynamical system. Let $T : X \to X$ be conservative and $A \in \mathcal{F}$ be such that $0 < m(A) < \infty$. Prove the following.

(a) If T is ergodic, then T_A is ergodic.

(b) If T_A is ergodic and A is a sweep-out set, then T is ergodic.

Exercise 11.3.5. Consider the rotation from Example 1.3.2, i.e., $Tx = x + \theta \pmod 1$ for some $\theta \in (0,1)$. Assume that θ irrational. Determine explicitly the induced transformation T_A of T on the interval $A = [0, \theta)$.

Exercise 11.3.6. Let $G = \frac{1+\sqrt{5}}{2}$ be the golden mean, so that $G^2 = G+1$. Consider the set

$$X = \Big[0, \frac{1}{G}\Big) \times [0,1) \cup \Big[\frac{1}{G}, 1\Big) \times \Big[0, \frac{1}{G}\Big),$$

endowed with the product Lebesgue σ-algebra and the normalized Lebesgue measure λ^2. Define the transformation

$$T(x,y) = \begin{cases} (Gx, \frac{y}{G}), & \text{if } (x,y) \in [0, \frac{1}{G}) \times [0,1], \\ (Gx - 1, \frac{1+y}{G}), & \text{if } (x,y) \in [\frac{1}{G}, 1) \times [0, \frac{1}{G}). \end{cases}$$

(a) Show that T is measure preserving with respect to λ^2.

(b) Determine explicitly the induced transformation U of T on the set $[0,1) \times [0, \frac{1}{G})$.

Now let $S : [0,1)^2 \to [0,1)^2$ be given by

$$S(x,y) = \begin{cases} (Gx, \frac{y}{G}), & \text{if } (x,y) \in [0, \frac{1}{G}) \times [0,1), \\ & \\ (G^2 x - G, \frac{G+y}{G^2}), & \text{if } (x,y) \in [\frac{1}{G}, 1) \times [0,1). \end{cases}$$

(c) Show that S is measure preserving with respect to Lebesgue measure on $[0,1)^2$.

(d) Show that the map $\phi : [0,1)^2 \to [0,1) \times [0, \frac{1}{G})$ given by

$$\phi(x,y) = \Big(x, \frac{y}{G}\Big)$$

defines an isomorphism from S to U, where $[0,1) \times [0, \frac{1}{G})$ has the induced measure structure.

The next proposition is one of the reasons why inducing is so useful for infinite measure systems.

Proposition 11.3.3. *Let $(X, \mathcal{F}, m, T)$ be a dynamical system with $T : X \to X$ conservative. Let $A \in \mathcal{F}$ be a sweep-out set. Assume that the induced map T_A preserves some probability measure ν on $(A, \mathcal{F} \cap A)$ that is absolutely continuous with respect to the measure m_A.*

(i) *The measure μ on $(X, \mathcal{F})$ given by*

$$\mu(B) = \sum_{n \geq 0} \nu(T^{-n}B \cap \{x \in A : r(x) > n\}) \tag{11.6}$$

for all $B \in \mathcal{F}$ is a T-invariant measure on $(X, \mathcal{F})$ and is absolutely continuous with respect to m.

(ii) *The system $(X, \mathcal{F}, \mu, T)$ is conservative.*

Proof. For (i) it is straightforward to check that μ is a measure. To prove that it is T-invariant, let $B \in \mathcal{F}$ be given and write $A_k = \{x \in A : r(x) = k\}$ as before. Then $B \cap A \in \mathcal{F} \cap A$, so by (11.3)

$$\nu(B \cap A) = \nu(T_A^{-1}(B \cap A)) = \sum_{n \geq 1} \nu(A_n \cap T^{-n}B).$$

Then

$$\begin{aligned}
\mu(T^{-1}B) &= \sum_{n \geq 0} \nu(T^{-(n+1)}B \cap \{x \in A : r(x) > n\}) \\
&= \sum_{n \geq 1} \nu(A_n \cap T^{-n}B) + \sum_{n \geq 1} \nu(T^{-n}B \cap \{x \in A : r(x) > n\}) \\
&= \nu(B \cap A) + \sum_{n \geq 1} \nu(T^{-n}B \cap \{x \in A : r(x) > n\}) = \mu(B).
\end{aligned}$$

So μ is T-invariant. Finally, suppose $m(B) = 0$. The non-singularity of T then gives $m(T^{-n}B) = 0$ for all n. Hence $m_A(T^{-n}B \cap \{x \in A : r(X) > n\}) = 0$ for all n, and since ν is absolutely continuous with respect to m_A, also $\nu(T^{-n}B \cap \{x \in A : r(X) > n\}) = 0$ for all n. This shows that $\mu(A) = 0$ and μ is absolutely continuous with respect to m.

To prove (ii), we first note that

$$\begin{aligned}
\mu(A) &= \sum_{n \geq 0} \nu(T^{-n}A \cap \{x \in A : r(x) > n\}) \\
&= \nu(A) + \sum_{n \geq 1} \nu(T^{-n}A \cap \{x \in A : r(x) > n\}) = \nu(A) = 1.
\end{aligned}$$

By T-invariance we have $\mu(T^{-k}A) = 1$ for all k. Since $m(X \setminus \bigcup_{k\geq 0} T^{-k}A) = 0$ it follows by absolute continuity that $\mu(X \setminus \bigcup_{k\geq 0} T^{-k}A) = 0$, so $(X, \mathcal{F}, \mu)$ is a σ-finite measure space and A is a sweep-out set with respect to μ. The conservativity of μ then follows from Maharam's Recurrence Theorem. □

Under the hypotheses of Proposition 11.3.3 the measure ν equals the induced measure μ_A. To see this, let $C \in \mathcal{F} \cap A$. Then, since $C \subseteq A$,

$$\mu_A(C) = \frac{\mu(C)}{\mu(A)} = \sum_{n\geq 0} \nu(T^{-n}C \cap \{x \in A : r(x) > n\}) = \nu(C).$$

The proof of Proposition 11.3.3 shows that A is a sweep-out set for μ as well as for m. It then follows from Exercise 11.3.4 that if ν is ergodic, then also μ is ergodic.

Exercise 11.3.7 (Kac's Lemma). Let $(X, \mathcal{F}, \mu, T)$ be a dynamical system with $T : X \to X$ a measure preserving, conservative and ergodic transformation. Let $A \in \mathcal{F}$ satisfy $0 < \mu(A) < \infty$. Prove that

$$\int_A r \, d\mu = \mu(X).$$

Conclude that if μ is a probability measure, then $r \in L^1(A, \mathcal{F} \cap A, \mu_A)$, and

$$\lim_{n\to\infty} \frac{1}{n} \sum_{i=0}^{n-1} r(T_A^i x) = \frac{1}{\mu(A)},$$

for μ_A-almost every x.

From the previous exercise we immediately deduce the following.

Corollary 11.3.1. *Under the conditions of Proposition 11.3.3, suppose that T is ergodic with respect to the measure μ from* (11.6). *Then μ is infinite if and only if $\int_A r \, d\mu = \infty$.*

Example 11.3.2. Define the transformation $T : [0, \infty) \to [0, \infty)$ by

$$Tx = \begin{cases} 0, & \text{if } x = 0, \\ \frac{1}{x} - 1, & \text{if } 0 < x \leq 1, \\ x - 1, & \text{if } x > 1. \end{cases}$$

See Figure 11.4(a) for the graph. Let $A = [0, 1)$ and consider the induced transformation $T_A : [0,1) \to [0,1)$. Note that $r(x) = 1$ for all $x \in (\frac{1}{2}, 1] \cup \{0\}$. For any $n \geq 2$ and $x \in (\frac{1}{n+1}, \frac{1}{n}]$ we have $Tx = \frac{1}{x} - 1 \in [n-1, n)$ and thus $r(x) = n$. We then immediately see that T_A becomes a very familiar map: $T_A 0 = 0$ and $T_A x = \frac{1}{x} \pmod 1$ for $x \neq 0$. The induced map therefore has the Gauss measure (see (1.6)) as an invariant probability measure ν and we can use Proposition 11.3.3 to obtain an invariant measure μ for T. The ergodicity of T for μ follows immediately from the ergodicity of T_A for ν (see Theorem 8.2.1). Proposition 11.3.3 also gives the conservativity of T with respect to μ. We will compute μ.

It is easily verified that $\{x \in [0,1) : r(x) > n\} = (0, \frac{1}{n+1})$ for any $n \geq 1$. Moreover, $T^n|_{(0,\frac{1}{n+1})} x = \frac{1}{x} - n$. Let $B \subseteq (0, \infty)$ be an arbitrary Lebesgue set. If $B \subseteq (0,1)$ it holds that

$$\mu(B) = \nu(B) = \frac{1}{\log 2} \int_B \frac{1}{x+1}\, d\lambda(x).$$

If, on the other hand, $B \subseteq (1, \infty)$, then

$$\begin{aligned} \mu(B) &= \sum_{n \geq 0} \nu\left(T^{-n} B \cap \left(0, \frac{1}{n+1}\right)\right) \\ &= \sum_{n \geq 0} \nu\left(\frac{1}{B+n}\right). \end{aligned}$$

The same computation that shows ν is the invariant measure for the Gauss map yields that

$$\mu(B) = \frac{1}{\log 2} \int_B \frac{1}{x+1}\, d\lambda(x) \tag{11.7}$$

also in this case. So, (11.7) gives $\mu(B)$ for any Lebesgue measurable subset $B \subseteq [0, \infty)$. It is now immediately clear that $\mu([0,\infty)) = \infty$, but we could also deduce this from Kac's formula.

Exercise 11.3.8. Show that in the above example,

$$\int_{[0,\infty)} r \, d\mu = \infty.$$

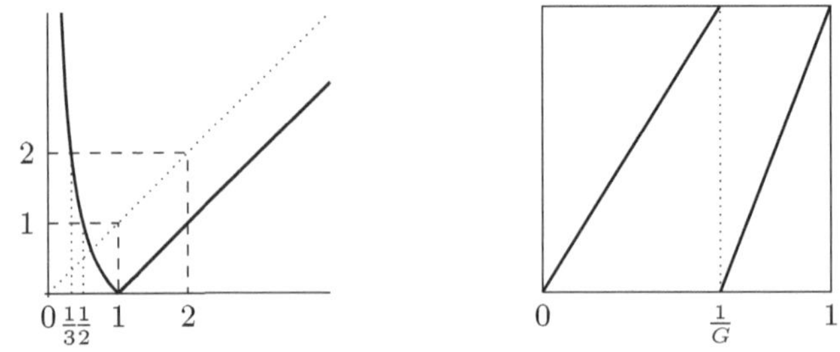

(a) T from Example 11.3.2

(b) J_A

Figure 11.4 The graph of the map T from Example 11.3.2 on the left. The dashed lines indicate that the interval $(\frac{1}{3}, \frac{1}{2}]$ is first mapped to $[1, 2)$ and then back to $[0, 1)$ giving return time 2 on this interval. On the right we see the jump transformation J_A from Example 11.4.1.

11.4 JUMP TRANSFORMATIONS

A construction very similar to the induced transformation is the jump transformation which we introduce next. Let $(X, \mathcal{F}, m)$ be a measure space and $T : X \to X$ a transformation. Let $A \in \mathcal{F}$ be a sweep-out set with the property that $TA \in \mathcal{F}$ and $m(X \setminus TA) = 0$. Define the *first passage time* $p : X \to \mathbb{N}$ by

$$p(x) = 1 + \inf\{n \geq 0 \,:\, T^n x \in A\},$$

so it is 1 plus the number of iterates needed to map x inside A. The *jump transformation* $J_A : X \to X$ is defined by

$$J_A x = T^{p(x)} x.$$

In contrast to an induced transformation, a jump transformation is defined on the whole space X. A jump transformation can be used for systems that are slow on some part of the space X (often the reason that an invariant measure is infinite). The system is then accelerated to the point that it admits a finite invariant measure. This is formulated in the next proposition, the proof of which is to a large extent equal to the proof of Proposition 11.3.3.

Proposition 11.4.1. *Let $(X, \mathcal{F}, m, T)$ be a dynamical system. Let $A \in \mathcal{F}$ be a sweep-out set with the properties that $TA \in \mathcal{F}$ and $m(X \setminus TA) =$*

0. *Assume that* J_A *preserves a probability measure* ν *on* $(X, \mathcal{F})$. *Then the measure* μ *on* $(X, \mathcal{F})$ *given by*

$$\mu(B) = \sum_{n \geq 0} \nu(T^{-n}B \cap \{x \in X : p(x) > n\}) \quad \text{for all } B \in \mathcal{F} \tag{11.8}$$

is T*-invariant. Moreover, if* ν *is ergodic for* J_A*, then the measure* μ *from* (11.8) *is ergodic for* T.

Exercise 11.4.1. Prove Proposition 11.4.1.

Exercise 11.4.2. Let $T : X \to X$ be a measure preserving transformation on a measure space $(X, \mathcal{F}, \mu)$ and let A be a sweep-out set. Prove that

$$\mu(\{x \in A : r(x) > k\}) = \mu(\{x \in X : p(x) = k + 1\})$$

for any $k \geq 0$.

Example 11.4.1. Let $G = \frac{1+\sqrt{5}}{2}$ be the golden mean (so $G^2 = G + 1$) and let T be the β-transformation from Example 1.3.6 with $\beta = G$. The formula for the invariant density from (1.4) can easily be deduced from Proposition 11.4.1. Let $A = [0, \frac{1}{G})$. Then $p(x) = 1$ for $x \in A$ and $p(x) = 2$ for $x \notin A$. So,

$$J_A x = \begin{cases} Gx, & \text{if } x \in A, \\ G^2 x - G, & \text{if } x \notin A. \end{cases}$$

See Figure 11.4(b) for the graph. One immediately sees that Lebesgue measure λ is invariant for J_A, so by Proposition 11.4.1 the measure $\hat{\mu}$ given by

$$\begin{aligned} \hat{\mu}(B) &= \sum_{n \geq 0} \lambda(T^{-n}B \cap \{x \in [0,1) : p(x) > n\}) \\ &= \lambda(B) + \lambda\Big(T^{-1}B \cap \Big[\frac{1}{G}, 1\Big)\Big) \\ &= \lambda(B) + \frac{1}{G}\lambda\Big(B \cap \Big[0, \frac{1}{G}\Big)\Big) \end{aligned}$$

for any Lebesgue set B is T-invariant. Note that $\hat{\mu}([0,1)) = 1 + \frac{1}{G^2} < \infty$, so if we normalize $\hat{\mu}$ we get the invariant probability measure μ for T given by

$$\mu(B) = \frac{G^2}{1+G^2}\Big(\lambda\Big(B \cap \Big[\frac{1}{G}, 1\Big)\Big) + \Big(1 + \frac{1}{G}\Big)\lambda\Big(B \cap \Big[0, \frac{1}{G}\Big)\Big)\Big),$$

which is precisely the measure from (1.4).

Example 11.4.2. Consider again the Farey map $T : [0,1] \to [0,1]$ from Example 11.1.1 with its σ-finite, infinite invariant measure μ on $([0,1], \mathcal{B})$ given by

$$\mu(A) = \int_A \frac{1}{x}\, d\lambda(x) \qquad \text{for all } A \in \mathcal{B}.$$

Consider the jump transformation of T to the interval $A = \left(\frac{1}{2}, 1\right]$. Obviously $TA = [0,1)$. As in Example 11.2.3 for any $n \geq 2$,

$$T^{n-1}\left(\frac{1}{n+1}, \frac{1}{n}\right] = A.$$

Hence, $p(x) = n$ for $x \in \left(\frac{1}{n+1}, \frac{1}{n}\right]$ and $\bigcup_{n\geq 0} T^{-n}A = (0,1]$, so A is a sweep-out set with respect to λ. By (11.4) the jump transformation $J_A : [0,1] \to [0,1]$ is given on $\left(\frac{1}{n+1}, \frac{1}{n}\right]$ by

$$J_A x = \frac{1 - T^{n-1}x}{T^{n-1}x} = \frac{1 - nx}{x} = \frac{1}{x} - n.$$

In other words, J_A is the Gauss map from Example 1.3.7: $J_A x = \frac{1}{x}$ (mod 1). Let ν denote the Gauss measure from (1.6). One can check that

$$\mu(B) = \log 2 \sum_{n \geq 0} \nu(T^{-n}B \cap \{x \in [0,1] \,:\, p(x) > n\}) \tag{11.9}$$

for all $B \in \mathcal{B}$. Since J_A is ergodic, it follows from Proposition 11.4.1 that also T is ergodic.

Exercise 11.4.3. Prove (11.9).

11.5 INFINITE ERGODIC THEOREMS

While the Pointwise Ergodic Theorem was one of the main results we discussed, the statement is not true for infinite measure systems. In fact we have the following result.

Theorem 11.5.1. *Let $(X, \mathcal{F}, \mu)$ be a measure space with $\mu(X) = \infty$ and let $T : X \to X$ be a conservative, measure preserving and ergodic transformation. Then for all $f \in L^1(X, \mathcal{F}, \mu)$,*

$$\lim_{n\to\infty} \frac{1}{n} \sum_{i=0}^{n-1} f(T^i x) = 0 \quad \mu - a.e.$$

The proof of this result is a consequence of the following more general theorem.

Theorem 11.5.2 (Hopf's Ratio Ergodic Theorem). *Let $(X, \mathcal{F}, \mu)$ be a measure space with $\mu(X) = \infty$ and let $T : X \to X$ be a conservative, measure preserving and ergodic transformation. Let $f, g \in L^1(X, \mathcal{F}, \mu)$ with $g \geq 0$ and $\int_X g \, d\mu > 0$. Then*

$$\lim_{n\to\infty} \frac{\sum_{i=0}^{n-1} f(T^i x)}{\sum_{i=0}^{n-1} g(T^i x)} = \frac{\int_X f \, d\mu}{\int_X g \, d\mu} \quad \mu - a.e.$$

Before we prove this theorem, we first show how to derive Theorem 11.5.1 from it.

Proof of Theorem 11.5.1. By the σ-finiteness of the measure space, we can find for each $k \geq 1$ a set $B_k \in \mathcal{F}$ with $k \leq \mu(B_k) < \infty$. Let $f \in L^1(X, \mathcal{F}, \mu)$ be given and assume without loss of generality that $f \geq 0$. We apply Theorem 11.5.2 on the functions f and 1_{B_k}. Note that by Proposition 2.2.2 the set B_k is a sweep-out set, so that for μ-a.e. $x \in X$ for all n large enough it holds that $\sum_{i=0}^{n-1} 1_{B_k}(T^i x) \geq 1$. Therefore, for μ-almost every $x \in X$,

$$\begin{aligned} 0 \leq \limsup_{n\to\infty} \frac{1}{n} \sum_{i=0}^{n-1} f(T^i x) &\leq \limsup_{n\to\infty} \frac{\sum_{i=0}^{n-1} f(T^i x)}{\sum_{i=0}^{n-1} 1_{B_k}(T^i x)} \\ &= \lim_{n\to\infty} \frac{\sum_{i=0}^{n-1} f(T^i x)}{\sum_{i=0}^{n-1} 1_{B_k}(T^i x)} = \frac{\int_X f \, d\mu}{\mu(B_k)}. \end{aligned}$$

Let

$$Y_k = \Big\{ x \in X \, : \, \limsup_{n\to\infty} \frac{1}{n} \sum_{i=0}^{n-1} f(T^i x) > \frac{\int_X f \, d\mu}{\mu(B_k)} \Big\}$$

denote the exceptional set for B_k and put $Y = \bigcup_{k\geq 1} Y_k$. Then $\mu(Y) = 0$ and for all $x \in X \setminus Y$ and $k \geq 1$,

$$0 \leq \limsup_{n\to\infty} \frac{1}{n} \sum_{i=0}^{n-1} f(T^i x) \leq \frac{1}{k} \int_X f \, d\mu.$$

Since $f \geq 0$ the statement follows. □

The proof of Theorem 11.5.2 we give here is basically the same proof as the one given for the Pointwise Ergodic Theorem in Chapter 3. We give a sketch of the proof with a description of the necessary modifications.

Sketch of the proof of Theorem 11.5.2. Assume without loss of generality that $f \geq 0$. Define

$$\overline{R}(x) = \limsup_{n\to\infty} \frac{\sum_{i=0}^{n-1} f(T^i x)}{\sum_{i=0}^{n-1} g(T^i x)} \quad \text{and} \quad \underline{R}(x) = \liminf_{n\to\infty} \frac{\sum_{i=0}^{n-1} f(T^i x)}{\sum_{i=0}^{n-1} g(T^i x)}.$$

From Proposition 2.2.3 it follows that $\sum_{i\geq 0} g(T^i x) = \infty$ for μ-almost every $x \in X$, so that $\overline{R}$ and $\underline{R}$ are well-defined. Then $\overline{R}$ and $\underline{R}$ are T-invariant functions (this is an exercise), so that by Theorem 2.2.2(ii) $\overline{R}$ and $\underline{R}$ are constants μ-almost everywhere. We claim that to prove the theorem it is enough to prove that

$$\int_X \overline{R} g \, d\mu \leq \int_X f \, d\mu \leq \int_X \underline{R} g \, d\mu. \tag{11.10}$$

Since $\overline{R} \geq \underline{R}$ and $g \geq 0$ from (11.10) it would follow that

$$\int_X g \cdot (\overline{R} - \underline{R}) d\mu = 0$$

and hence $g \cdot (\overline{R} - \underline{R}) = 0$ μ-a.e. Since $\overline{R}$ and $\underline{R}$ are both constant and $\int_X g \, d\mu > 0$ we see that $\overline{R} - \underline{R} = 0$. Hence,

$$\overline{R} = \underline{R} = \lim_{n\to\infty} \frac{\sum_{i=0}^{n-1} f(T^i x)}{\sum_{i=0}^{n-1} g(T^i x)}.$$

The proofs of the inequalities $\int_X \overline{R} g \, d\mu \leq \int_X f \, d\mu$ and $\int_X f \, d\mu \leq \int_X \underline{R} g \, d\mu$ are the same as the corresponding estimates in the proof of Theorem 3.1.1 with the some small modifications. We treat the estimate $\int_X \overline{R} g \, d\mu \leq \int_X f \, d\mu$ and leave the other one to the reader. For any $L > 0$ the measure ν defined by $\nu(A) = \int_A L \cdot g \, d\mu$ for all $A \in \mathcal{F}$ is absolutely continuous with respect to μ. So for any given $\delta > 0$ there is a $\delta' < \delta$ such that if $\mu(A) < \delta'$, then $\int_A L \cdot g \, d\mu < \delta$. By the definition of limsup, given $0 < \varepsilon < 1$ and $L > 0$ there is an $M \geq 1$, such that if

$$X_0 = \left\{ x \in X : \exists m \leq M \text{ with } \frac{\sum_{i=0}^{m-1} f(T^i x)}{\sum_{i=0}^{m-1} g(T^i x)} \geq \min\{\overline{R}(x), L\}(1-\varepsilon) \right\},$$

then $\mu(X \setminus X_0) < \delta' < \delta$. The function F is modified to

$$F(x) = \begin{cases} f(x), & \text{if } x \in X_0, \\ Lg(x), & \text{if } x \notin X_0. \end{cases}$$

The sequences $(a_n(x))$ and $(b_n(x))$ are now defined by $a_n(x) = F(T^n x)$ and $b_n(x) = g(T^n x) \min(\overline{R}(x), L)(1 - \epsilon)$. The rest of the proof follows the same steps as in Theorem 3.1.1. □

Exercise 11.5.1. Prove that the functions $\underline{R}$ and $\overline{R}$ from the proof of Theorem 11.5.2 are T-invariant, and then finish the proof of this theorem.

The statement from Theorem 11.5.1 is bad news, but one could argue that the sequence $(\frac{1}{n})_{n\geq 1}$ used in front of the ergodic sums is just converging to 0 too fast and that the result could maybe be salvaged by replacing this sequence by another sequence (c_n) with slower convergence to 0. The following result by J. Aaronson, the proof of which can be found in [1, Section 2.4], shows that this will not work.

Theorem 11.5.3 (Aaronson's Theorem). *Let $(X, \mathcal{F}, \mu)$ be a measure space with $\mu(X) = \infty$ and $T : X \to X$ a conservative, measure preserving and ergodic transformation. Let $(c_n)_{n\geq 1} \subseteq (0, \infty)$ be an arbitrary sequence. Then either*

(i) $\liminf_{n\to\infty} \frac{1}{c_n} \sum_{i=0}^{n-1} f(T^i x) = 0$ *μ-a.e. for all $f \in L^1(X, \mathcal{F}, \mu)$ with $f \geq 0$,*

or

(ii) *there is a subsequence $(c_{n_k})_{k\geq 1}$ such that* $\lim_{k\to\infty} \frac{1}{c_{n_k}} \sum_{i=0}^{n_k-1} f(T^i x) = \infty$ *μ-a.e. for all $f \in L^1(X, \mathcal{F}, \mu)$ with $f \geq 0$.*

To end on a positive note, there exist many other tools to describe the long term behavior of infinite measure dynamical systems, but they fall beyond the scope of this book. We refer the reader to e.g. [1] or [29]. Here we give a small example, showing that even with Hopf's Ratio Ergodic Theorem we can still obtain a little bit of information on relative digit frequencies as we did in Section 3.2.

Example 11.5.1. Recall the definition of the Rényi map $T : [0, 1) \to [0, 1)$ by

$$Tx = \frac{x}{1-x} \pmod 1.$$

from Exercise 11.1.1. It can be used to produce *backwards continued fraction expansions* of numbers in the interval $[0, 1)$ of the form

$$x = 1 - \cfrac{1}{b_1 - \cfrac{1}{b_2 - \cfrac{1}{b_3 - \ddots}}}, \quad b_n \geq 2. \tag{11.11}$$

The digit sequence (b_n) is obtained from setting

$$b_n = b_n(x) = k + 1 \quad \text{if } T^{n-1}x \in \Big[\frac{k-1}{k}, \frac{k}{k+1}\Big), \ k \geq 1.$$

Then $Tx = \frac{1}{1-x} - b_1 + 1$ and rewriting and iterating gives

$$\begin{aligned} x = 1 - \frac{1}{b_1 - 1 + Tx} &= 1 - \cfrac{1}{b_1 - \cfrac{1}{b_2 - 1 + T^2x}} \\ = \cdots = 1 - &\cfrac{1}{b_1 - \cfrac{1}{b_2 - \ddots - \cfrac{1}{b_n - 1 + T^n x}}} \end{aligned} \tag{11.12}$$

for any $n \geq 1$. From Exercise 11.1.1(a) we see how T relates to the α-continued fraction map with $\alpha = 0$. It then follows from Proposition 8.4.1 that the process from (11.12) converges.

The measure μ on $([0,1), \mathcal{B})$ given by (11.1) is invariant for T. It is a straightforward application of Theorem 11.5.1 that for Lebesgue almost all $x \in [0,1)$ and all $k \geq 3$,

$$\lim_{n\to\infty} \frac{\#\{1 \leq j \leq n : b_j(x) = k\}}{n} = \lim_{n\to\infty} \frac{1}{n} \sum_{i=0}^{n-1} 1_{[\frac{k-1}{k}, \frac{k}{k+1})}(T^i x) = 0.$$

So for Lebesgue almost all x any digit $k > 2$ occurs with frequency 0 in the digit sequence (b_n). As a consequence, the frequency of the digit 2 is one. Using Hopf's Ratio Ergodic Theorem we can compare the frequency of any two digits $k, \ell \neq 2$. For any $k \geq 2$ the interval $[\frac{k-1}{k}, \frac{k}{k+1})$ has

$$\mu\Big(\Big[\frac{k-1}{k}, \frac{k}{k+1}\Big)\Big) = \log\Big(\frac{k^2}{k^2-1}\Big).$$

Hence, for Lebesgue almost all $x \in [0,1)$ and any $k, \ell > 2$,

$$\lim_{n\to\infty} \frac{\#\{1 \leq j \leq n : b_n(x) = k\}}{\#\{1 \leq j \leq n : b_n(x) = \ell\}} = \frac{\log\big(\frac{k^2}{k^2-1}\big)}{\log\big(\frac{\ell^2}{\ell^2-1}\big)}.$$

Using this expression it is easy to calculate, for example, that for any $k \geq 3$ and any $m \geq 2$

$$\lim_{n\to\infty} \frac{\#\{1 \leq j \leq n : b_n(x) = k\}}{\#\{1 \leq j \leq n : b_n(x) = mk\}} = m^2.$$

So in typical digit sequences (b_n) the digit k appears m^2 times as often as the digit mk.

CHAPTER 12

Appendix

12.1 TOPOLOGY

We will assume that the reader has some familiarity with basic (analytic) topology, but for future reference we start by recalling some concepts over compact metric spaces that are needed in some parts of the book.

For any subset $A \subseteq X$ we write A^o for the *interior* of A, $\overline{A}$ for the *closure* of A and $\partial A = \overline{A} \setminus A^o$ for the *boundary* of A.

Lemma 12.1.1. *Let (X, d) be a compact metric space and $T : X \to X$ a continuous function. Then for any subset A of X we have*

$$\partial(T^{-1}A) \subseteq T^{-1}(\partial A).$$

Proof. Let $y \in \partial(T^{-1}A)$, then there exists a sequence $(x_n) \in T^{-1}A$ converging to y. By continuity we have $Tx_n \to Ty$ with $Tx_n \in A \subseteq \overline{A}$. Since X is compact, there exists a subsequence (x_{n_j}) such that (Tx_{n_j}) converges to some $a \in \overline{A}$. However, (Tx_{n_j}) converges to Ty so that $Ty = a$ and $Ty \in \overline{A} = \partial A \cup A^o$. Since $\partial(T^{-1}A) = \overline{T^{-1}A} \setminus (T^{-1}A)^o \subseteq T^{-1}\overline{A} \setminus T^{-1}A^o$ and $y \in \partial(T^{-1}A)$, we see that $Ty \notin A^o$. This implies that $Ty \in \partial A$ or that $y \in T^{-1}(\partial A)$. □

Theorem 12.1.1. *Let X be a compact metric space, Y a topological space and $f : X \to Y$ continuous. Then all the following hold.*

(i) *X is a Hausdorff space.*

(ii) *Every closed subspace of X is compact.*

(iii) *Every compact subspace of X is closed.*

(iv) *X has a countable basis for its topology.*

(v) *X is* normal, *i.e., for any pair of disjoint closed sets A and B of X there are disjoint open sets U, V containing A and B, respectively.*

(vi) *If $\mathcal{O}$ is an open cover of X, then there is a $\delta > 0$ such that every subset A of X with diam$(A) < \delta$ is contained in an element of $\mathcal{O}$. We call $\delta > 0$ a Lebesgue number for $\mathcal{O}$;*

(vii) *X is a Baire space, i.e., every intersection of a countable collection of open, dense subsets of X is itself dense.*

(viii) *f is uniformly continuous.*

(ix) *If Y is an ordered space, then f attains a maximum and minimum on X.*

(x) *If A and B are closed sets in X and $[a, b] \subset \mathbb{R}$, then there exists a continuous map $g : X \to [a, b]$ such that $g(x) = a$ for all $x \in A$ and $g(x) = b$ for all $x \in B$.*

The reader may recognize that this theorem contains some of the most important results in analytic topology, most notably the Lebesgue Number Lemma, Baire Category Theorem, the Extreme Value Theorem and Urysohn's Lemma.

12.2 MEASURE THEORY

We also assume a basic knowledge of measure theory, since the underlying state space of our dynamical systems is always (with the exception of Section 7.3) a measure space. We denote this by $(X, \mathcal{F}, \mu)$, where $\mathcal{F}$ is a σ-algebra on X and μ is a measure on the measurable space $(X, \mathcal{F})$. In this section we quickly recall some basic concepts and list some famous results that are used in the main text for easy reference. We do not provide proofs of these results, since they can be found in any textbook on measure theory. This section also includes some less well known results, some are provided with a proof.

The measure spaces we consider are either a *probability space* with $\mu(X) = 1$ or a *σ-finite, infinite* measure space in which case $\mu(X) = \infty$ and there exists a countable collection $(A_n)_{n\geq 1} \subseteq \mathcal{F}$ with $\mu(A_n) < \infty$ for all $n \geq 1$ and $\bigcup_{n\geq 1} A_n = X$. A set $A \in \mathcal{F}$ is a *μ-null set* if $\mu(A) = 0$. A result is said to hold *μ-almost everywhere* if the set of points for which the result does not hold is a μ-null set. Often the results in this book hold μ-almost everywhere and sets of measure zero can be neglected. Two sets are said to be *equal modulo sets of μ-measure zero* if they are equal up to adding or subtracting μ-null sets. If μ and ν are two measures on the same measurable space $(X, \mathcal{F})$, then μ is called *absolutely continuous* with respect to ν if for each $A \in \mathcal{F}$ with $\nu(A) = 0$ it holds that also $\mu(A) = 0$. We denote this by $\mu \ll \nu$. The measures μ and ν are called *equivalent* if they have the same null sets.

Besides σ-algebras there are three other types of collections of subsets of X that can sometimes make life easier.

A collection $\mathcal{S}$ of subsets of X is called a *semi-algebra* if it satisfies

(i) $\emptyset \in \mathcal{S}$,

(ii) $A \cap B \in \mathcal{S}$ whenever $A, B \in \mathcal{S}$, and

(iii) if $A \in \mathcal{S}$, then $A^c = \cup_{i=1}^n E_i$ is a disjoint union of elements of $\mathcal{S}$.

An *algebra* $\mathcal{A}$ is a collection of subsets of X satisfying

(i) $\emptyset \in \mathcal{A}$,

(ii) if $A, B \in \mathcal{A}$, then $A \cap B \in \mathcal{A}$, and finally

(iii) if $A \in \mathcal{A}$, then $A^c \in \mathcal{A}$.

Clearly an algebra is a semi-algebra. Furthermore, given a semi-algebra $\mathcal{S}$ one can form an algebra by taking all finite disjoint unions of elements of $\mathcal{S}$. We denote this algebra by $\mathcal{A}(\mathcal{S})$, and we call it the *algebra generated* by $\mathcal{S}$. It is in fact the smallest algebra containing $\mathcal{S}$. Likewise, given a semi-algebra $\mathcal{S}$ (or an algebra $\mathcal{A}$), the σ-algebra generated by $\mathcal{S}$ (or $\mathcal{A}$) is denoted by $\sigma(\mathcal{S})$ (or $\sigma(\mathcal{A})$), and is the smallest σ-algebra containing $\mathcal{S}$ (or $\mathcal{A}$).

A *monotone class* $\mathcal{M}$ is a collection of subsets of X with the following two properties

(i) if $E_1 \subseteq E_2 \subseteq \ldots$ are elements of $\mathcal{M}$, then $\cup_{i=1}^\infty E_i \in \mathcal{M}$,

(ii) if $F_1 \supseteq F_2 \supseteq \ldots$ are elements of $\mathcal{M}$, then $\cap_{i=1}^\infty F_i \in \mathcal{M}$.

The *monotone class generated* by a collection $\mathcal{S}$ of subsets of X is the smallest monotone class containing $\mathcal{S}$. The following theorem relates these concepts and will be used in the proof of Theorem 12.3.1 further on.

Theorem 12.2.1 (Monotone Class Theorem). *Let $\mathcal{A}$ be an algebra of X, then the σ-algebra $\sigma(\mathcal{A})$ generated by $\mathcal{A}$ equals the monotone class generated by $\mathcal{A}$.*

The next useful lemma states that any measurable set can be approximated arbitrarily well by an element in a generating algebra.

Lemma 12.2.1 (Approximation Lemma). *Let $(X, \mathcal{F}, \mu)$ be a finite measure space, and $\mathcal{A}$ an algebra generating $\mathcal{F}$. Then, for any $A \in \mathcal{F}$ and any $\epsilon > 0$, there exists a $C \in \mathcal{A}$ such that $\mu(A\Delta C) < \epsilon$.*

Proof. Let $\mathcal{D}$ be the collection of all sets $A \in \mathcal{F}$ satisfying the property that for any $\epsilon > 0$, there exists a $C \in \mathcal{A}$ such that $\mu(A\Delta C) < \epsilon$. We will show that $\mathcal{D} = \mathcal{F}$. First note that since $X \in \mathcal{A}$, then $X \in \mathcal{D}$. Now let $A \in \mathcal{D}$ and $\varepsilon > 0$. There exists $C \in \mathcal{A}$ such that $\mu(A\Delta C) < \varepsilon$. Since $C^c \in \mathcal{A}$ and $A\Delta C = A^c \Delta C^c$, we have $\mu(A^c \Delta C^c) < \varepsilon$ and hence $A^c \in \mathcal{D}$. Finally, suppose $(A_n)_n \subset \mathcal{D}$ and $\varepsilon > 0$. For each n, there exists $C_n \in \mathcal{A}$ such that $\mu(A_n \Delta C_n) < \frac{\varepsilon}{2^{n+1}}$. It is easy to check that

$$\bigcup_{n=1}^{\infty} A_n \Delta \bigcup_{n=1}^{\infty} C_n \subseteq \bigcup_{n=1}^{\infty} (A_n \Delta C_n),$$

so that

$$\mu\left(\bigcup_{n=1}^{\infty} A_n \Delta \bigcup_{n=1}^{\infty} C_n\right) \leq \sum_{n=1}^{\infty} \mu(A_n \Delta C_n) < \frac{\varepsilon}{2}.$$

Since $\mathcal{A}$ is closed under finite unions we do not know at this point if $\bigcup_{n=1}^{\infty} C_n$ is an element of $\mathcal{A}$. To solve this problem, we proceed as follows. First note that $\bigcap_{n=1}^{m} C_n^c \searrow \bigcap_{n=1}^{\infty} C_n^c$, hence

$$\mu\left(\bigcup_{n=1}^{\infty} A_n \cap \bigcap_{n=1}^{\infty} C_n^c\right) = \lim_{m\to\infty} \mu\left(\bigcup_{n=1}^{\infty} A_n \cap \bigcap_{n=1}^{m} C_n^c\right),$$

and therefore,

$$\mu\Big(\bigcup_{n=1}^{\infty} A_n \Delta \bigcup_{n=1}^{\infty} C_n\Big) = \lim_{m\to\infty} \mu\left(\left(\bigcup_{n=1}^{\infty} A_n \cap \bigcap_{n=1}^{m} C_n^c\right) \cup \left(\bigcap_{n=1}^{\infty} A_n^c \cap \bigcup_{n=1}^{\infty} C_n\right)\right).$$

Hence there exists an m sufficiently large so that

$$\mu\Big(\Big(\bigcup_{n=1}^{\infty} A_n \cap \bigcap_{n=1}^{m} C_n^c\Big) \cup \Big(\bigcap_{n=1}^{\infty} A_n^c \cap \bigcup_{n=1}^{\infty} C_n\Big)\Big) < \mu\left(\bigcup_{n=1}^{\infty} A_n \Delta \bigcup_{n=1}^{\infty} C_n\right) + \frac{\varepsilon}{2}.$$

Since $\bigcap_{n=1}^{\infty} A_n^c \cap \bigcup_{n=1}^{m} C_n \subseteq \bigcap_{n=1}^{\infty} A_n^c \cap \bigcup_{n=1}^{\infty} C_n$, we get

$$\mu\left(\left(\bigcup_{n=1}^{\infty} A_n \cap \bigcap_{n=1}^{m} C_n^c\right) \cup \left(\bigcap_{n=1}^{\infty} A_n^c \cap \bigcup_{n=1}^{m} C_n\right)\right)$$
$$< \mu\left(\bigcup_{n=1}^{\infty} A_n \Delta \bigcup_{n=1}^{\infty} C_n\right) + \frac{\varepsilon}{2}.$$

Thus,

$$\mu\left(\left(\bigcup_{n=1}^{\infty} A_n \Delta \bigcup_{n=1}^{m} C_n\right)\right) < \varepsilon,$$

and $\bigcup_{n=1}^{m} C_n \in \mathcal{A}$ since $\mathcal{A}$ is closed under finite unions. This shows that $\bigcup_{n=1}^{\infty} A_n \in \mathcal{D}$. Thus, $\mathcal{D}$ is a σ-algebra.

By definition of $\mathcal{D}$ we have $\mathcal{D} \subseteq \mathcal{F}$. Since $\mathcal{A} \subseteq \mathcal{D}$, and $\mathcal{F}$ is the smallest σ-algebra containing $\mathcal{A}$ we have $\mathcal{F} \subseteq \mathcal{D}$. Therefore, $\mathcal{F} = \mathcal{D}$. □

Given a collection of measurable sets $\mathcal{G} \subseteq \mathcal{F}$ one can wonder what the smallest measurable set containing every element from $\mathcal{G}$ is. In case $\mathcal{G}$ is a *hereditary* collection, so if $C \in \mathcal{G}$, $B \subseteq C$ and $B \in \mathcal{F}$ imply that $B \in \mathcal{G}$, then such a set is given by the measurable union. This is a set $U \subseteq \mathcal{F}$ with the following two properties:

(i) U is a *cover* of $\mathcal{G}$, i.e., $\mu(A \setminus U) = 0$ for all $A \in \mathcal{G}$.

(ii) $\mathcal{G}$ *saturates* U, i.e., for all $B \in \mathcal{F}$ with $B \subseteq U$ and $\mu(B) > 0$ there is a $C \in \mathcal{G}$ with $C \subseteq B$ and $\mu(C) > 0$.

Proposition 12.2.1. *Let $(X, \mathcal{F}, \mu)$ be a σ-finite measure space and $\mathcal{G} \subseteq \mathcal{F}$ a hereditary collection. Then a measurable union U for $\mathcal{G}$ exists. In fact, $U = \bigcup_{k \geq 1} A_k$ for a countable collection (A_k) of disjoint set in $\mathcal{G}$.*

Proof. Let $\{A_n\}_n \subseteq \mathcal{F}$ be a countable collection of pairwise disjoint sets with the properties that $0 < \mu(A_n) < \infty$ for each n and $\mu(X \setminus \bigcup_n A_n) = 0$. Note that each of the collections $\mathcal{G} \cap A_n := \{G \cap A_n : G \in \mathcal{G}\}$ is hereditary. We first prove that for each $\mathcal{G} \cap A_n$ a measurable union exists that is the countable union of disjoint sets in $\mathcal{G} \cap A_n$. Fix n and let

$$\varepsilon_1 = \sup\{\mu(A) : A \in \mathcal{G} \cap A_n\} < \infty.$$

Let $A_{n,1}$ be any set in $\mathcal{G} \cap A_n$ with $\mu(A_{n,1}) \geq \frac{\varepsilon_1}{2}$. Then inductively define a sequence $(\varepsilon_j)_j$ and a sequence of sets $(A_{n,j})_j$ with the properties that for each j,

$$\varepsilon_j = \sup\{\mu(A) : A \in \mathcal{G} \cap A_n \text{ and } A \cap A_{n,i} = \emptyset \text{ for all } 1 \leq i < j\}$$

and $A_{n,j} \in \mathcal{G} \cap A_n$ satisfies $\mu(A_{n,j}) \geq \frac{\varepsilon_j}{2}$ and $A_{n,j} \cap A_{n,i} = \emptyset$ for all $1 \leq i < j$. Note that

$$\sum_{j\geq 1} \varepsilon_j \leq \sum_{j\geq 1} 2\mu(A_{n,j}) = 2\mu\Big(\bigcup_j A_{n,j}\Big) \leq 2\mu(A_n) < \infty.$$

Hence, $\lim_{j\to\infty} \varepsilon_j = 0$. We claim that $U_n := \bigcup_j A_{n,j}$ is a measurable union of $\mathcal{G} \cap A_n$. Suppose U_n does not cover $\mathcal{G} \cap A_n$. Then there is a set $A \in \mathcal{G} \cap A_n$ with $\mu(A) > 0$ and $A \cap A_{n,j} = \emptyset$ for each j. But this would imply that $\mu(A) < \varepsilon_j$ for each j, contradicting that $\mu(A) > 0$. Hence U_n covers $\mathcal{G} \cap A_n$. To see that $\mathcal{G} \cap A_n$ saturates U_n, let $B \in \mathcal{F}$ with $B \subseteq U_n$ and $\mu(B) > 0$. Then there is a j, such that $\mu(B \cap A_{n,j}) > 0$. Since $A_{n,j} \in \mathcal{G} \cap A_n$ and $\mathcal{G} \cap A_n$ is hereditary, also $B \cap A_{n,j} \in \mathcal{G} \cap A_n$. Thus U_n is a measurable union of $\mathcal{G} \cap A_n$.

Now let $U = \bigcup_n U_n = \bigcup_n \bigcup_j A_{n,j}$. We show that U is a measurable union of $\mathcal{G}$. To see that U covers $\mathcal{G}$, let $A \in \mathcal{G}$. Then for each n, $A \cap A_n \in \mathcal{G} \cap A_n$. Since U_n covers $\mathcal{G} \cap A_n$, $\mu(A \cap A_n \setminus U_n) = 0$. Then

$$\mu(A \setminus U) = \sum_n \mu(A \cap A_n \setminus U) \leq \sum_n \mu(A \cap A_n \setminus U_n) = 0.$$

To see that $\mathcal{G}$ saturates U, let $B \in \mathcal{F}$ satisfy $B \subseteq U$ and $\mu(B) > 0$. Then there is an n, such that $\mu(B \cap U_n) > 0$. Since $\mathcal{G} \cap A_n$ saturates U_n, there is a set $G \in \mathcal{G}$, such that $G \cap A_n \subseteq B \cap U_n$ and $\mu(G \cap A_n) > 0$. Since $G \cap A_n \subseteq G$ and $\mathcal{G}$ is hereditary, it follows that $G \cap A_n \in \mathcal{G}$. Hence, $\mathcal{G}$ saturates U and U is a measurable union of $\mathcal{G}$. □

Theorem 12.2.2 (Carathéodory Extension Theorem). *Let X be a non-empty set, $\mathcal{S}$ a semi-algebra on X and $\mu_0 : \mathcal{S} \to [0,\infty]$ a countably additive function, i.e., $\mu(\bigcup_{i\geq 1} A_i) = \sum_{i\geq 1} \mu(A_i)$ for any countable collection $(A_i)_{i\geq 1} \subseteq \mathcal{S}$ of pairwise disjoint sets with $\bigcup_{i\geq 1} A_i \in \mathcal{S}$. Then there exists a measure μ on $(X, \sigma(\mathcal{S}))$ that satisfies $\mu(A) = \mu_0(A)$ for all $A \in \mathcal{S}$. This measure μ is unique if there is a countable sequence $(S_n)_{n\geq 1} \subseteq \mathcal{S}$ with $\mu_0(S_n) < \infty$ for all $n \geq 1$ and $\bigcup_{n\geq 1} S_n = X$.*

Note that the uniqueness of μ from the previous theorem holds in particular in case $\mu_0 : \mathcal{S} \to [0,1]$.

Theorem 12.2.3 (Kolmogorov 0-1 Law). *Let $(X, \mathcal{F}, \mu)$ be a probability space. Let $(\mathcal{A}_n) \subseteq \mathcal{F}$ be a sequence of independent σ-algebras, so for any $n \geq 1$ and any sets $A_i \in \mathcal{A}_i$, $1 \leq i \leq n$, it holds that*

$$\mu(A_1 \cap A_2 \cap \cdots \cap A_n) = \mu(A_1)\mu(A_2)\cdots\mu(A_n).$$

For each $n \geq 1$ *let* $\mathcal{T}_n = \sigma(\bigcup_{m\geq n} \mathcal{A}_m)$. *Set* $\mathcal{T}_\infty = \bigcap_{n\geq 1} \mathcal{T}_n$, *the* tail σ-algebra. *Then for any* $A \in \mathcal{T}_\infty$ *either* $\mu(A) = 0$ *or* $\mu(A) = 1$.

12.3 LEBESGUE SPACES

A function $f : (X_1, \mathcal{F}_1) \to (X_2, \mathcal{F}_2)$ between two measurable spaces is called *measurable* if $f^{-1}C \in \mathcal{F}_1$ for any $C \in \mathcal{F}_2$. We call it measure preserving if moreover the measure is preserved.

Definition 12.3.1. Let $(X_i, \mathcal{F}_i, \mu_i)$, $i = 1, 2$, be two measure space and $\phi : X_1 \to X_2$ measurable. The map ϕ is said to be *measure preserving* if $\mu_1(\phi^{-1}A) = \mu_2(A)$ for all $A \in \mathcal{F}_2$.

Two probability spaces $(X_i, \mathcal{F}_i, \mu_i)$, $i = 1, 2$, are called *measurably isomorphic* if there is a bijection $\phi : X_1 \to X_2$ that is measurable and measure preserving up to sets of measure zero.

The definitions of measurability and measure preservingness require one to verify the conditions for all measurable sets. The following theorem states that it is enough to check them on a generating semi-algebra only.

Theorem 12.3.1. *Let* $(X_i, \mathcal{F}_i, \mu_i)$ *be measure spaces and* $T : X_1 \to X_2$ *a map. Suppose* $\mathcal{S}_2$ *is a generating semi-algebra of* $\mathcal{F}_2$ *that contains an exhausting sequence* (S_n), *i.e., an increasing sequence with* $Y = \bigcup_{n=1}^{\infty} S_n$. *Suppose that for each* $A \in \mathcal{S}_2$ *one has* $T^{-1}A \in \mathcal{F}_1$ *and* $\mu_1(T^{-1}A) = \mu_2(A)$. *If furthermore,* $\mu_2(S_n) = \mu_1(T^{-1}S_n) < \infty$ *for all* n, *then* T *is measurable and measure preserving.*

Proof. Let $m \geq 1$ and consider the collection

$$\mathcal{D}_m = \{B \in \mathcal{F}_2 : T^{-1}(B \cap S_m) \in \mathcal{F}_1 \text{ and } \mu_1(T^{-1}(B \cap S_m)) = \mu_2(B \cap S_m)\},$$

then $\mathcal{S}_2 \subseteq \mathcal{D}_m \subseteq \mathcal{F}_2$. We show that $\mathcal{D}_m$ is a monotone class. Let $E_1 \subseteq E_2 \subseteq \ldots$ be elements of $\mathcal{D}_m$, and let $E = \bigcup_{i=1}^{\infty} E_i$. Then, $T^{-1}(E \cap S_m) = \bigcup_{i=1}^{\infty} T^{-1}(E_i \cap S_m) \in \mathcal{F}_1$, and

$$\begin{aligned}
\mu_1(T^{-1}(E \cap S_m)) = \mu_1\Big(\bigcup_{n=1}^{\infty} T^{-1}(E_n \cap S_m)\Big) &= \lim_{n\to\infty} \mu_1(T^{-1}(E_n \cap S_m)) \\
&= \lim_{n\to\infty} \mu_2(E_n \cap S_m) = \mu_2\Big(\bigcup_{n=1}^{\infty} (E_n \cap S_m\Big) \\
&= \mu_2(E \cap S_m).
\end{aligned}$$

Thus, $E \in \mathcal{D}_m$. A similar proof shows that if $F_1 \supseteq F_2 \supseteq \ldots$ are elements of $\mathcal{D}_m$, then $\bigcap_{i=1}^{\infty} F_i \in \mathcal{D}_m$. Hence, $\mathcal{D}_m$ is a monotone class containing the algebra $\mathcal{A}(\mathcal{S}_2)$. By the Monotone Class Theorem, $\mathcal{F}_2$ is the smallest monotone class containing $\mathcal{A}(\mathcal{S}_2)$, hence $\mathcal{F}_2 \subseteq \mathcal{D}_m$. This shows that $\mathcal{F}_2 = \mathcal{D}_m$ for all m. Now let $B \in \mathcal{F}_2$, then $B \in \mathcal{D}_m$ for all m. This implies that $T^{-1}(B \cap S_m) \in \mathcal{F}_1$ and $\mu_1(T^{-1}(B \cap S_m)) = \mu_2(B \cap S_m)$ for all m. Since the sequence $(B \cap S_m)$ increases to B, we have $T^{-1}B = \bigcup_{m=1}^{\infty} T^{-1}(B \cap S_m) \in \mathcal{F}_1$ and

$$\begin{aligned}
\mu_1(T^{-1}B) &= \lim_{m\to\infty} \mu_1(T^{-1}(B \cap S_m)) \\
&= \lim_{m\to\infty} \mu_2(B \cap S_m) \\
&= \mu_2\Big(\bigcup_{m=1}^{\infty} (B \cap S_m)\Big) = \mu_2(B).
\end{aligned}$$

This proves that T is measurable and measure preserving. □

On a topological space X we usually consider the *Borel σ-algebra* $\mathcal{B}$, which is the smallest σ-algebra containing all open sets, together with the Lebesgue measure λ. For some purposes it is more convenient to consider the completion of the Borel σ-algebra. A measure space is called *complete* if any subset of a null set is measurable. The *Lebesgue σ-algebra* is the completion of $\mathcal{B}$, containing all subsets of Lebesgue null sets as well.

Throughout this book, the spaces we are working with are assumed to be standard Lebesgue spaces. Let $(X, \mathcal{F}, \mu)$ be a measure space. A subset $A \in \mathcal{F}$ is called an *atom* if $\mu(A) > 0$ and if each measurable subset $B \subseteq A$ with $\mu(B) < \mu(A)$ has $\mu(B)=0$.

Definition 12.3.2. Let $(X, \mathcal{F}, \mu)$ be a complete finite measure space. It is called a *Lebesgue space* if we can remove from X an at most countable (possibly empty) set of atoms, such that the remainder $(X_0, \mathcal{F} \cap X_0, \mu|_{X_0})$ is measurably isomorphic to a measure space $([a,b), \mathcal{B}([a,b)), \lambda)$, where $[a,b) \subseteq \mathbb{R}$ is an interval and $\mathcal{B}([a,b))$ and λ are the corresponding Lebesgue σ-algebra and Lebesgue measure.

In case $(X, \mathcal{F}, \mu)$ is an infinite, σ-finite measure space with $\{A_n\} \subseteq \mathcal{F}$ such that $\mu(A_n) < \infty$ for all n and $\bigcup_{n \geq 1} A_n = X$, then we call it a *Lebesgue space* if each of the restrictions $(A_n, \mathcal{F} \cap A_n, \mu|_{A_n})$ is a Lebesgue space.

In [54], see also [60], the following equivalent characterization of Lebesgue spaces was given.

Theorem 12.3.2 (Rohlin). *Let $(X, \mathcal{F}, \mu)$ be a complete probability space. Then it is a Lebesgue space if and only if there is a countable collection $\{B_n : n \geq 0\} \subseteq \mathcal{F}$ with all the following properties:*

(i) $\sigma(\{B_n : n \geq 0\}) = \mathcal{F}$;

(ii) there is a full measure set X_0 such that for all $x, y \in X_0$ there is an $n \geq 0$ such that either $x \in B_n$ and $y \notin B_n$ or $x \notin B_n$ and $y \in B_n$;

(iii) the intersection $\bigcap_{n \geq 0} E_n \neq \emptyset$, where each E_n either equals B_n or B_n^c.

12.4 LEBESGUE INTEGRATION

For any $A \subseteq X$ we use $1_A : X \to \mathbb{C}$ to denote the *indicator function*, i.e.,

$$1_A(x) = \begin{cases} 1, & \text{if } x \in A, \\ 0, & \text{if } x \notin A. \end{cases}$$

This is a measurable function if $A \in \mathcal{F}$. A *simple function* is a linear combination of indicator functions of measurable sets, so a function $h : X \to \mathbb{C}$ of the form $h(x) = \sum_{i=1}^n a_i 1_{A_i}(x)$, where $a_i \in \mathbb{C}$ and all $A_i \in \mathcal{F}$ are pairwise disjoint.

Theorem 12.4.1 (Simple Function Approximation Theorem, part I). *Let $(X, \mathcal{F}, \mu)$ be a measure space and $f : X \to \mathbb{R}$ a measurable function with respect to the Lebesgue σ-algebra on $\mathbb{R}$. Then f is the pointwise limit of a sequence of simple functions.*

We denote by $L^0(X, \mathcal{F}, \mu)$ the space of all complex valued measurable functions on a σ-finite measure space $(X, \mathcal{F}, \mu)$. Let

$$L^p(X, \mathcal{F}, \mu) = \Big\{ f \in L^0(X, \mathcal{F}, \mu) : \int_X |f|^p \, d\mu(x) < \infty \Big\}.$$

On $L^p(X, \mathcal{F}, \mu)$, $p \in [1, \infty)$, the L^p-norm $\|\cdot\|_p$ is defined by

$$\|f\|_p = \Big(\int_X |f|^p \, d\mu \Big)^{\frac{1}{p}}.$$

We say that the sequence $(f_n)_{n \geq 1} \subseteq L^p(X, \mathcal{F}, \mu)$ *converges in L^p* to f if

$$\lim_{n \to \infty} \int_X |f_n - f|^p \, d\mu = 0.$$

Finally, $L^\infty(X,\mathcal{F},\mu)$ is the space of essentially bounded measurable functions on $(X,\mathcal{F},\mu)$, i.e.,

$$L^\infty(X,\mathcal{F},\mu) = \{f \in L^0(X,\mathcal{F},\mu) : \exists c > 0 \text{ s.t. } \mu(\{x \in X : |f(x)| \geq c\}) = 0\}.$$

The L^∞-norm is defined by

$$\|f\|_\infty = \inf\{c > 0 : \mu(\{x \in X : |f(x)| \geq c\}) = 0\}. \tag{12.1}$$

A sequence $(f_n)_{n\geq 1} \subseteq L^\infty(X,\mathcal{F},\mu)$ converges to $f \in L^\infty(X,\mathcal{F},\mu)$ if $\lim_{n\to\infty} \|f_n - f\|_\infty = 0$. This is equivalent to the statement that there exists a set $A \in \mathcal{F}$ with $\mu(X \setminus A) = 0$ such that on A the sequence $(f_n)_{n\geq 1}$ converges to f uniformly.

There are several well known convergence results for Lebesgue Integration that we will use throughout the book.

Theorem 12.4.2 (Monotone Convergence Theorem). *Let $(X,\mathcal{F},\mu)$ be a measure space. Let $f, f_n : X \to \mathbb{R}$ be non-negative, measurable functions such that $f_n \nearrow f$ pointwise. Then*

$$\int_X f\,d\mu = \lim_{n\to\infty} \int_X f_n\,d\mu.$$

Theorem 12.4.3 (Dominated Convergence Theorem). *Let $(X,\mathcal{F},\mu)$ be a measure space. Let $f, f_n : X \to \mathbb{R}$ be measurable functions such that $f_n \to f$ pointwise. If there is a Lebesgue integrable function $g : X \to \mathbb{R}$, such that $|f_n| \leq g$ for all n, then f_n and f are Lebesgue integrable as well and*

$$\int_X f\,d\mu = \lim_{n\to\infty} \int_X f_n\,d\mu.$$

Lemma 12.4.1 (Scheffé's Lemma). *Let $(X,\mathcal{F},\mu)$ be a measure space and (f_n) a sequence in $L^1(X,\mathcal{F},\mu)$ converging μ-a.e. to a function $f \in L^1(X,\mathcal{F},\mu)$. Then, the sequence (f_n) converges to f in L^1 if and only if $\lim_{n\to\infty} \int_X f_n\,d\mu = \int_X f\,d\mu$.*

Theorem 12.4.4 (Radon-Nikodym Theorem). *Let $(X,\mathcal{F})$ be a measurable space with two σ-finite measures μ and ν, such that $\mu \ll \nu$. Then there exists a measurable function $f : X \to [0,\infty)$, such that for any $A \in \mathcal{F}$,*

$$\mu(A) = \int_A f\,d\nu.$$

The function f is called the Radon-Nikodym derivative *and is denoted by $\frac{d\mu}{d\nu}$.*

Theorem 12.4.5 (Simple Function Approximation Theorem, part II). *Let $(X, \mathcal{F}, \mu)$ be a measure space and $f : X \to \mathbb{R}$ Lebesgue integrable. Then f is the L^1-limit of a sequence of simple functions.*

A last convergence result is the following.

Theorem 12.4.6 (Fubini's Theorem). *Let $(X_i, \mathcal{F}_i, \mu_i)$, $i = 1, 2$, be two σ-finite measure spaces. If $f \in L^1_{\mathbb{R}}(X_1 \times X_2, \mathcal{F}_1 \otimes \mathcal{F}_2, \mu_1 \times \mu_2)$, then*

$$\int_{X_1} \Big(\int_{X_2} f \, d\mu_2 \Big) d\mu_1 = \int_{X_1 \times X_2} f \, d\mu_1 \times \mu_2 = \int_{X_2} \Big(\int_{X_1} f \, d\mu_1 \Big) d\mu_2.$$

We now define the concept of conditional expectation.

Definition 12.4.1. Let $(X, \mathcal{F}, \mu)$ be a probability space, $\mathcal{G}$ a sub-σ-algebra of $\mathcal{F}$ and $f \in L^1(X, \mathcal{F}, \mu)$. The *conditional expectation of f given $\mathcal{G}$*, denoted by $\mathbb{E}_\mu(f|\mathcal{G})$ is the μ-a.e. unique integrable function satisfying the following two properties:

(i) $\mathbb{E}_\mu(f|\mathcal{G})$ is $\mathcal{G}$ measurable,

(ii) for any $A \in \mathcal{G}$,

$$\int_A f \, d\mu = \int_A \mathbb{E}_\mu(f|\mathcal{G}) \, d\mu.$$

The following version of the Martingale Convergence Theorem is adapted to our setting.

Theorem 12.4.7 (Martingale Convergence Theorem). *Let $(X, \mathcal{F}, \mu)$ be a probability space. Let $\mathcal{C}_1 \subseteq \mathcal{C}_2 \subseteq \ldots$ be a sequence of increasing σ-algebras, and let $\mathcal{C} = \sigma(\cup_n \mathcal{C}_n)$. If $f \in L^1(X, \mathcal{F}, \mu)$, then*

$$\mathbb{E}_\mu(f|\mathcal{C}) = \lim_{n \to \infty} \mathbb{E}_\mu(f|\mathcal{C}_n)$$

μ-a.e., and in L^1.

12.5 HILBERT SPACES

For the material treated in this book we need a few basic notions from functional analysis.

Definition 12.5.1. A *Banach space* X is a complete normed vector space, i.e., it is a vector space with a norm $\|\cdot\|$ defined on it such that X is complete with respect to the metric induced by the norm.

Definition 12.5.2. A *Hilbert space* X is a real or complex vector space with an inner product $(\cdot,\cdot)$ defined on it such that X is complete with respect to the metric induced by the inner product.

If H is a Hilbert space and S is a closed linear subspace of H, then the *orthogonal complement* $S^\perp$ of S is the space

$$S^\perp = \{h \in H : (h,g) = 0 \text{ for all } g \in S\}.$$

Theorem 12.5.1 (Decomposition Theorem). *Let S be a closed linear subspace of a Hilbert space H. Then for any element $h \in H$, there exists a unique element $g \in S$ satisfying*

$$\inf\{\|h-s\|_H : s \in S\} = \|h-g\|_H,$$

where $\|\cdot\|_H$ denotes the norm induced by the inner product on H. Furthermore, if we define $\Pi : H \to S$ by $\Pi(h) = g$, then every element $h \in H$ can be written uniquely as $h = \Pi(h) + t$, where $t \in S^\perp$. The transformation Π is called the orthogonal projection *of H onto S.*

Let $(X, \mathcal{F}, \mu)$ be a measure space. For $p \geq 1$ the space $L^p(X, \mathcal{F}, \mu)$ is a Banach space under the L^p-norm. The space $L^2(X, \mathcal{F}, \mu)$ equipped with the inner product $(f,g) = \int_X f\overline{g}\,d\mu$, where $\overline{g}$ is the complex conjugate of g, is a Hilbert space.

Let V be a normed vector space over $\mathbb{C}$ (or $\mathbb{R}$). The (topological) *dual space* V^* is the space of all continuous linear functionals $\phi : V \to \mathbb{C}$ (or $\mathbb{R}$). The Riesz Representation Theorem establishes a relation between a Hilbert space and its dual space.

Theorem 12.5.2 (Riesz Representation Theorem for Hilbert Spaces). *Let H be a Hilbert space with inner product $(\cdot,\cdot)$ and let $\phi \in H^*$. Then there exists an element $h \in H$, such that for any $g \in H$, $\phi(g) = (h,g)$ and $\|h\|_H = \|\phi\|_{H^*}$.*

Let $L : H \to G$ be a linear operator between two Hilbert spaces H and G and use $\|\cdot\|_H$ and $\|\cdot\|_G$ to denote the norms on H and G, respectively. Then L is called *bounded* if there is an $M \geq 0$, such that $\|Lh\|_H \leq M\|h\|_H$ for all $h \in H$. The *operator norm* is defined by

$$\|L\| = \inf\left\{M \geq 0 : \frac{\|Lh\|_G}{\|h\|_H} \leq M \text{ for all } h \in H, \text{ with } \|h\| \neq 0\right\}.$$

L is called *positive* if $(Lh,h) \geq 0$ for every $h \in H$. It is an *isometry* if it preserves distances, i.e., $\|Lh\|_G = \|h\|_H$ for all $h \in H$.

If $L : H \to H$ is a bounded linear operator from a Hilbert space H to itself, then the *adjoint* of L is the bounded linear operator $L^* : H \to H$ defined by

$$(Lh, g) = (h, L^* g) \quad \text{for all } h, g \in H.$$

Existence and uniqueness of the adjoint follow from Theorem 12.5.2.

12.6 BOREL MEASURES ON COMPACT METRIC SPACES

If (X, d) is a compact metric space, then we can say a bit more about the collection of Borel probability measures on X. Let $M(X)$ be the collection of all Borel probability measures on X. This set is non-empty if $X \neq \emptyset$, since for each $x \in X$ the Dirac measure δ_x concentrated at x, which is given by

$$\delta_x(A) = \begin{cases} 1, & \text{if } x \in A, \\ 0, & \text{if } x \notin A, \end{cases}$$

is a Borel probability measure on X. The space $M(X)$ is also *convex*, i.e., $p\mu + (1-p)\nu \in M(X)$ whenever $\mu, \nu \in M(X)$ and $0 \le p \le 1$.

We denote by $C(X)$ the Banach space of all complex valued continuous functions on X under the supremum norm $\|f\|_\infty$, see (12.1). Since X is a compact space and f is continuous we in fact have $\|f\|_\infty = \sup_{x\in X} |f(x)|$. This implies that a sequence (f_n) in $C(X)$ converges to f in the supremum norm if and only if (f_n) converges uniformly to f. Furthermore, $C(X)$ is a *separable* space, i.e., it contains a countable dense subset.

Theorem 12.6.1. *Every member of* $M(X)$ *is* regular, *i.e., if* $\mu \in M(X)$, *then for all* $A \in \mathcal{B}$ *and every* $\varepsilon > 0$ *there exist an open set* O_ε *and a closed set* C_ε *such that* $C_\varepsilon \subseteq A \subseteq O_\varepsilon$ *and* $\mu(O_\varepsilon \setminus C_\varepsilon) < \varepsilon$.

Idea of proof. Call a set $B \in \mathcal{B}$ with the above property a *regular* set. Let $\mathcal{R}$ be the collection of all regular sets $B \in \mathcal{B}$. Show that $\mathcal{R}$ is a σ-algebra containing all the closed sets. □

Corollary 12.6.1. *For any* $A \in \mathcal{B}$, *and any* $\mu \in M(X)$,

$$\mu(A) = \sup_{C \subseteq A : C \text{ closed}} \mu(C) = \inf_{A \subseteq O : O \text{ open}} \mu(O).$$

Theorem 12.6.2 below shows that a member of $M(X)$ is determined by how it integrates continuous functions.

Theorem 12.6.2. *Let $\mu, \nu \in M(X)$. If for all $f \in C(X)$*

$$\int_X f\, d\mu = \int_X f\, d\nu,$$

then $\mu = \nu$.

Proof. Let $\mu, \nu \in M(X)$ be such that $\int_X f\, d\mu = \int_X f\, d\nu$ for all $f \in C(X)$. We want to show that $\mu(A) = \nu(A)$ for all $A \in \mathcal{B}$. By Corollary 12.6.1 it is enough to show that $\mu(C) = \nu(C)$ for all closed subsets C of X. Let $\varepsilon > 0$. By regularity of the measure ν there exists an open set O_ε such that $C \subseteq O_\varepsilon$ and $\nu(O_\varepsilon \setminus C) < \varepsilon$. Define $f \in C(X)$ by

$$f(x) = \frac{d(x, X \setminus O_\varepsilon)}{d(x, X \setminus O_\varepsilon) + d(x, C)}.$$

Notice that $1_C \le f \le 1_{O_\varepsilon}$, thus

$$\mu(C) \le \int_X f\, d\mu = \int_X f\, d\nu \le \nu(O_\varepsilon) \le \nu(C) + \varepsilon.$$

Using a similar argument, one can show that $\nu(C) \le \mu(C) + \varepsilon$. Therefore, $\mu(C) = \nu(C)$ for all closed sets, and hence for all Borel sets. □

This allows us to define a metric structure on $M(X)$ as follows. A sequence (μ_n) in $M(X)$ is said to *converge weakly* to $\mu \in M(X)$ if and only if

$$\lim_{n\to\infty} \int_X f\, d\mu_n = \int_X f\, d\mu$$

for all $f \in C(X)$. We will show that under this notion of convergence the space $M(X)$ is compact, but first we need a second version of the Riesz Representation Theorem.

Theorem 12.6.3 (Riesz Representation Theorem for Compact Metric Spaces). *Let X be a compact metric space and $J : C(X) \to \mathbb{C}$ a continuous linear map such that J is a positive operator and $J(1) = 1$. Then there exists a $\mu \in M(X)$ such that $J(f) = \int_X f\, d\mu$.*

Theorem 12.6.4. *The space $M(X)$ is compact.*

Idea of proof. Let (ν_n) be a sequence in $M(X)$, and choose a countable dense subset of (g_n) of $C(X)$. The sequence $(\int_X g_1\, d\nu_n)$ is a bounded sequence of complex numbers, hence one can find a subsequence $(\nu_{1,n})$

of (ν_n) such that the sequence $(\int_X g_1\, d\nu_{1,n})$ is convergent. Now, the sequence $(\int_X g_2\, d\nu_{1,n})$ is bounded, and hence one can find a subsequence $(\nu_{2,n})$ of $(\nu_{1,n})$ such that the sequence $(\int_X g_2\, d\nu_{2,n})$ is convergent. Notice that $(\int_X g_1\, d\nu_{2,n})$ is also convergent. We continue in this manner, to get for each i a subsequence $(\nu_{i,n})$ of (ν_n) such that for all $j \leq i$, $(\nu_{i,n})$ is a subsequence of $(\nu_{j,n})$ and $(\int_X g_j\, d\nu_{i,n})$ converges. Consider the diagonal sequence $(\nu_{n,n})$, then $(\int_X g_j\, d\nu_{n,n})$ converges for all j. Since (g_n) is dense in C(X), $(\int_X f\, d\nu_{n,n})$ converges for all $f \in C(X)$. Now define $L : C(X) \to \mathbb{C}$ by $L(f) = \lim_{n\to\infty} \int_X f\, d\nu_{n,n}$. Then L is linear, continuous ($|L(f)| \leq \sup_{x\in X} |f(x)| = \|f\|_\infty$), positive and $L(1) = 1$. Thus, by Theorem 12.6.3, there exists a measure $\nu \in M(X)$ such that $L(f) = \lim_{n\to\infty} \int_X f\, d\nu_{n,n} = \int_X f\, d\nu$. Therefore, $\lim_{n\to\infty} \nu_{n,n} = \nu$, and $M(X)$ is compact. □

The following result on equivalent definitions for weak convergence of measures is part of a longer list of equivalences sometimes known as the Portmanteau Theorem.

Theorem 12.6.5. *Let (X, d) be a metric space with Borel σ-algebra $\mathcal{B}$. Let $(\mu_n) \subseteq M(X)$ and $\mu \in M(X)$. Then (μ_n) converges weakly to μ if and only if* $\lim_{n\to\infty} \mu_n(A) = \mu(A)$ *for all sets $A \in \mathcal{F}$ with $\mu(\partial A) = 0$.*

The next theorem is an infinite dimensional version of the statement that each element of a compact convex subset of a finite dimensional vector space can be written as a finite convex combination of the extreme points.

Theorem 12.6.6 (Choquet). *Let Y be a metrizable compact convex subset of a locally convex space V and let $y \in Y$. Let $E \subseteq Y$ denote the collection of extreme points of Y. Then there is a probability measure ν on Y with $\nu(Y \setminus E) = 0$ and such that for each continuous linear functional ϕ on V one has $\phi(y) = \int_X \phi\, d\nu$.*

12.7 FUNCTIONS OF BOUNDED VARIATION

As in Chapter 6 we define the total variation as follows. Let $a, b \in \mathbb{R}$ with $a < b$ and fix the measure space $([a,b], \mathcal{B}, \lambda)$, where $\mathcal{B}$ is the Lebesgue σ-algebra on the interval $[a,b]$ and λ is the one-dimensional Lebesgue measure. The *total variation* of a function $g : [a,b] \to \mathbb{R}$ is defined by

$$Var_{[a,b]}(g) = \sup_{x_0=a<x_1<\cdots<x_n=b} \sum_{i=1}^{n} |g(x_i) - g(x_{i-1})|,$$

where the supremum is taken over all possible n and all possible collections of points $x_0 = a < x_1 < \cdots < x_n = b$. The function f is said to be of *bounded variation* if $Var_{[a,b]}(g) < \infty$ and we let $BV([a,b])$ denote the set of functions $g : [a,b] \to \mathbb{R}$ that are of bounded variation. The first proposition collects some properties of the total variation that are easy to verify.

Proposition 12.7.1. *Let $g : [a,b] \to \mathbb{R}$, $h : [a,b] \to \mathbb{R}$ be two functions. Then the following hold.*

(i) For each $x \in [a,b]$, $|g(x)| \leq |g(a)| + Var_{[a,b]}(g)$.

(ii) $Var_{[a,b]}(g+h) \leq Var_{[a,b]}(g) + Var_{[a,b]}(h)$.

(iii) $Var_{[a,b]}(g \cdot h) \leq \|g\|_\infty Var_{[a,b]}(h) + \|h\|_\infty Var_{[a,b]}(g)$.

Note that Proposition 12.7.1 implies that any $g \in BV([a,b])$ is bounded. A function of bounded variation is also Lebesgue integrable.

Theorem 12.7.1. *If $g \in BV([a,b])$, then $g \in L^1([a,b], \mathcal{B}, \lambda)$.*

The following proposition gives two more properties.

Proposition 12.7.2. *(i) Let $c, d \in \mathbb{R}$ with $c < d$ be given. Let $g : [a,b] \to [c,d]$ and $h : [c,d] \to \mathbb{R}$ be two functions and assume that g is monotone. Then*

$$Var_{[a,b]}(h \circ g) \leq Var_{[c,d]}(h).$$

(ii) Let $g \in BV([a,b])$ and $h \in C^1([a,b])$. Then

$$Var_{[a,b]}(g \cdot h) \leq \|h\|_\infty Var_{[a,b]}(g) + \int_{[a,b]} |g \cdot h'| \, d\lambda.$$

(iii) Let $g \in L^1([a,b], \mathcal{B}, \lambda)$. Then

$$\|g\|_\infty \leq Var_{[a,b]}(g) + \frac{1}{\lambda([a,b])} \int_{[a,b]} |g| \, d\lambda.$$

We mention two other results.

Theorem 12.7.2 (Yorke's Inequality). *Let $a, b, c, d \in \mathbb{R}$ with $a \leq c < d \leq b$. For $g \in BV([a,b])$ it holds that*

$$Var_{[a,b]}(g \cdot 1_{[c,d]}) \leq 2Var_{[c,d]}(g) + \frac{2}{\lambda([c,d])} \int_{[c,d]} |g| \, d\lambda.$$

Theorem 12.7.3 (Helly's First Theorem). *Let $a, b \in \mathbb{R}$ with $a < b$ and let $G = \{g\} \subseteq BV([a, b])$ be an infinite collection of functions with the property that there exists a constant $C > 0$, such that for each $g \in G$,*

$$Var_{[a,b]}(g) < C \quad \text{and} \quad \|g\|_\infty < C.$$

Then there is a sequence $(g_n) \subseteq G$ and a function $h : [a, b] \to \mathbb{R}$ with $Var_{[a,b]}(h) \leq C$, such that (g_n) converges pointwise to h.

Bibliography

[1] J. Aaronson. *An introduction to infinite ergodic theory*, volume 50 of *Mathematical Surveys and Monographs*. American Mathematical Society, Providence, RI, 1997.

[2] R. L. Adler, A. G. Konheim, and M. H. McAndrew. Topological entropy. *Trans. Amer. Math. Soc.*, 114:309–319, 1965.

[3] V. I. Arnol′d and A. Avez. *Ergodic problems of classical mechanics.* Translated from the French by A. Avez. W. A. Benjamin, Inc., New York-Amsterdam, 1968.

[4] V. Baladi. *Positive transfer operators and decay of correlations*, volume 16 of *Advanced Series in Nonlinear Dynamics*. World Scientific Publishing Co., Inc., River Edge, NJ, 2000.

[5] G. D. Birkhoff. Proof of the ergodic theorem. *Proc. Nat. Acad. Sci. USA*, 17:656–660, 1931.

[6] É. Borel. Les probabilités dénombrables et leurs applications arithmétiques. *Rend. Circ. Mat. Palermo*, 27:247–271, 1909.

[7] R. Bowen. Entropy for group endomorphisms and homogeneous spaces. *Trans. Amer. Math. Soc.*, 153:401–414, 1971.

[8] A. Boyarsky and P. Góra. *Laws of chaos. Invariant measures and dynamical systems in one dimension.* Probability and Its Applications. Birkhäuser Boston, Inc., Boston, MA, 1997.

[9] L. Breiman. The individual ergodic theorem of information theory. *Ann. Math. Statist.*, 28:809–811, 1957.

[10] M. Brin and G. Stuck. *Introduction to dynamical systems.* Cambridge University Press, Cambridge, 2002.

[11] L. Carleson. On convergence and growth of partial sums of Fourier series. *Acta Math.*, 116:135–157, 1966.

[12] I. P. Cornfeld, S. V. Fomin, and Ya. G. Sinaĭ. *Ergodic theory*, volume 245 of *Grundlehren der Mathematischen Wissenschaften [Fundamental Principles of Mathematical Sciences]*. Springer-Verlag, New York, 1982. Translated from the Russian by A. B. Sosinskiĭ.

[13] K. Dajani and A. Fieldsteel. Equipartition of interval partitions and an application to number theory. *Proc. Amer. Math. Soc.*, 129(12):3453–3460, 2001.

[14] K. Dajani and C. Kraaikamp. *Ergodic theory of numbers*, volume 29 of *Carus Mathematical Monographs*. Mathematical Association of America, Washington, DC, 2002.

[15] E. I. Dinaburg. A connection between various entropy characterizations of dynamical systems. *Izv. Akad. Nauk SSSR Ser. Mat.*, 35:324–366, 1971.

[16] W. Doeblin. Remarques sur la théorie métrique des fractions continues. *Compositio Math.*, 7:353–371, 1940.

[17] M. Einsiedler and T. Ward. *Ergodic theory with a view towards number theory*, volume 259 of *Graduate Texts in Mathematics*. Springer-Verlag London, Ltd., London, 2011.

[18] T. N. T. Goodman. Relating topological entropy and measure entropy. *Bull. London Math. Soc.*, 3:176–180, 1971.

[19] L. W. Goodwyn. Topological entropy bounds measure-theoretic entropy. *Proc. Amer. Math. Soc.*, 23:679–688, 1969.

[20] P. R. Halmos. *Measure Theory*. D. Van Nostrand Company, Inc., New York, 1950.

[21] F. Hofbauer. On intrinsic ergodicity of piecewise monotonic transformations with positive entropy. *Israel J. Math.*, 34(3):213–237 (1980), 1979.

[22] F. Hofbauer. On intrinsic ergodicity of piecewise monotonic transformations with positive entropy. II. *Israel J. Math.*, 38(1–2):107–115, 1981.

[23] F. Hofbauer and G. Keller. Ergodic properties of invariant measures for piecewise monotonic transformations. *Math. Z.*, 180(1):119–140, 1982.

[24] W. Hurewicz. Ergodic theorem without invariant measure. *Ann. of Math. (2)*, 45:192–206, 1944.

[25] H. Jager. The distribution of certain sequences connected with the continued fraction. *Nederl. Akad. Wetensch. Indag. Math.*, 48(1):61–69, 1986.

[26] C. Kalle. Isomorphisms between positive and negative β-transformations. *Ergodic Theory Dynam. Systems*, 34(1):153–170, 2014.

[27] C. Kalle, T. Kempton, and E. Verbitskiy. The random continued fraction transformation. *Nonlinearity*, 30(3):1182–1203, 2017.

[28] T. Kamae and M. Keane. A simple proof of the ratio ergodic theorem. *Osaka J. Math.*, 34(3):653–657, 1997.

[29] M. Kesseböhmer, S. Munday, and B. O. Stratmann. *Infinite ergodic theory of numbers.* De Gruyter Graduate. De Gruyter, Berlin, 2016.

[30] J. F. C. Kingman and S. J. Taylor. *Introduction to measure and probability.* Cambridge University Press, London-New York-Ibadan, 1966.

[31] K. Knopp. Mengentheoretische Behandlung einiger Probleme der diophantischen Approximationen und der transfiniten Wahrscheinlichkeiten. *Math. Ann.*, 95(1):409–426, 1926.

[32] W. Krieger. On entropy and generators of measure-preserving transformations. *Trans. Amer. Math. Soc.*, 149:453–464, 1970.

[33] A. Lasota and M. C. Mackey. *Chaos, fractals, and noise*, volume 97 of *Applied Mathematical Sciences.* Springer-Verlag, New York, second edition, 1994. Stochastic aspects of dynamics.

[34] A. Lasota and J. A. Yorke. On the existence of invariant measures for piecewise monotonic transformations. *Trans. Amer. Math. Soc.*, 186:481–488 (1974), 1973.

[35] P. Lévy. Sur le loi de probabilité dont dependent les quotients complets wet incompletes d'une fraction continue. *Bull. Soc. Math. de France*, 57:178–194, 1929.

[36] C. Liverani, B. Saussol, and S. Vaienti. A probabilistic approach to intermittency. *Ergodic Theory Dynam. Systems*, 19(3):671–685, 1999.

[37] G. Lochs. Vergleich der Genauigkeit von Dezimalbruch und Kettenbruch. *Abh. Math. Sem. Univ. Hamburg*, 27:142–144, 1964.

[38] J. Lüroth. Ueber eine eindeutige Entwickelung von Zahlen in eine unendliche Reihe. *Math. Ann.*, 21(3):411–423, 1883.

[39] L. Luzzi and S. Marmi. On the entropy of Japanese continued fractions. *Discrete Contin. Dyn. Syst.*, 20(3):673–711, 2008.

[40] D. Maharam. Incompressible transformations. *Fund. Math.*, 56:35–50, 1964.

[41] B. McMillan. The basic theorems of information theory. *Ann. Math. Statistics*, 24:196–219, 1953.

[42] M. Misiurewicz. A short proof of the variational principle for a $\mathbf{Z}_+^N$ action on a compact space. In *International Conference on Dynamical Systems in Mathematical Physics (Rennes, 1975)*, pages 147–157. Astérisque, No. 40. 1976.

[43] M. Misiurewicz and W. Szlenk. Entropy of piecewise monotone mappings. *Studia Math.*, 67(1):45–63, 1980.

[44] J. R. Munkres. *Topology.* Prentice Hall, Inc., Upper Saddle River, NJ, 2000.

[45] H. Nakada. Metrical theory for a class of continued fraction transformations and their natural extensions. *Tokyo J. Math.*, 4(2):399–426, 1981.

[46] H. Nakada, S. Ito, and S. Tanaka. On the invariant measure for the transformations associated with some real continued-fractions. *Keio Engrg. Rep.*, 30(13):159–175, 1977.

[47] D. Ornstein. Bernoulli shifts with the same entropy are isomorphic. *Advances in Math.*, 4:337–352, 1970.

[48] W. Parry. *Topics in ergodic theory*, volume 75 of *Cambridge Tracts in Mathematics.* Cambridge University Press, Cambridge, 2004. Reprint of the 1981 original.

[49] O. Perron. *Die Lehre von den Kettenbrüchen. Bd I. Elementare Kettenbrüche.* B. G. Teubner Verlagsgesellschaft, Stuttgart, 1954. 3te Aufl.

[50] K. Petersen. *Ergodic theory*, volume 2 of *Cambridge Studies in Advanced Mathematics.* Cambridge University Press, Cambridge, 1983.

[51] H. Poincaré. Sur le problème des trois corps et les équations de la dynamique. *Acta Math.*, 13:1–270, 1890.

[52] M. Pollicott and H. Weiss. Multifractal analysis of Lyapunov exponent for continued fraction and Manneville-Pomeau transformations and applications to Diophantine approximation. *Comm. Math. Phys.*, 207(1):145–171, 1999.

[53] Y. Pomeau and P. Manneville. Intermittent transition to turbulence in dissipative dynamical systems. *Comm. Math. Phys.*, 74(2):189–197, 1980.

[54] V. A. Rohlin. On the fundamental ideas of measure theory. *Amer. Math. Soc. Translation*, 1952(71):55, 1952.

[55] V. A. Rohlin. Exact endomorphisms of a Lebesgue space. *Izv. Akad. Nauk SSSR Ser. Mat.*, 25:499–530, 1961.

[56] W. Rudin. *Functional analysis.* International Series in Pure and Applied Mathematics. McGraw-Hill, Inc., New York, second edition, 1991.

[57] M. Rychlik. Bounded variation and invariant measures. *Studia Math.*, 76(1):69–80, 1983.

[58] R. L. Schilling. *Measures, integrals and martingales.* Cambridge University Press, Cambridge, second edition, 2017.

[59] C. E. Shannon. A mathematical theory of communication. *Bell System Tech. J.*, 27:379–423, 623–656, 1948.

[60] P. Shields. *The theory of Bernoulli shifts.* The University of Chicago Press, Chicago, Ill.-London, 1973. Chicago Lectures in Mathematics.

[61] Ya. Sinaĭ. On the concept of entropy for a dynamic system. *Dokl. Akad. Nauk SSSR*, 124:768–771, 1959.

[62] M. Thaler. Transformations on [0, 1] with infinite invariant measures. *Israel J. Math.*, 46(1–2):67–96, 1983.

[63] M. Viana and K. Oliveira. *Foundations of ergodic theory*, volume 151 of *Cambridge Studies in Advanced Mathematics*. Cambridge University Press, Cambridge, 2016.

[64] J. Von Neuman. Proof of the quasi-ergodic hypothesis. *Proc. Nat. Acad. Sci. USA*, 18:70–82, 1931.

[65] P. Walters. *An introduction to ergodic theory*, volume 79 of *Graduate Texts in Mathematics*. Springer-Verlag, New York-Berlin, 1982.

[66] R. Zweimüller. Ergodic properties of infinite measure-preserving interval maps with indifferent fixed points. *Ergodic Theory Dynam. Systems*, 20(5):1519–1549, 2000.

Index

(n, ε)-covering, 177
(n, ε)-separated, 178, 193
(n, ε)-spanning, 177
$B(x, r)$, 178
$B_n(x, r)$, 178
$Co(n, \varepsilon, T)$, 177
$H_\mu(\alpha)$, 150
$H_{top}(\alpha)$, 174
L^∞ norm, 232, 235
L^p norm, 107, 231
L^p space, 231
L^p-convergence, 231
$M_{max}(X, T)$, 195
$N(\alpha)$, 174
$Se(\varepsilon, T)$, 192
$Se(n, \varepsilon, T)$, 178
$Sp(n, \varepsilon, T)$, 177
T-expansion, 166
α-continued fraction map, 100, 140, 222
β-expansion, 11, 166
β-transformation, 11, 32, 39, 68, 75, 89, 98, 156, 165, 217
    entropy, 167
∂A, 191
σ-algebra
    Borel, 5
    Lebesgue, 5
$\times 10$ transformation, 168
$\times N$ transformation, 44, 67, 172
d_{disc}, 117
d_n, 176
$diam_d(A)$, 174
$diam_d(\alpha)$, 174
$h_\mu(T)$, 154
$h_\mu(\alpha, T)$, 154
$h_{top}(T)$, 179
$h_{top}(\alpha, T)$, 175

Aaronson's Theorem, 221
absolutely continuous, 225
absolutely continuous measure, 42
absolutely normal, 45
adjoint, 54, 83
algebra, 225
algebra generated, 225
almost everywhere, 224
analytic topology, 224
approximation coefficient, 136
arithmetic mean, 135
Arnold's cat map, 13, 77, 102, 121
atom, 147, 204, 230

backwards continued fraction transformation, 201
Baire Category Theorem, 224
Baire space, 224
baker's transformation, 8, 69, 71, 90
Banach space, 105, 233, 235
base N expansion, 44, 67, 166
Bernoulli measure, 14, 15, 63
Bernoulli shift, 14, 28, 50, 67, 69, 71, 72, 78, 84, 85, 102, 121, 148, 156, 158, 184, 196, 197
    topological entropy, 184
binary expansion, 7, 43, 71, 78, 84

binary odometer, 15, 63, 84, 102, 115, 117
Boltzmann Hypothesis, 22
Boole's transformation, 8, 199, 206
Borel σ-algebra, 5, 230
Borel-Cantelli Lemma, 171
boundary, 191, 223, 237
bounded operator, 234
bounded variation, 94, 238

Carathéodory Extension Theorem, 73, 228
Cauchy Schwartz inequality, 54
change of variables, 90, 92, 95, 132, 136
Choquet's Theorem, 108, 237
closure, 223
common refinement, 150
 open cover, 174
commuting diagram, 66
compact disc, 148
compact metric space, 101, 235
complement, 22
complete measure space, 4, 230
completion, 4
composition, 79
concavity, 188
conditional entropy
 of a partition, 151
conditional expectation, 38, 86, 160, 233
conditional information, 162
 function, 160
conjugacy, 181, 194
conservativity, 20, 24, 25, 57, 206, 207
 Boole's transformation, 207
 Farey map, 207
 odometer, 63
 Rényi map, 207
continued fraction expansion, 11, 100, 125, 165, 168
contraction, 79
convergence
 in L^p, 231
 pointwise, 110
 uniform, 109, 232, 235
 weak, 101
convergents, 12, 126, 142
convex combination, 101, 104, 191
convexity, 101, 104, 195, 235
countable additivity, 228
cover, 227
cylinder set, 14, 31, 46, 73, 102, 148, 158, 168

decimal expansion, 168
Decomposition Theorem, 234
dense orbit, 115
density zero, 50
diameter
 of a set, 174
 open cover, 174
diffeomorphism, 166
Dirac measure, 101, 104, 114, 193, 235
dissipative part, 203
Dominated Convergence Theorem, 97, 107, 111, 159, 164, 166, 232
doubling map, 7, 30, 43, 66–68, 70, 77, 93
dyadic intervals, 30
dyadic rationals, 30, 44, 93
dynamical system, 2

eigenvalue, 47, 121
entropy
 β-transformation, 165, 167
 of a transformation w.r.t. a partition, 154

Bernoulli shift, 158
Gauss map, 165
Markov shift, 159
metric, 154, 173
of the partition, 150
of the transformation, 154
sequence of interval partitions, 169
topological, 173, 175
equivalence, 29, 38, 79, 130, 225
topological, 180
ergodicity, 22, 40, 66, 72, 74, 78, 81, 83, 84, 104, 106, 130, 195, 204
β-transformation, 32
Bernoulli shift, 28
binary odometer, 63, 84
doubling map, 30
Gauss map, 130
irrational rotation, 27
Lüroth transformation, 30
Markov shift, 46
Euclid's algorithm, 126
even continued fraction expansion, 139
exactness, 84
β-transformation, 89
Bernoulli shift, 84
Gauss map, 132
expanding, 94, 121
expansion
T-, 166
β-, 11, 166
base N, 44, 166
binary, 7, 43, 71, 78, 84
continued fraction, 11, 100, 125, 168
decimal, 168
even continued fraction, 139
Lüroth, 9, 45, 166
nearest integer continued fraction, 140
odd continued fraction, 139
regular continued fraction, 125
semi-regular continued fraction, 139
expansiveness, 196
expected value, 149
extreme point, 104, 109, 195
Extreme Value Theorem, 224

factor, 71
map, 71
topological, 122
factor map, 71
Farey map, 200, 218
Fibonacci sequence, 133
first digit sequence, 112
first passage time, 216
first return time, 207
fixed point, 80
forward measurable, 85–87
forward orbit, 3, 115
Fourier series, 27, 120
Fubini's Theorem, 162, 233
fundamental interval, 31, 89, 94, 129, 166, 185
partition, 185

Gauss map, 11, 100, 125, 165, 168, 201, 218
Gauss measure, 12, 130, 218
generator, 157
geometric mean, 135
golden mean, 11, 32, 68, 89, 134, 156, 165, 212, 217

Hölder's Inequality, 93, 107
Haar measure, 6, 66, 112, 195
Halmos Recurrence Theorem, 21, 202

Helly's First Theorem, 97, 239
hereditary, 202, 227
Hilbert space, 53, 234
homeomorphism, 102, 108, 181
    conjugate, 123
Hopf decomposition, 203
Hopf's Ratio Ergodic Theorem, 219
Hurewicz Ergodic Theorem, 58
hyperbolic toral automorphism, 13

independence, 148, 151
indicator function, 56, 84, 90, 108, 231
induced operator, 5
induced system, 209
induced transformation, 208
information function, 160
    conditional, 160
interior, 92, 115, 191, 223
interval partition, 91, 169, 185
invariant measure, 3
invertibility, 3, 56, 72, 86, 90, 203, 204
irrational rotation, 6, 27, 51, 111, 119, 183, 195, 212
irreducibility, 22, 118
    Markov chain, 47
    matrix, 47
isometric operator, 234
isometry, 53, 112, 117, 120, 183, 195
isomorphism, 65, 66, 74, 78, 122, 148, 155, 194
    of probability spaces, 229

Jacobian, 90, 136
Jensen's Inequality, 149, 188, 189
jump transformation, 216

Kac's Lemma, 214
Knopp's Lemma, 29, 130
Kolmogorov 0-1 Law, 84, 228
Kolmogorov-Sinai Theorem, 157, 166
Koopman operator, 5, 26, 53, 82
Krieger's Generator Theorem, 157

Lüroth expansion, 9, 45, 166
Lüroth transformation, 9, 30, 45, 74
Lasota-Yorke inequality, 94
Lebesgue
    σ-algebra, 5, 230
    measure, 230
    number, 176, 224
    Number Lemma, 224
    space, 2, 230
left shift, 14
Lenstra's conjecture, 137
Lipschitz continuity, 195
local inverse, 92
Lochs' Theorem, 168, 169
logistic map, 92, 187
    topological entropy, 187
LSV map, 202

Möbius transformation, 126
Markov
    measure, 15, 46
    shift, 15, 46, 159
Martingale Convergence Theorem, 159, 163, 233
Mean Ergodic Theorem, 55
measurability, 2, 229
measurable union, 202, 227
measure
    σ-finite, 224
    absolutely continuous, 42, 81, 97, 225
    Bernoulli, 14, 15, 63

- complete, 4, 230
- Dirac, 77, 101, 104, 114, 193, 235
- equivalent, 29, 38, 130, 225
- Haar, 6, 66, 112, 195
- Lebesgue, 230
- Markov, 15, 46
- of maximal entropy, 195
- probability, 34, 65, 74, 81, 224
- regular, 101, 235
- singular, 42, 78

measure preservingness, 3, 229
measure theoretic entropy, 154, 173
metric entropy, 154, 173
minimal set, 118
minimality, 117
- binary odometer, 117
- irrational rotation, 119

mixing, 49
- Bernoulli shift, 50
- irrational rotation, 51
- strongly, 49, 66, 70, 72, 74, 75, 83, 87, 91
- topological, 121
- weakly, 49, 66, 72, 74, 83

modulo sets of measure zero, 224
monotone class, 225, 230
Monotone Class Theorem, 225
Monotone Convergence Theorem, 58, 61, 232
Morse code, 148
multinacci number, 69

natural extension, 72, 75
- Gauss map, 135

nearest integer continued fraction expansion, 140
negative β-transformation, 69, 156
neutral fixed point, 199
non-singularity, 2, 56, 63, 79, 90
normal
- absolutely, 45
- in base N, 45
- in base 2, 44
- number, 43, 44

NTFM, 165
null set, 224
number theoretic fibred map, 165

odd continued fraction expansion, 139
open cover, 174
operator
- induced, 5
- Koopman, 5, 26, 53, 82
- norm, 234
- Perron-Frobenius operator, 79

orbit, 3, 23, 66, 68, 115, 141
- dense, 115
- forward, 3, 115
- two-sided, 3

orthogonal complement, 54, 234
orthogonal projection, 55, 234

partial quotients, 126
partition
- atom, 147
- finite, 147
- independent, 151
- interval, 91, 169

periodic point, 27, 77, 108, 114
Perron-Frobenius operator, 77, 79, 90, 167
piecewise monotone, 91, 165, 185
Poincaré Recurrence Theorem, 19, 22
pointwise convergence, 110
Pointwise Ergodic Theorem, 34, 44, 46, 53, 58, 105, 132, 137, 139, 163, 218

Pomeau Manneville map, 202
Portmanteau Theorem, 237
positive operator, 234
probability vector, 46
product topology, 183

quality of approximation, 128

Rényi map, 145, 201, 221
Radon-Nikodym derivative, 56, 79, 90, 91, 232
Radon-Nikodym Theorem, 232
random continued fraction map, 141
random shift, 17
rational independence, 119
recurrence, 19
refinement, 150
    common, 150
    open cover, 174
regular continued fraction expansion, 125
regular measure, 101, 235
Riesz Representation Theorem, 110, 234, 236
Rohlin's formula, 166
rotation, 6, 27, 195
    topological entropy, 184

saturate, 227
Scheffé's Lemma, 56, 232
semi-algebra, 4, 41, 50, 225, 229
semi-regular continued fraction expansion, 139
separable
    set, 105
    space, 235
Shannon-McMillan-Breiman Theorem, 163, 166
shift
    Bernoulli, 14, 28, 50, 67, 69, 71, 72, 78, 84, 85, 121, 148, 156, 158, 184, 196
    Markov, 15, 46, 159
    random, 17
simple function, 231
Simple Function Approximation Theorem, 231, 233
singular measure, 42
stable
    foliation, 121
    manifold, 121
stationary process, 4
    stochastic, 16
stochastic matrix, 46, 47
strongly mixing, 49, 66, 70, 72, 74, 75, 83, 87, 91
subadditivity, 95, 150, 153, 175, 177, 186
support, 109
    of a function, 81
supremum norm, 105, 111, 112, 235
sweep-out set, 206
symmetric difference, 22

tail σ-algebra, 84, 85, 229
tent map, 68
time-zero partition, 158
topological conjugacy, 123
topological entropy, 173, 175, 179
    Bernoulli shift, 184
    logistic map, 187
    of open cover, 174
    open cover, 175
    properties, 181
    rotation, 184
topological factor, 123
topological semi-conjugacy, 123
topologically mixing, 121
    Arnold's cat map, 121

Bernoulli shift, 121
topologically transitive, 115
one-sided, 115
topology, 101, 116, 180, 181
total variation, 94, 237
transformation, 2
α-continued fraction, 100, 140, 222
β-, 11, 32, 39, 68, 75, 89, 98, 156, 165, 167, 217
$\times 10$, 168
$\times N$, 44, 67, 172
Arnold's cat, 13, 102, 121
baker's, 8, 69, 71, 90
Boole's, 8, 199, 206
conservative, 57
doubling, 7, 30, 43, 67, 68, 70, 77, 93
Farey, 200, 218
Gauss, 11, 100, 125, 165, 168, 201, 218
invertible, 3, 56, 72, 86, 90, 157, 203, 204, 212
Lüroth, 9, 30, 45, 74
logistic, 92, 187
Möbius, 126
measurable, 2
negative β-, 69, 156
non-invertible, 72, 74
non-singular, 2
Rényi, 145, 201, 221
random continued fraction, 141
tent, 68
translation, 6, 20
tribonacci number, 75
trigonometric polynomials, 112
two-sided orbit, 3

uniform continuity, 181, 183
uniform convergence, 109, 232, 235
unique ergodicity, 109, 195
irrational rotation, 112
odometer, 115
unique measure of maximal entropy, 195
unstable
foliation, 121
manifold, 121
Urysohn's Lemma, 224

Variational Principle, 173, 194

wandering set, 20, 203
weak convergence, 101, 103, 193, 236
weakly mixing, 49, 66, 72, 74, 83
Weyl's Theorem, 112

Yorke's Inequality, 95, 238

Zorn's Lemma, 118